Handbook of Embryogenesis

Handbook of Embryogenesis

Edited by **Leonard Roosevelt**

New York

Published by Callisto Reference,
106 Park Avenue, Suite 200,
New York, NY 10016, USA
www.callistoreference.com

Handbook of Embryogenesis
Edited by Leonard Roosevelt

International Standard Book Number: 978-1-63239-390-6 (Hardback)

Printed in the United States of America.

Contents

Preface

The purpose of the book is to provide a glimpse into the dynamics and to present opinions and studies of some of the scientists engaged in the development of new ideas in the field from very different standpoints. This book will prove useful to students and researchers owing to its high content quality.

This book is a collection of views on contemporary trends in modern biology, emphasizing on gametogenesis, fertilization, early and/or delayed embryogenesis in animals, plants and other small organisms. Written by international experts, this book provides an introduction as well as in-depth review on classical as well as contemporary problems that pose a challenge in understanding how living organisms - take birth, grow, and reproduce at levels varying from molecular and cellular levels to individual level. This book covers extensive topics such as somatic embryogenesis in forest plants, microspore embryogenesis, liquid-crystal in embryogenesis and pathogenesis of human diseases, etc.

At the end, I would like to appreciate all the efforts made by the authors in completing their chapters professionally. I express my deepest gratitude to all of them for contributing to this book by sharing their valuable works. A special thanks to my family and friends for their constant support in this journey.

Editor

Bamboo Regeneration via
Embryogenesis and Organogenesis

Xinchun Lin[1], Lichun Huang[2] and Wei Fang[1]
[1]The Nurturing Station for the State Key Laboratory of Subtropical Silviculture,
Zhejiang Agriculture and Forestry University, Zhejiang,
[2]Institute of Plant and Microbial Biology, Academia Sinica, Taibei, Taiwan,
China

1. Introduction

Bamboo is a member of grass family (Poaceae: Bambusoideae) (Wu & Raven, 2006). With the characteristics of short rotation, marketability of culms every year and immediate returns, bamboos are the fast growing multipurpose plants of high economic and environmental value that converts solar radiation into useful goods and services better than most tree species (Franklin, 2006; Kassahun, 2000). Besides producing fresh edible shoots and culms for timber, furniture and handicraft or as a raw material for pulping, bamboo serves as an efficient agent for conservation of water and soil (Christanty et al., 1996, 1997; Kassahun, 2003; Kleinhenz & Midmore, 2001; Mailly et al., 1997). Additionally, new products such as bamboo charcoal, bamboo vinegar, bamboo juice, bamboo healthy food, bamboo fiber product have been developed. World-wide interest in bamboo as a source of biofuel or bioenergy has also increased rapidly in recent years (Fu, 2001; Scurlock, 2000).

There are about 88 genera and 1400 recorded species of bamboo in the world, 34 genera and 534 species of which are in China (Wu & Raven, 2006). Bamboo is found in an area of more than 14 million ha throughout the tropics, subtropics and temperate zones of the world. Eighty percent of the species and area are confined to South and Southeast Asia, and largely in China, India, and Myanmar. China, with the richest resources and largest bamboo industry worldwide, possesses 5 million ha of bamboo forests (Bystriakova et al., 2003; McNeely, 1999).

Because bamboos flowering is unpredictable and has a long interval, the manipulation of bamboo crossbreeding is difficult (Lin et al, 2010a, 2010b; Lu et al., 2009). Gene transformation is another efficient approach to increase productivity and quality in plants. However, there have been no successful reports on bamboo gene transformation till now, because a stable and efficient regeneration system, the prerequisite to bamboo gene transformation, is still not completely established yet (Huang et al., 1989; Zhang et al., 2010; Zuo & Liu, 2004).

2. Process of bamboo regeneration and influencing factor

2.1 Regeneration process

Shoot, seeds, mature zygotic embryo, immature embryo, anther and inflorescence (Figure 1) can be used as explants of bamboo regeneration (Huang & Murashige, 1983; Pei et al.,

2011; Saxena & Dhawan, 1999; Zhang et al., 2010). After inoculation, calli are induced in the medium with a high concentration of 2,4-dichlorophenoxyacetic acid (2,4-D) usually. The quantity and quality of calli differ significantly between different concentration of 2,4-D in the medium. Three kinds of calli are often observed after treated by 2,4-D: (1) Yellowish, granular, and compact calli, with good potential regeneration ability (Fig.2a); (2) Pale-yellow, translucent, watery, and sticky calli, unable to regenerate generally (Fig.2b); (3) Creamy-yellow, compact, and non-embryogenic calli (Fig.2c), unable to regenerate. The yellowish, granular, and compact calli will proliferate after treatment in the callus growth maintenance medium with no or lower levels of 2,4-D compared with callus initiation medium. Then the calli will develop into adventitious shoots, embryo or adventitious roots, after subjected to the differentiated medium, and the callus with adventitious roots (Fig.2d) will not usually continue to differentiate. The destiny of calli is determined by different kinds of plant growth regulators (PGRs) such as 2-isopentenyladenine (2iP), thidiazuron (TDZ), zeatin (ZT), 6-benzyladenine (BA), kinetin (KT), naphthaleneacetic acid (NAA), and indole-3-butyricacid (IBA), etc., and high cytokinin/low auxin will result in adventitious shoot formation. Embryoids have radicles (Fig.2e-g), while adventitious shoot need to root before transplantation. Some shoots will produce roots naturally without any treatment, but others need to be treated with high auxin and minimal or no cytokinin during root development (Fig.2h-i). After rooting, the plantlet can be transferred to potting soil in the greenhouse (Fig.2j). (Huang et al., 1989; Yeh & Chang, 1986a, 1986b, 1987; Zhang et al., 2010).

Fig. 1. Explants of bamboos.

a) Seeds of *Melocalamus compactiflorus*; b) Seeds of *Qiongzhurea tumidinoda*; c) Seeds of *Phyllostachys edulis*; d) The seed of *Melocanna baccifera*; e) The inflorescence of *Phyllostachys violascens*; f) Flower buds proliferation of *Bambusa oldhamii*; g) The globular embryo of *Phyllostachys violascens*; h) The cotyledon embryo of *Phyllostachys violascens*; i) The mature embryo of *Dendrocalamus hamiltonii*.

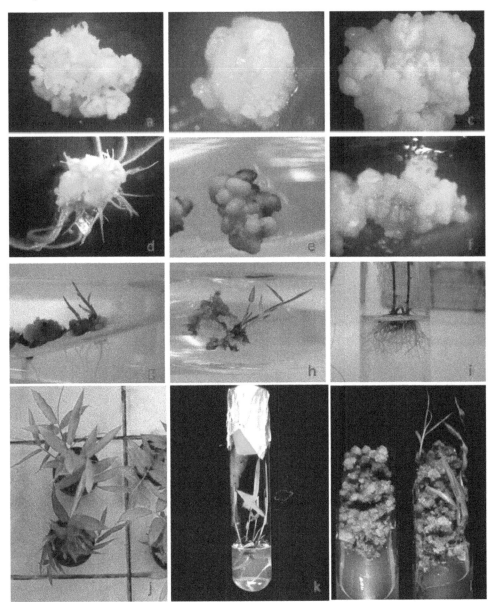

Fig. 2. The somatic embryogenesis and organogenesis of Bamboos.

a-c) Three kinds of calli of *Dendrocalamus hamiltonii*; d) Adventitious roots differentiation of *Dendrocalamus hamiltonii*; e-g) Embryogenesis of *Dendrocalamus hamiltonii*; h-i) Organogenesis of *Dendrocalamus hamiltonii*; j) Transplanted plantlets of *Dendrocalamus hamiltonii*; k) Organogenesis of *Bambusa oldhamii*; l) Organogenesis of *Phyllostachys aurea*.

2.2 Regeneration type

Bamboo can regenerate via embryogenesis and organogenesis (Fig.2e-k), and the frequency of embryogenesis is lower than that of organogenesis (Ramanayake & Wanniarachchi, 2003; Woods et al., 1992; Zhang et al., 2010).

During the bamboo embryogenesis, the embryoids initiate and develop from somatic cells. Compared with organogenesis, somatic embryogenesis is characterized by the formation of a bipolar structure, which will develop into plumule and radicle (Fig.2e-g). Histological analysis reveals that embryogenic cells are small in size, isodiametric with dense cytoplasm, generally locate along the periphery of calli, distribute in clusters, and intersperse with large parenchymal cells. Somatic embryos gradually developed from the granular onsite to heart-shaped, torpedo-shaped, and final cotyledons (Zhang et al., 2010).

In contrasted to the embryogenesis, organogenesis occurs via apparent shoot meristem or leaf primordial (Fig.2h-i). Histological analysis shows that non-embryogenic cells are large, and vacuolated parenchymal cells contain few plastids (Zhang et al., 2010). Callus will differentiate into adventitious shoots, and develop with subsequent formation of adventitious roots.

2.3 Factors affecting regeneration

The bamboo embryogenesis and organogenesis will be affected by many factors such as bamboo species (including cultivars, genotypes and ecotypes), type and age of explants, type of basal media, type and concentration of plant growth regulators, etc. (Godbole et al., 2002; Hassan & Debergh, 1987). Serious browning and difficult differentiation are popular in bamboo regeneration experiments (Huang & Murashige, 1983).

3. Innovative approaches about bamboo regeneration

To establish a stable and efficient regeneration system of bamboo, some efficient measures are proposed as follows:

3.1 Screening of bamboo species

About 20 bamboo species are used for regeneration system establishment in our lab, and we find that the bamboo regeneration ability differs significantly in different kinds of bamboo species, and the sympodial bamboo is easier to regenerate than monopodial bamboo. There are about 1400 kinds of bamboo species in the world, the successful reports about bamboo regeneration mainly focus on the species of *Bambusa* and *Dendrocalamus* (*Sinocalamus* is the anonymus of *Dendrocalamus*), so we can select the bamboo species which are easy to regenerate for overcoming the obstacle of bamboo gene transformation at first.

In addition to the difference among different kinds of species, genotype has distinct influence on the efficiency of plant regeneration via organogenesis or embryogenesis of various plant species such as soybean (*Glycine max*), rapeseed (*Brassica napus*), rice (*Oryza sativa*), etc. (Akasaka-Kennedy et al., 2005; Bailey et al., 1993; Hoque & Mansfield, 2004). Screening of bamboo genotypes or cultivars with strong regeneration ability may be a good choose for setting up an efficient system for bamboo gene transformation.

3.2 Selection of explants

Mature embryos, immature embryos, shoot tips, leaves, young inflorescences, hypocotyls, flower stalks, cotyledons, anthers and nodal segments are the common explants in plant regeneration. Most of those explants (Fig.1) are also efficient during the bamboo regeneration experiments (Lin et al, 2003, 2004; Rout & Das, 1994; Yuan et al., 2009). Within a species, the age of explants is important, loss of competence is correlated with maturation extent of explants, i.e. extent of differentiation, developmentally immature organs (or less differentiated cells) are most likely to contain morphogenetically competent cells. The regeneration ability of mature embryos, immature embryos, shoot tips and flower buds as explants of bamboo regeneration are tested in our lab. We find that the induction and differentiation of shoot tips of young bamboo seedling are easier than those of adult bamboo plants, and embryos, especially for immature embryos, are more efficient than other kinds of explants. Using the immature embryos (Fig.1g-h) as the explants, we have succeeded in setting up a regeneration system of *Phyllostachys violascens*, a species of monopodial bamboo (Pei et al., 2011). However, the materials related to bamboo flowers and seeds such as embryos, inflorescences and anthers are difficult to get for bamboo seldom flowering, shoots which are easier to obtain would be a better choice as the explants.

3.3 Selection of media

The components of media are also important during bamboo callus induction, callus growth maintenance, shoot differentiation, and root development. MS, NB(including N6's macrosalts, B5's microsalts and organic compounds), N6, HB and B5 basal media have been used in bamboo regeneration, and MS basal medium is the most common (Rao et al., 1985; Sun et al., 1999; Tsay et al., 1990; Wu & Chen, 1987; Zhang et al., 2010).

Different kinds of plant growth regulators are needed in different stage of bamboo regeneration, and auxin and cytokinin are the critical components for the morphogenesis in vitro. Besides, ethylene, abscisic acid (ABA) and brassinosteroid (BR) will also have different effects on embryogenesis and organogenesis (Aydin et al., 2006; Torrizo & Zapata, 1986; Vain et al., 1989).

3.4 Complementary approaches for advancing regeneration ability

3.4.1 *NiR* gene

Nitrate assimilation is an important process in rice regeneration. A quantitative trait loci gene encoding the ferrodoxin-nitrite reductase (*NiR*), an enzyme that catalyzes the reduction of nitrite to ammonium leading to the accumulation of toxic nitrite in culture media, was isolated from the high-regeneration rice strain Kasalath. The level of *NiR* expression in the

Koshihikari, a notorious poor rice line for genetic transformation, may result in lower enzymatic activity, and the enzymatic activity is correlated with the regeneration ability. With the introduction of Kasalath *NiR* gene encoding high enzymatic activity, the regeneration ability of low-regeneration rice strain Koshihikari had been improved (Nishimura et al., 2005). The *NiR* gene cloned from the high-regeneration rice strain Konansou has the similar function with that of Kasalath (Ozawa & Kawahigashi, 2006). However, *NiR* gene based improvement method will be suitable for the major Japonica rice varieties but not Indica rice varieties for having high *NiR* activity (Nishimura et al., 2006). In addition, the *NiR* gene can be used as a selection marker for rice gene transformation (Nishimura et al., 2005; Ozawa & Kawahigashi, 2006). The *NiR* gene isolated from the high-regeneration rice line may be useful in promoting the regeneration capacity during bamboo regeneration and gene transformation.

3.4.2 *ipt* gene

Cytokinins play a vital role in the differentiation process of plants. The isopentenyl transferase (*ipt*) gene, isolated from *Agrobacterium tumefaciens*, encodes for isopentenyltransferase which catalyzes the condensation of adenosine - 5' - monophosphate and isopentenylpyrophosphate to isopentenyladenosine - 5' - monophosphate (Akiyoshi et al. 1984). Integrating with the *ipt* gene, a cytokinin biosynthetic gene, the transformed cells present higher concentrations of endogenous cytokinins, and lead to higher frequencies of differentiation and transformation than untransformed control cells (Endo et al., 2002; Lopez-Noguera et al., 2009; Smigocki & Owens, 1988). Overexpression of *ipt* gene driven by a strong constitutive promoter favors plant regeneration, and the transformed plants exhibit abnormal morphogenetic variations such as an increased rate of branching, shorter stem internodes and little or no root formation. These morphogenetic changes can be used as selective markers during gene transformation, but these changes, especially for rooting difficultly, also disturb the normal development of transformed plants (Endo et al., 2001; Molinier et al., 2002; Smigocki & Owens, 1988). These defects derived from overexpression of *ipt* gene can be amended by gene deletor technology to delete the exogenous *ipt* gene. (Luo et al., 2007, 2008). It may be a good choose integrating those genes advancing regeneration frequency such as the *ipt* gene with objective genes during bamboo gene transformation.

In addition to *ipt* gene, a number of genes involved in hormone signal transduction have been indentified to influence the regenerative competence of plant cells for somatic embryogenesis and/or adventitious shoot formation (Sakamoto et al., 2006; Srinivasan et al., 2007, 2011).

*KNOX1*genes regulate the shoot apical meristem differentiation, and upregulate cytokinin biosynthesis and decrease gibberellin accumulation in plants, ectopic expression of *KNOX1* genes induces adventitious shoot regeneration in vitro-cultured explants (Sakamoto et al., 2006; Srinivasan et al., 2011).

Heterologous expression of the *BABY BOOM* (*BBM*) AP2/ERF transcription factor enhances the competence of tissues to undergo organogenesis and somatic embryogenesis (Srinivasan et al., 2007).

Identification and application of those genes regulating plant development and regeneration may lead to new approaches for plant regeneration in vitro.

3.4.3 Chemical additives

Many chemical additives including osmoticums, antioxidants, ethylene inhibitors, etc. have the good effects on plant regeneration, and may be used for increasing the efficiency of bamboo regeneration.

Osmotic pressure is correlated with plant development and differentiation, appropriate treatment by the common osmoticums such as sucrose, mannitol and polyethylene glycol (PEG) enhances the embryogenesis and organogenesis in *Solanum melongena* (Mukherjee et al.,1991), *Brassica napus* (Ferrie & Keller, 2007) and white spruce (*Picea glauca*) (Misra et al., 1993).

Tissue browning is the major problem of bamboo regeneration. Oxidized phenolic compounds produced from the damaged explants will suppress enzyme activity, darken the culture medium, and lead to the death of the explants. Treated with antioxidants and absorbents such as cysteine and ascorbic acid, polyvinylpyrrolidone (PVP) and activated carbon, will alleviate the phenolic oxidation and favor the plant regeneration (Abdelwahd et al., 2008; Sanyal et al., 2005; Toth et al., 1994).

Ethylene plays an important role in plant morphogenesis, and has a negative effect on plant regeneration, ethylene inhibitors such as AgNO3 and AVG promote the callus initiation and plant regeneration in cabbage, tobacco, maize and wheat accordingly. The silver ion is reported to be an inhibitor of ethylene action by competitively binding to ethylene receptors which are located predominantly at the intracellular membrane, while AVG inhibits ethylene biosynthesis directly (Vain et al., 1989; Zhang et al., 1998). In addition to silver ion, other heavy metals including Cu and other ethylene inhibitors (Co and Ni), also significantly facilitate the regeneration and somatic embryogenesis (Purnhauser & Gyulai, 1993; Roustan et al. 1989).

3.4.4 Partial desiccation

Partial desiccation has been reported to accelerate plant organogenesis or embryogenesis and results in high regeneration ability significantly in grape (Gray, 1989), wheat (Cheng et al., 2003), *Brassica napus* (Kott & Beversdorf, 1990), rice (Rance et al., 1994; Saharan et al., 2004; Tsukahara & Hirosawa, 1992) and cassava (Mathews et al., 1993). The possible mechanism about its promotion on plant regeneration capacity may be that partial desiccation terminates the developmental mode and "switches" the embryo into a germination mode (Attree et al., 1991). Partial desiccation not only enhances the plant regeneration efficiency, but also benefits the plant organogenesis or embryogenesis and subsequent differentiated stage, and thus reduces the whole time in plant tissue culture (Rance et al., 1994). In addition to the positive effect on wheat organogenesis or embryogenesis, partial desiccation during co-culture greatly enhances the transformation efficiency through inhibiting the growth of Agrobacterium which will suppress the recovery of wheat tissue, and favoring the transfer DNA (T-DNA) delivery (Cheng et al., 2003). Partial desiccation may be also beneficial in bamboo differentiation.

4. Somaclonal variation

Somaclonal variation is the common phenomenon during plant tissue culture. Being high efficient, time-saving and low cost, somaclonal variation has become a useful tool to creating new germplasms with beneficial economic traits of the breeding process in rice, potato, maize, barley and sugar cane, etc. (Karp et al., 1995; Larkin & Scowcroft, 1981). Somaclonal variation, such as mosaic leaf, albino, etiolated shoots, polyploidization and early flowering in vitro etc., also occurs during bamboo embryogenesis and organogenesis (Fig.3), and the frequency of variation will increase after continuous subculture. Most of those bamboo variants can grow normally, but is generally difficult in root formation (Zhang et al., 2010). Compared with poplar and other economic tree species, the process of bamboo breeding is slower for its peculiar flowering characteristics, screening of somaclonal variants with stable and valuable traits may be an alternative choose for bamboo improvement.

Fig. 3. Somaclonal variation during regeneration of *Dendrocalamus hamiltonii.*

a) Albino plantlets; b) The plantlet with mosaic leaf; c-d) Plantlets flowering in vitro, as indicated with red arrows.

5. Genetic transformation

Using the regeneration system of *Dendrocalamus hamiltonii* of our lab, we tried to establish a bamboo genetic transformation system. After pre-culture for 4 days, good calli were infected by Agrobacterium strains *EHA105* habouring the *pCAMBIA 1301* vector. After co-culture, for 3 days, the calli were transferred to the recovery medium for 8 days, then the calli were transferred to the selection medium with hygromycin selection. The resistant calli produced and then differentiate shoots and plantlets. To determine the genetic transformation frequency, 10 resistant plantlets were examined by PCR, and all of them are positive, which shows that *Hygromycin B phosphotransferase* (*HPT*) gene of *pCAMBIA 1301* was successfully integrated into the genome of *Dendrocalamus hamiltonii* via agrobacterium, the result was further verified through sequencing the PCR products.

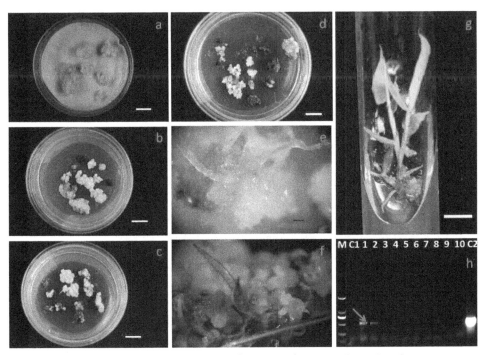

Fig. 4. The genetic transformation and PCR detection of *Dendrocalamus hamiltonii*.

a) Co-culture; b) First selection culture; c) Second selection culture; d) Third selection culture; e-f) Resistant calli differentiate shoots; g) Resistant plantlet; h) PCR detection of plantlets. M: marker; C1: Untransferred plantlet; 1-10: Resistant plantlets; C2: Positive control. Obiective bands were indicated with red arrows.6. Conclusions and prospects

6. Conclusions and prospects

Gene transformation has been proved to be efficient in plant breeding (Nishimura et al., 2006; Varshney et al., 2007). Being one of the most challenging aspects of the gene transformation protocol, a stable and efficient regeneration system must be developed. Lack of a well established regeneration system is the main obstacle for bamboo gene transformation. Understanding bamboo regeneration process and adopting innovative approaches about it will help to enhance the regeneration ability of bamboos and make breakthrough in bamboo gene transformation. The application of gene transformation will open up a new field for bamboo breeding.

Although we have succeeded in establishing a genetic transformation for *Dendrocalamus hamiltonii* at the first time, there are still many problems during the bamboo genetic transformation process such as serious browning, low differentiate frequency, etc (Zuo & Liu, 2004). More effort is needed to establish a stable and efficient genetic transformation system about bamboos.

7. Acknowledgements

This study is supported by research grants from the Natural Science Foundation of Zhejiang Province [Grant No. Z3100366] and the Ministry of Science and Technology of China (Grant No. 2012CB723008).

8. References

Akasaka-Kennedy, Y., Yoshida, H. & Takahata, Y. (2005). Efficient plant regeneration from leaves of rapeseed (*Brassica napus* L.): the influence of AgNO3 and genotype. Plant Cell Reports, 24: 649-654

Akiyoshi, D.E., Klee, H., Amasino, R.M., Nester, E.W. & Gordon, M.P. (1984). T-DNA of *Agrobacterium tumefaciens* encodes an enzyme of cytokinin biosynthesis. Proceedings of the National Academy of Science USA, 81:5994-5998

Attree, S.M., Moore, D., Sawhney, V.K. & Fowke, L.C. (1991). Enhanced maturation and desiccation tolerance of white spruce [*Picea glauca* (Moench.) Voss] somatic embryos: Effects of a non-plasmolysing water stress and abscisic acid. Annals of Botany, 68: 519-525

Aydin, Y., Talas-Ogras, T., Ipekçi-Altas, Z. & Gözükirmizi, N. (2006). Effects of brassinosteroid on cotton regeneration via somatic embryogenesis. Biologia, 61(3): 289-293

Bailey, M.A., Boerma H.R. & Parrot D.W.A. (1993). Genotype effects on proliferative embryogenesis and plant regeneration of soybean. In Vitro Cellular & Developmental Biology - Plant, 29(3): 102-108

Bystriakova, N., Kapos, V., Lysenko, I. & Stapleton, C.M.A. (2003). Distribution and conservation status of forest bamboo biodiversity in the Asia-Pacific Region. Biodiversity and Conservation, 12: 1833-1841

Cheng, M., Hu, T., Layton, J., Liu, C.N. & Fry, J. E. (2003). Desiccation of plant tissues post-Agrobacterium infection enhances T-DNA delivery and increases stable transformation efficiency in wheat. In Vitro Cellular & Developmental Biology-Plant, 39: 595-604

Christanty, L., Mailly, D. & Kimmins, J. P. (1996). "Without bamboo, the land dies": Biomass, litterfall, and soil organic matter dynamics of a Javanese bamboo talun-kebun system. Forest Ecology and Management, 87: 75-88.

Christanty, L., Kimmins, J. P. & Mailly, D. (1997). "Without bamboo, the land dies": A conceptual model of the biogeochemical role of bamboo in an Indonesian agroforestry system. Forest Ecology and Management, 91: 83-91.

Endo, S., Kasahara, T., Sugita, K., Matsunaga, E. & Ebinuma, H. (2001). The isopentenl-transferase gene is effective as a selectable marker gene for plant transformation in tobacco (*Nicotiana tabacum* cv. Petite Havana SRI). Plant Cell Reports, 20(1): 60-66

Endo, S., Kasahara, T., Sugita, K. & Ebinuma, H. (2002). A new GST-MAT vector containing both *ipt* and *iaaM/H* genes can produce marker-free transgenic tobacco plants with high frequency. Plant Cell Reports, 20(10): 923-928

Ferrie, A. M. R. & Keller, W. A. (2007). Optimization of methods for using polyethylene glycol as a non-permeating osmoticum for the induction of microspore embryogenesis in the Brassicaceae. In Vitro Cellular & Developmental Biology-Plant, 43: 348–355

Franklin, D.C. (2006). Wild bamboo stands fail to compensate for a heavy 1-year harvest of culm shoots. Forest Ecology and Management, 237:115-118

Fu, J. (2001). Bamboo Juice, Beer and Medicine. The Magazine of The American Bamboo Society, 22(5): 16

Godbole, S., Sood, A., Thakur, R., Sharma, M. & Ahuja, P.S. (2002). Somatic embryogenesis and its conversion into plantlets in a multipurpose bamboo, *Dendrocalamus hamiltonu* Nees et Arn. Ex Munro. Current Science, 83: 885-889

Gray, D.J. (1989). Effects of dehydration and exogenous growth regulators on dormancy, quiescence and germination of grape somatic embryos. In Vitro Cellular & Developmental Biology-Plant, 25: 1173-1178

Hassan, A.E. & Debergh, P. (1987). Embryogenesis and plantlet development in the bamboo *Phyllostachys viridis* (Young) McClure. Plant cell Tissue and Organ Culture, 10: 73-77

Hoque, M.E. & Mansfield, J.W. (2004). Effect of genotype and explant age on callus induction and subsequent plant regeneration from root-derived callus of Indica rice genotypes. Plant Cell, Tissue and Organ Culture, 78: 217–223

Huang, L.C. & Murashige, T. (1983). Tissue culture investigation of bamboo I: Callus culture of *Bambusa, Phyllostachys* and *Sasa*. Academia Sinica Institute of Botany Bulletin, 24:31-52

Huang, L.C., Huang, B.L. & Chen, W.L. (1989). Tissue culture investigations of bamboo IV: Organogenesis leading to adventitious shoots and plants in excised shoot apices. Environmental and Experimental Botany, 29: 307-315.

Karp, A. (1995). Somaclonal variation as a tool for crop improvement. Euphytica, 85: 295-302

Kassahun, E. (2000). The indigenous bamboo forests of Ethiopia: an overview. Ambio, 29: 518-521.

Kassahun, E. (2003). Ecological aspects and resource management of bamboo forests in Ethiopia. PhD Dissertation. Swedish University of Agricultural Sciences, Uppsala

Kleinhenz, V. & Midmore, D.J. (2001). Aspects of bamboo agronomy. Advances in Agronomy, 74: 99-145.

Kott, L.S. & Beversdorf, W.D. (1990). Enhanced plant regeneration from microspore-derived embryos of *Brassica napus* by chilling, partial desiccation and age selection. Plant Cell, Tissue and Organ Culture, 23: 187-192

Larkin, P.J. & Scowcroft, W.P. (1981). Somaclonal variation-novel source of variability from cell culture for plant improvement. Theoretical and Applied Genetics, 60: 197-214

Lin, C.S., Lin, C.C. & Chang, W.C. (2003). In vitro flowering of *Bambusa edulis* and subsequent plantlet survival. Plant Cell, Tissue and Organ Culture, 72: 71-78

Lin, C.S., Lin, C.C. & Chang, W.C. (2004). Effect of thidiazuron on vegetative tissue-derived somatic embryogenesis and flowering of bamboo *Bambusa edulis*. Plant Cell, Tissue and Organ Culture, 76: 75-82

Lin, X.C., Chow, T.Y., Chen H.H., Liu, C.C., Chou, S.J., Huang, B.L., Kuo, C.I., Wen, C.K., Huang, L.C. & Fang, W. (2010a). Understanding bamboo flowering based on large-scale analysis of expressed sequence tags. Genetics and Molecular Research, 9(2): 1085-1093

Lin, X.C., Lou, Y.F., Liu, J., Peng, J.S., Liao, G.L. & Fang, W. (2010b). Crossbreeding of *Phyllostachys* species (Poaceae) and identification of their hybrids using ISSR markers. Genetics and Molecular Research, 9(3): 1398-1404

Lopez-Noguera, S., Petri, C. & Burgos, L. (2009). Combining a regeneration-promoting *ipt* gene and site-specific recombination allows a more efficient apricot transformation and the elimination of marker genes. Plant Cell Reports, 28:1781-1790

Luo, K., Duan, H., Zhao, D., Zheng, X., Deng, W., Chen, Y., Jr, C.N.S., McAvoy, R., Jiang, X., Wu, Y., He, A., Pei, Y. & Li, Y. (2007). 'GM-gene-deletor' for transgenic plants: fused *loxP-FRT* recognition sequences dramatically improve the efficiency of *FLP* or *CRE* recombinase on transgene excision from pollen and seed of tobacco plants. Plant Biotechnology Journal, 5: 263-274

Luo, K., Sun, M., Deng, W. & Xu, S. (2008). Excision of selectable marker gene from transgenic tobacco using the GM-gene-deletor system regulated by a heat-inducible promoter. Biotechnology Letters, 30(7): 1295-1302

Lu, J.J., Yoshinaga, K., Fang, W. & Tang, D.Q. (2009). Identification of the hybrid bamboo F1 by SSR markers, Scientia Silvae Sinicae, 45(3): 29-34.

Mailly, D., Christanty, L. & Kimmins, J. P. (1997). "Without bamboo, the land dies": nutrient cycling and biogeochemistry of a Javanese bamboo talun-kebun system. Forest Ecology and Management, 91: 155-173.

Mathews, H., Schopke, C., Carcamo, R., Chavarriaga, P., Fauquet, C. & Beachy R.N. (1993). Improvement of somatic embryogenesis and plant recovery in cassava. Plant Cell Reports, 12: 328-333

McNeely, J.A. (1999). Biodiversity and bamboo genetic resources in Asia: in situ, community-based and *ex situ* approaches to conservation. Chinese biodiversity, 7(1): 38-51

Misra, S., Attree, S. M., Leal, I. & Fowke, L.C. (1993). Effect of abscisic acid, osmoticum, and desiccation on synthesis of storage proteins during the development of white spruce somatic embryos. Annals of Botany, 71:11-22

Molinier, J., Thomas, C., Brignou, M. & Hahne, G. (2002). Transient expression of *ipt* gene enhances regeneration and transformation rates of sunflower shoot apices (*Helianthus annuus* L.). Plant Cell Reports, 21:251-256

Mukherjee, S. K., Rathinasabapathi, & Gupta, N. (1991). Low sugar and osmotic requirements for shoot regeneration from leaf pieces of *Solanum melongena* L.. Plant Cell, Tissue and Organ Culture, 25: 13-16

Nishimura, A., Aichi, I. & Matsuoka, M. (2006). A protocol for Agrobacterium-mediated transformation in rice. Nature Protocols, 1(6): 2796-2802

Nishimura, A., Ashikari, M., Lin, S.Y., Takashi, T., Angeles, E.R., Yamamoto, T. & Matsuoka, M. (2005). Isolation of a rice regeneration quantitative trait loci gene and its application to transformation systems. Proceedings of the National Academy of Science USA, 102(33): 11940-11944

Ozawa, K. & Kawahigashi, H. (2006). Positional cloning of the nitrite reductase gene associated with good growth and regeneration ability of calli and establishment of a new selection system for Agrobacterium-mediated transformation in rice (*Oryza sativa* L.). Plant Science, 170: 384-393

Pei, H.Y., Lin, X.C., Fang, W. & Huang, L.C. (2011). A preliminary study of somatic embryogenesis of *Phyllostachys violascens* in vitro. Chinese Bulletin of Botany, 46 (2): 170-178

Purnhauser, L. & Gyulai, G. (1993). Effect of copper on shoot and root regeneration in wheat, triticale, rape and tobacco tissue cultures. Plant Cell, Tissue and Organ Culture, 35: 131-139

Ramanayake, S.M.S.D. & Wanniarachchi, W.A.V.R. (2003). Organogenesis in callus derived from an adult giant bamboo (*Dendrocalamus giganteus* Wall. ex Munro). Scientia Horticulturae, 98:195-200

Rance, I.M., Tian, W., Mathews, H., Kochko, A. d., Beachy, R.N. & Fauquet, C. (1994). Partial desiccation of mature embryo-derived caili, a simple treatment that dramatically enhances the regeneration ability of indica rice. Plant Cell Reports, 13: 647-651

Rao, I.U., Rao, I.V.R. & Narang, V. (1985). Somatic embryogenesis and regeneration of plants in the bamboo *Dendrocalamus strictus*. Plant Cell Reports, 4:191-194

Roustan, J.P., Latche, A. & Fallot, J. (1989). Stimulation of *Daucus carota* somatic embryogenesis by inhibitors of ethylene synthesis: cobalt and nickel. Plant Cell Reports, 8:182-185

Rout, G.R. & Das, P. (1994). Somatic embryogenesis and *in vitro* flowering of 3 species of bamboo. Plant Cell Reports, 13:683-686

Saharan, V., Yadav, R.C., Yadav, N.R. & Chapagain, B.P. (2004). High frequency plant regeneration from desiccated calli of indica rice (*Oryza sativa* L.). African Journal of Biotechnology, 3(5): 256-259

Sakamoto, T., Sakakibara, H., Kojima, M., Yamamoto, Y., Nagasaki, H., Inukai, Y., Sato, Y. & Matsuoka, M. (2006). Ectopic expression of *KNOTTED1*-like homeobox protein induces expression of cytokinin biosynthesis genes in rice. Plant Physiology, 142: 54–62

Sanyal, I., Singh, A.K., Kaushik, M. & Amla, D.V. (2005). Agrobacterium mediated transformation of chickpea (*Cicer arietinum* L.) with Bacillus thuringiensis *cry1Ac* gene for resistance against pod borer insect *Helicoverpa armigera*. Plant Science, 168: 1135-1146.

Saxena, S. & Dhawan, V. (1999). Regeneration and large-scale propagation of bamboo (*Dendrocalamus strictus* Nees) through somatic embryogenesis. Plant Cell Reports, 18:438-443

Scurlock, J.M.O., Dayton, D.C. & Hames B. (2000). Bamboo: an overlooked biomass resource? Biomass and Bioenergy, 19: 229-244

Smigocki, A. & Owens, L. (1988). Cytokinin gene fuses with a strong promoter enhances shoot organogenesis and zeatin level in transformed plant cells. Proceedings of the National Academy of Science USA, 85: 5131-5135

Srinivasan, C., Liu, Z., Heidmann, I., Supena, E.D.J., Fukuoka H., Joosen, R., Lambalk, J., Angenent, G., Scorza, R., Custers, J.B.M. & Boutilier, K. (2007) Heterologous expression of the *BABY BOOM* AP2/ERF transcription factor enhances the regeneration capacity of tobacco (Nicotiana tabacum L.). Planta, 225: 341-351

Srinivasan, C., Liu, Z., & Scorza, R. (2011) Ectopic expression of class 1 *KNOX* genes induce adventitious shoot regeneration and alter growth and development of tobacco

(*Nicotiana tabacum* L) and European plum (*Prunus domestica* L). Plant Cell Reports, 30: 655–664

Sun, G.Z., Ma, M.Q., Zhang, Y.Q., Xie, X.L., Cai, J.F., Li, X.P., Yin, Z.M., Qin, J., Zhang, R., Xu, D., Yang, J.Y. & Wang, H.B. (1999). A suitable medium for callus induction and maintenance for wheat. Journal of Hebei Agriculture Science, 3: 224-226

Toth, K., Haapala, T. & Hohtola, A. (1994). Alleviation of browning in oak explants by chemical pretreatments. Biologia Plantarum, 36: 511-517

Torrizo, L. B. & Zapata, F. J. (1986). Anther culture in rice: IV. The effect of abscisic acid on plant regeneration. Plant Cell Reports, 5(2): 136-139

Tsay, H.S., Yeh, C.C. & Hsu, J.Y. (1990). Embryogenesis and plant regeneration from anther culture of bamboo (*Sinocalamus latiflora* (Munro) McClure). Plant Cell Reports, 9:349-351

Tsukahara, M. & Hirosawa, T. (1992). Simple dehydration treatment promotes plantlet regeneration of rice (*Oryza sativa* L.) callus. Plant Cell Reports, 11: 550-553

Vain, P., Yean, H. & Flament, P. (1989). Enhancement of production and regeneration of embryogenic type II callus in *Zea mays* L. by AgNO3. Plant Cell, Tissue and Organ Culture, 18:143-151

Varshney, R. K., Langridge, P. & Graner, A. (2007). Application of genomics to molecular breeding of wheat and barley. Advances in Genetics, 58: 121-155

Woods, S.H., Phillips, G.C., Woods, J.E. & Collins, G.B. (1992). Somatic embryogenesis and plant regeneration from zygotic embryo explants in Mexican weeping bamboo, *Otatea acuminate aztecorum*. Plant Cell Reports, 11:257-261

Wu, C.Y. & Chen, Y. (1987). A study on the genotypical differences in anther culture of Keng rice (*Oryza sativa* subsp. *keng*). Acta Genetica Sinica, 14: 168-174

Wu, Z.Y. & Raven, P.H., eds. (2006). Flora of China: Poaceae. Vol. 22. Beijing: Science Press, St. Louis: MBG Press.

Yeh, M. & Chang, W.C. (1986a). Plant regeneration through somatic embryogenesis in callus culture of green bamboo (*Bambusa oldhamii* Munro). Theoretical and Applied Genetics, 73:161-163

Yeh, M. & Chang, W.C. (1986b). Somatic embryogenesis and subsequent plant regeneration from inflorescence callus of *Bambusa beecheyana* Munro var. *beecheyana*. Plant Cell Reports, 5:409-411

Yeh, M. & Chang, W.C. (1987). Plant regeneration via somatic embryogenesis in mature embryo-derived callus culture of *Sinocalamus latiflora* (Munro) McClure. Plant Science,51: 93-96

Yuan, J.L., Gu, X.P., Li, L.B., Yue, J.J., Yao, N. & Guo, G.P. (2009). Induction and plantlet regeneration of *Bambusa multiplex*. Scientia Silvae Sinicae, 3:35- 40

Zhang, F.L., Takahata, Y. & Xu, J.B. (1998). Medium and genotype factors influencing shoot regeneration from cotyledonary explants of Chinese cabbage (*Brassica campestris* L. ssp. *pekinensis*). Plant Cell Reports, 17: 780–786

Zhang, N., Fang, W., Shi, Y., Liu, Q.Q., Yang, H.Y., Gui, R.Y. & Lin, X.C. (2010). Somatic embryogenesis and organogenesis in *Dendrocalamus hamiltonii*. Plant Cell, Tissue and Organ Culture, 103: 325–332.

Zhuo, R.Y. & Liu, X.G. (2004). Factors effecting transgenic breeding of *Dendrocalamus latiflorus*. Acta Agriculturae Universitatis Jiangxiensis, 26: 551-554.

Role of Polyamines in Efficiency of Norway Spruce (Hurst Ecotype) Somatic Embryogenesis

J. Malá[1], M. Cvikrová[2], P. Máchová[1] and L. Gemperlová[2]
[1]*Forestry and Game Management Research Institute, Jíloviště*
[2]*Institute of Experimental Botany, Academy of Sciences of the Czech Republic, Praha*
Czech Republic

1. Introduction

Somatic embryogenesis is considered as to be an advantageous methodology for plant micropropagation *in vitro*, particularly for conifers. It offers, moreover, a row of possibilities to study developmental processes during the embryo differentiation, and it enables also more detailed analyses of the mechanisms of embryo conversion.

Somatic embryogenesis can be divided into several stages, which are comparable to the stages in zygotic embryogenesis (Atree & Fowke, 1993). However, since somatic embryos develop without the protective environment of the surrounding maternal tissue, there is a need to supply developing embryos both the nutrients and the regulatory compounds exogenously. Induction and continuous proliferation requires the auxins and the cytokinins, whereas the further growth and maturation of embryos depends on the abscisic acid (Attree et al., 1991). By the end of the maturation stage, all structures of the embryo are morphologically fully developed (Find, 1997) but the embryo becomes biochemically mature after a desiccation processing (Flinn et al., 1993).

Somatic embryogenesis in some coniferous species provides sufficient numbers of fully developed embryos usable for propagation. On the other hand, the yield of converted somatic embryos in other species is often too low for practical applications (Igasaki et al., 2003). In some instances, the successful somatic embryogenesis is a question of the selection of responsible clones within the range of a coniferous species. Research of somatic embryogenesis in conifers has increased rapidly since the eighties of the last century. Promising results were achieved especially with Norway spruce (*Picea abies* (L.) Karst.), in which the successful regeneration of complete plants was obtained (Attree & Fowke, 1993; Bornman, 1985; Chalupa, 1985; Chalupa et al., 1990; Hakman et al., 1985; Malá et al., 1995). The process of somatic embryogenesis in Norway spruce can be divided into four stages characterized by the different degree of embryonic tissue differentiation: the induction of embryogenic tissue, the proliferation of somatic embryos, maturation, and finally, the conversion of mature embryos into complete plants.

The induction of embryogenic tissues can be achieved by applying phytohormones on mature or immature zygotic embryos. The initiation rate is higher when immature zygotic embryos are used; however, it is difficult to determine an optimal cone harvest time (Chalupa, 1985). The transfer of embryogenic tissue from proliferation onto maturation medium leads to the induction of embryo development. Despite of the successful protocol for establishment of Norway spruce somatic embryogenesis technique, there is a lack of data concerning the endogenous composition of biologically active compounds both in somatic and zygotic embryos. Generally, the development of embryos as well as their conversion into complete plantlets is closely associated with changes in endogenous phytohormone levels. Changes in endogenous hormone levels (IAA, ABA, and ethylene) during Norway spruce somatic embryo development and maturation have been recently reported (Vágner et al., 2005).

Beside the key roles of auxins and cytokinins, very important function during growth and differentiation of plant tissues belongs to polyamines (PAs) (Mattoo et al., 2010; Vera-Sirera et al., 2010). PAs have a wide spectrum of action with some similarities with plant phytohormones. In cooperation with auxins and cytokinins, PAs modulate morphogenetic processes (Altamura et al., 1993). From this point of view PAs could be considered as new category of plant hormones acting particularly as regulators of such processes as gene expression, cell proliferation, cell wall formation, etc. (Cohen, 1998). PAs are involved also in the transmission of cellular signals. They regulate the synthesis of nitric oxide, which is known as a plant signaling molecule (Besson-Bard et al., 2008).

PAs are small polycationic molecules with several amino groups, which are ubiquitous in both prokaryotes and eukaryotes (Cohen, 1998; Tiburcio et al., 1997). Also the mechanisms of PAs synthesis, transport, and catabolism are conserved from bacteria to animals and plants (Kusano et al., 2008; Tabor & Tabor, 1984). Most of the biological functions of PAs can be explained by their polycationic nature, which allows interactions with anionic macromolecules such as DNA, RNA, and with negative groups of cellular membrane components. The synthesis of PAs within various plant tissues depends on a great variety of physiological growth regulatory stimuli as well as on various external influences as periodicity or stressing conditions (humidity, droughts), and environmental damaging factors, too.

PAs are essential compounds for life. Decrease or arrest of internal PA production inhibits a row of cellular functions, e. g. cell growth. Therefore, in the case of PAs deficiency, plant cells are equipped with a high efficient system for transfer and functional utilization of PA molecules from external sources (Cohen, 1998, Hanfrey et al., 2001, Kusano et al., 2008).

Three commonly occurring PAs in plants are diamine putrescine (Put), triamine spermidine (Spd) and tetramine spermine (Spm). All these compounds are present in the free form or as conjugates with other low molecular substances (e. g. phenolic acids) or macromolecules (proteins, nucleoproteins). They are found in cell walls, vacuoles, mitochondria, chloroplasts and cytoplasm (Kaur-Sawhney et al., 2003). PAs are detected in increased amounts within actively differentiating and growing plant tissues. Also their activity increases mainly in growing plant tissues, during embryogenesis, root formation and stem elongation, fruit development and ripening, and during response to abiotic and biotic stress factors, too (Kumar et al., 1997).

Polyamines play a fundamental role in the regulation of somatic and zygotic embryogenesis (Kong et al., 1998; Silveira et al., 2004). The role of PAs during *in vivo* and *in vitro* development, including somatic embryogenesis, were recently reviewed (Baron & Stasolla, 2008). The accumulation of high levels of PAs in somatic embryos contributes to their reserve consisting predominately of proteins and triglycerides, which are utilized during embryo germination. PAs changes were studied in embryogenic cultures of *Picea abies* (R. Minocha et al., 1993; Serapiglia et al., 2008; Vondráková et al., 2010), *Picea rubens* (R. Minocha et al., 1993), *Pinus taeda* (Silveira et al., 2004), and *P. radiata* (R. Minocha et al., 1999). Polyamine profiles in germinating somatic embryos derived from long term cultivated embryogenic mass and germinating zygotic embryos of Norway spruce were studied by Gemperlová et al. (2009). There are also several reports indicating the participation of PAs in somatic embryo development of some coniferous species (R. Minocha et al., 1999; Santanen & Simola, 1992; Silveira et al., 2004) but the mechanism of how polyamines regulate cell differentiation processes is not fully elucidated up to the present.

Norway spruce is the most important forest tree species in the Czech Republic, both economically and due to its representation. However, only few populations can be regarded autochthonous. These populations are irreplaceable in the future, not only for maintenance and natural regeneration of the biotope, but also as an important source of genetic material needed for the breeding programs which are aimed to preserve the valuable forest sources for the future generations. The Hurst ecotype of Norway spruce is considered to be autochthonous and it is rarely preserved at the altitudes over 700 m. Reproduction of these populations is really difficult due to the high age of the trees and longer and lengthening flushing intervals. For conifers and mainly for Norway spruce, micropropagation technologies, mainly somatic embryogenesis, represent very suitable methods of reproduction and preservation of valuable genotypes.

The aim of this study was to compare PA levels during development of somatic embryos of high responsible AFO 541 cell line with less responsible cell lines derived from the Hurst ecotype of *Picea abies*. The results obtained could help to a better understanding of PAs function in regulation of plant tissue differentiation processes and contribute to improving the micropropagation of less responsible Hurst Norway spruce by somatic embryogenesis.

2. Experimental part

2.1 Materials and methods

2.1.1 Cell lines

Picea abies (L.) Karst. embryogenic cultures of high responsible cell line AFO 541 (AFOCEL, Nangis, France) and five less responsible cell lines (L10, L13, L16, L17, L28) derived from a *P. abies* Hurst ecotype were used.

2.1.2 Hurst cell lines initiation

Immature cones of 140 years old elite open pollinated Hurst Norway Spruce growing in the conservation area Labské Pískovce (Northern Bohemia, CR) were collected in late July 2006 and stored at 4 °C. To induce embryogenic tissue differentiation, the immature embryos extirpated from sterilized seeds (1% NaClO, Savo, Biochemie, CR) were cultivated in darkness at 24 °C on the modification of solid E medium (Gupta & Durzan,

1986) with 0.2 mg.l⁻¹ gelerit (Sigma-Aldrich, Germany) and phytohormones (0.5 mg.l⁻¹ of BAP, 1.0 mg.l⁻¹of 2,4-D, and 0.5 mg.l⁻¹ of Kin), pH adjusted to 5.8 prior to the medium autoclaving (Malá, 1991).

In a preliminary experiment the embryogenic capacity (i.e. the ability of the ESM to produce mature somatic embryos) were determined in 45 cell lines of Hurst Norway Spruce ESM. Five of them, with diverse characteristics, were further selected for the subsequent experiments.

2.1.3 Induction of proliferation

The cultures of embryogenic mass were after 4 wks of induction transferred onto fresh E medium of the same composition and cultured in the same conditions as previously. The embryogenic cultures were maintained by subculturing weekly.

2.1.4 Maturation of somatic embryos

After 12 wks of cultivation on proliferation medium were the cultures transferred onto maturation medium (solid E medium without phytohormones, supplemented with 8 mg.l⁻¹ of ABA (Sigma, Chemical Co., USA) and 20 mg.l⁻¹ of PEG (m. w. 3350, Sigma, Chemical Co., USA). Cultures were kept in same conditions as described above and subcultured every week. After 2 wks, the somatic embryo cultures were transferred onto solid E medium containing 0.1mg.l⁻¹ of IBA (Sigma Chemical Co., USA) and 20 mg.l⁻¹ of PEG and cultured under white fluorescent light (30 μmol.m⁻².s⁻¹) and 16 h photoperiod. After 3-4 wks the somatic embryos were harvested for desiccation.

2.1.5 Desiccation

Only the fully developed embryos were desiccated. The embryos were carefully transferred on the dry paper in small Petri dishes (3cm in diameter). Open dishes were placed into large Petri dishes (18cm in diameter) with several paper layers wetted by sterile water Large Petri dishes were covered by lids and sealed by parafilm. They were kept under the light regime of 12 hours of light and 12 hours of darkness, at 20±1⁰C for 2 wks.

2.1.6 Material for biochemical analyses

The contents of PAs were determined in the course of somatic embryo development. The samples were immediately frozen in liquid nitrogen and stored at −80 °C until determination.

2.1.7 Polyamine analysis

The cells were ground in liquid nitrogen and extracted overnight at 4 °C with 1 ml of 5% perchloric acid (PCA) per 100 mg fresh weight tissue. 1,7-Diaminoheptane was added as an internal standard, and the extracts were centrifuged at 21 000 x g for 15 min. PCA-soluble free PAs were determined in one-half volume of the supernatant. The remaining supernatant and pellet were acid hydrolysed in 6 M HCl for 18 h at 110 °C to obtain PCA-soluble and PCA-insoluble conjugates of PAs as described by Slocum et al. (1989). Standards (Sigma-Aldrich, Prague, Czech Republic) and PCA-soluble free PAs, and acid hydrolysed PA conjugates were benzoylated according to the method of Slocum et al. (1989), and the resulting benzoyl-amines were analyzed by HPLC using a Beckman chromatographic system equipped with a 125S Gradient Solvent Delivery Module, 507 Variable Mode Injection Autosampler, and 168 Diode Array Detector (Beckman Instruments, Inc., Fullerton, CA, USA). A Gold Nouveau software data system was used to collect, integrate and analyse the chromatographic data. A C18 column (Phenomenex Aqua, 5 μm, 125A, 250x4.6 mm, Phenomenex, Utrecht, NL) was used for the separation of polyamines. Elution was carried out at a flow rate of 0.4 mL min^{-1} at 45 °C. Standard sample (5 μl and 10 μl) was injected for a single run. The mobile phase consisted of solvent A (10% v/v methanol) and solvent B (80% v/v methanol). The gradient program (expressed as percentages of solvent A) was as follows: 0–10 min, 45% to 0%; 10-30 min, isocratic 0%; 30-40 min, 0% to 45%. Column was washed with 45% solvent A for 30 min between samples. Eluted polyamines were detected by UV detector at 254 nm by comparing of their R$_t$ values with those of standards (Sigma-Aldrich, Prague, the Czech Republic).

2.1.8 Statistical analyses

Two independent experiments were carried out. Analogous results were obtained in both experiments. Means ± S.E. of one experiment (with 3 replicates) are shown in the figures. Data were analyzed using the Student's t distribution criteria.

2.2 Results and discussion

Tissue culture approaches, in particular somatic embryogenesis, is considered as the advantageous technique for *in vitro* propagation and gene conservation of conifers. Generally, the development of embryos and their conversion into plantlets is closely associated with changes in endogenous phytohormone levels, including polyamines. Positive correlation between embryogenic capacity and total content of free PAs confirmed to a crucial role of PAs (with Spd predominating, Figs 1, 2, 3) during somatic embryo development in *Picea* as previously described (Gemperlová et al., 2009; S.C. Minocha & Long, 2004). On transfer from proliferation to maturation medium the levels of Put and Spd in the culture of highly responsible cell line AFO 541 were almost equal while in less responsible cultures of Hurst ecotype significantly higher levels of Put than Spd were determined as is apparent from the Put/Spd ratios in the embryogenic suspensor masses (ESM) of six cell lines of Norway spruce (AFO 541 and five of Hurst ecotype) grown on solid proliferation medium (Figs 1, 2). The contents of Put, Spd and Spm in highly responsible cell line AFO 541 steadily increased during maturation from ESM until early cotyledonary stage (Fig. 3). This stage of embryo development was characterized by very high Spd contents. On the contrary to Put and Spd, the level of Spm significantly increased during the desiccation phase (Fig. 3). The increase in Spm level in this phase might result from „certain" abiotic stress in the course of desiccation. The decline in PA contents in mature embryos was probably due to the increased catabolism of Put and Spd during the later stages of embryo development, as previously described in *Picea abies* (Santanen & Simola, 1992).

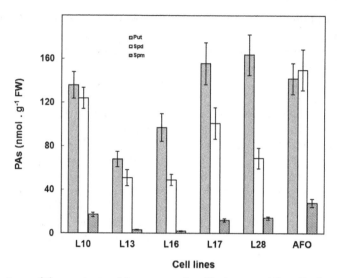

Fig. 1. Comparison of the contents of free putrescine (Put), spermidine (Spd) and spermine (Spm) in the ESMs of six cell lines of Norway spruce (AFO 541 and five of Hurst ecotype) grown on solid proliferation medium (before transfer to maturation medium). Bars represent SE of three replicates. ESM, embryogenic suspensor mass.

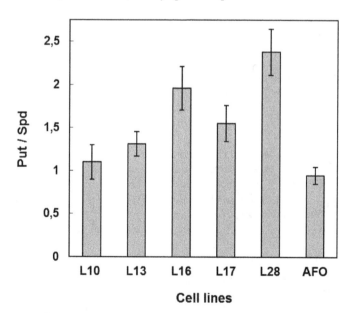

Fig. 2. Comparison of putrescine/spermidine (Put/Spd) ratios in the ESMs of six cell lines of Norway spruce (AFO 541 and five of Hurst ecotype) grown on solid proliferation medium (before transfer to maturation medium). Bars represent SE of three replicates. ESM, embryogenic suspensor mass.

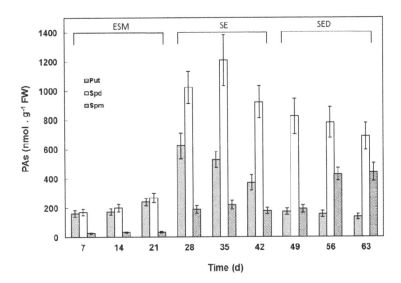

Fig. 3. Changes in cellular levels of free putrescine (Put), spermidine (Spd) and spermine (Spm) during Norway spruce (AFO 541) somatic embryo development from proliferation to desiccation. Bars represent SE of three replicates. ESM, embryogenic suspensor mass; SE, somatic embryo; SED, somatic embryo during desiccation; d, days.

Parallel rises in the content of Spd occurred during the 6 weeks of embryo development of Hurst ecotype cell lines. At this stage of somatic embryo development the predominant PA was Spd followed by Put (Fig. 4), although the mature embryos of highly responsible cell line AFO 541 contained still significantly higher level of Spd. Cell line AFO 514 and Hurst ecotype cell line L10 were characterized by high content of Spd and represented plant material with stable rapid growth during proliferation and a huge yield of somatic embryos was obtained at the end of maturation of cell line L10 (Fig. 5). On the contrary, rather low yield of less matured embryos was found in the remaining studied cell lines of less responsible ecotypes (L16 and L28) which contained lower level of Spd (Fig. 6). However, a high level of free PAs is not the only important PA-related factor in somatic embryogenesis, and (for instance) it has been proposed that an inadequate Spd/Put ratio may be causally linked to abnormal growth and disorganized cell proliferation of grape somatic embryos with high free PA contents (Faure et al., 1991). Less matured somatic embryos and lower yield of embryos in the remaining studied cell lines L13 and L17 (in comparison with the yield of embryos in the cell line L10) might coincide with the inadequate Put/Spd ratio determined in the embryos.

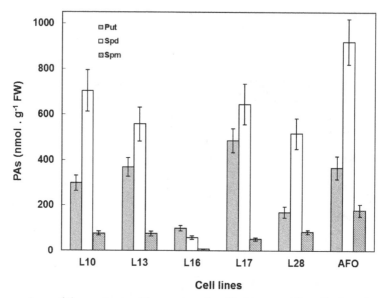

Fig. 4. Comparison of the contents of free putrescine (Put), spermidine (Spd) and spermine (Spm) in 6 week-old somatic embryos of six cell lines of Norway spruce (AFO 541 and five of Hurst ecotype) grown on solid maturation medium.

Fig. 5. Mature somatic embryos (6 week-old) of Hurst ecotype of Norway spruce L10.

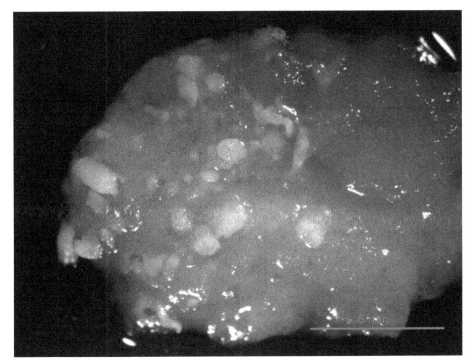

Fig. 6. Mature somatic embryos (6 week-old) of Hurst ecotype of Norway spruce L28.

Similarly, contents of PAs (higher Spd than Put levels) could be used as criteria for the physiological characterization of somatic embryogenesis in *Pinus nigra* Arn. (Noceda et al., 2009). Furthermore, the important role of cellular levels of metabolically active free PAs, especially contents of Spd, was shown to be essential for preservation of embryogenic potential of Norway spruce cultures after cryopreservation (Vondráková et al., 2010). The formation of somatic embryos in tissue cultures of wild carrot seemed to be also associated with high Spd level and much more Spd than Put was found in torpedo stage of these embryos (Mengoli et al., 1989). Especially Spd was implicated in somatic embryogenesis in tissue cultures of *Vigna aconitifolia* (Kaur-Sawhney et al., 1985), *Hevea brasiliensis* (El Hadrami & D'Auzac, 1992) and in the development of globular pro-embryos in alfalfa (Cvikrová et al., 1999). However, it is not always Spd, which is the dominant polyamine in somatic embryos of conifers. Putrescine was the most abundant PA in the embryogenic suspension culture of *Pinus taeda* (Silveira et al., 2004), whereas the development of both somatic and zygotic embryos of *Pinus radiata* was characterized by high level of Spd and its concentration positively correlated with the embryo development (R. Minocha et al., 1999). High level of Put was determined in pro-embryogenic tissue of *Picea rubens*, while Spd was predominant during embryo development in this culture (S.C. Minocha & Long, 2004).

As we have already mentioned in the Introduction, embryo maturation and low germination frequencies are main limitations for a broader use of somatic embryogenesis in practice. Requirement of exogenous phytohormones for efficient somatic embryogenesis is well established. Exogenously supplied polyamines might therefore affect and improve the induction and somatic embryo development in less responsible plant genotypes. It has been found that the exogenous application of Spd in the initiation phase significantly increased the production of embryos in *Panax ginseng* cultures (Kevers et al., 2000). This knowledge led as to try to improve the efficiency of somatic embryogenesis of less responsible genotypes by application of PAs into the growth medium. The studies of the possibility of improving the method of somatic embryogenesis in less responsible Hurst ecotype of Norway spruce are now in progress in our laboratory. Preliminary experiments (results not shown) reveal that low exogenous application of Put (0.1 and 0.01 mM) increased the number of early forms of embryos. However, on the bases of these results the effect of PA application into the growth medium in order to improve the efficiency of somatic embryogenesis of Norway spruce can not be generalized and further experiments are necessary.

The results presented here indicate a direct role for Spd and adequate Put/Spd ratio in somatic embryogenesis. Cell lines AFO 514 and L10, characterized by high content of Spd, represented plant material with stable rapid growth during proliferation and a huge yield of somatic embryos was obtained at the end of maturation. Less matured somatic embryos and rather low yield of embryos in the remaining studied cell lines (L13, L16, L17 and L28) might coincide with the lower level of Spd and/or inadequate Put/Spd ratio determined in these embryos.

3. Conclusion

Micropropagation technologies represent a powerful tool for improvement and acceleration of tree breeding programs in forestry. Tissue culture approaches, in particular somatic embryogenesis, hold considerable promise for breeding programs of coniferous trees. Utilization of somatic embryogenesis could facilitate reproduction of rare or selected coniferous genotypes. Despite the availability of a successful protocol for generating Norway spruce plants using a somatic embryogenesis technique, there is a lack of data concerning the endogenous composition of biologically active compounds in somatic embryos. Beside the key roles of auxins and cytokinins, very important function in differentiation processes belongs to polyamines although the mechanism of their action is still not fully cleared. Positive correlation between embryogenic capacity and total content of free PAs confirmed crucial role of PAs during somatic embryo development.

It was shown that the predominant PAs in somatic embryos of highly responsible Hurst ecotype of Norway spruce (L10) was spermidine, while embryos of less responsible ecotype (L13, L16, L17 and L28) contained lower level of Spd and/or inadequate Put/Spd ratio. Exogenously supplied PAs might therefore affect and improve the induction and somatic embryo development in less responsible plant genotypes. The studies related to the improving of somatic embryogenesis method in less responsible Hurst ecotype of Norway spruce are now in progress in our laboratory.

4. Acknowledgments

The authors acknowledge the funding of the Ministry for Agriculture (Project number QH 82303) and Institutional Grant AV0Z 50380511.

5. Abbreviations

ABA - abscisic acid, BAP - 6-benzylaminopurine, 2,4-D - 2,4 dichlorfenoxy acid, ESM – embryogenic suspensor mass, IBA - indolylbutyric acid, Kin - kinetin, PAs - polyamines, Put - putrescine, Spd - spermidine, Spm - spermine

6. References

Altamura, M.M.; Torrigiani, P.; Falasca, G.; Rossini, P. & Bagni, N. (1993). Morpho-functional gradients in superficial and deep tissues along tobacco stem: polyamine levels, biosynthesis and oxidation, and organogenesis in vitro. *Journal of Plant Physiology*, Vol. 142, pp. 543–551, ISSN 0176-1617.

Attree, S.M. & Fowke, L.C. (1993). Embryogeny of gymnosperms:advances in synthetic seed technology of conifers. *Plant Cell, Tissue and Organ Culture*, Vol. 35, No. 1, pp 1–35, ISSN: 0167 6857.

Attree, S.M.; Moore, D.; Sawhney, V.K. & Fowke, L.C. (1991). Enhanced maturation and desiccation tolerance of white spruce (*Picea glauca* [Moench.] Voss.) somatic embryos: effects of a nonplasmolysing water stress and abscisic acid. *Annals of Botany*, Vol. 68, No. 6, pp 519–525, ISSN 0305-7364.

Baron, K. & Stasolla, C. (2008). The role of polyamines during *in vivo* and *in vitro* development. *In Vitro Cellular & Developmental Biology – Plant*, Vol. 44, No. 5, pp 384-395, ISSN 1054-5476.

Besson-Bard, A.; Courtois C.; Gauthier, A.; Dahan, J.; Dobrowolska, G.; Jeandroz, S.; Pugin, A. & Wendehenne D. (2008). Nitric oxide in plants: production and cross-talk with Ca^{2+} signaling. *Molecular Plant*, Vol. 1, No. 2, pp 218-228, ISSN 1674-2052.

Bornman, C.H. (1985). Hormonal control of growth and differentiation in conifer tissues in vitro. *Biologia Plantarum*, Vol. 27, No. 4-5, pp 249–256, ISSN 006-3134.

Chalupa, V. (1985). Somatic embryogenesis and plantlet regeneration from cultured immature and mature embryos of *Picea abies*/L./Karst. *Communicationes Instituti Forestalis Cechoslovaca*, Vol. 14, pp 65–90, ISSN: 1211-2992.

Chalupa, V.; Malá, J. & Dujíčková, M. (1990). Somatic embryogenesis and regeneration of spruce *(Picea abies* /L./Karst.) and oak *(Quercus robur* L.). In: „Manipulation *in vitro* in higher plants", Proc. Conf. Inst. Exp. Botany, Prague, CSAS, Olomouc, CR: p. 98.

Cohen, S.S. (1998). A guide to the polyamines. pp. 1-595, Oxford university Press, New York, NY.

Cvikrová, M.; Binarová, P.; Eder, J.; Vágner, M.; Hrubcová, M.; Zon, J. & Macháčková, I. (1999). Effect of inhibition of phenylalanine ammonia-lyase activity on growth of alfalfa cell suspension culture: alterations in mitotic index, ethylene production, and contents of phenolics, cytokinins, and polyamines. *Physiologia Plantarum*, Vol 107, No. 3, pp 329–337, ISSN 1399-3054.

El Hadrami, M.I. & D'Auzac, J. (1992). Effects of polyamine biosynthetic inhibitors on somatic embryogenesis and cellular polyamines in *Hevea brasiliensis*. *Journal of Plant Physiology*. Vol. 140, pp 33–36, ISSN 0176-1617.

Find, J.I. (1997). Changes in endogenous ABA levels in developing somatic embryos of Norway spruce (*Picea abies* (L.) Karst.) in relation to maturation medium, dessication and germination. *Plant Science*, Vol. 128, No. 1, pp 75-83, , ISSN 0168-9452.

Flinn, B.S.; Roberts D.R.; Newton C.H.; Cyr D.R.; Webster F.B. & Taylor I.E.P. (1993). Storage protein gene expression in zygotic and somatic embryos of interior spruce. *Physiologia Plantarum*. Vol. 89, No. 4, pp 719-730, ISSN 1399-3054.

Faure, O.; Mengoli, M.; Nougarede A. & Bagni. N.(1991). Polyamine pattern and biosynthesis in zygotic and somatic embryo stages of *Vitis vinifera*. *Journal of Plant Physiology*, Vol. 138, pp 545-549, ISSN 0176-1617.

Gemperlová, L.; Fisherová, L.; Cvikrová, M.; Malá, J.; Vondráková, Z.; Martincová, O. & Vágner, M. (2009). Polyamine profiles and biosynthesis in somatic embryo development and comparison of germinating somatic and zygotic ambryos of Norway spruce. *Tree Physiology*, Vol. 29, No. 10, pp 1287-1298, ISSN 0829-318X.

Gupta, P. K. & Durzan, D. J. (1986). Somatic polyembryogenesis from callus of mature sugar pine embryos. *Bio/technology*, Vol. 4, pp 643-645.

Hakman, I.; Fowke, L.C.; Von Arnold, S. & Eriksson, T. (1985). The development of somatic embryos of *Picea abies* (Norway spruce). *Plant Science*, Vol. 38, No. 1, pp 53-59, ISSN 0168-9452.

Hanfrey, C.; Sommer, S.; Mayer, M.J.; Burtin, D. & Michael, A.J. (2001). Arabidopsis polyamine biosynthesis: absence of ornithine decarboxylase and the mechanism of arginine decarboxylase activity. The Plant Journal, Vol. 27, No. 6, pp 551-560, ISSN 1365-313X.

Igasaki, T.; Sato, T.; Akashi, N.; Mohri, T.; Maruyama, E.; Kinoshita, I.; Walter C. & Shinohara, K. 2003. Somatic embryogenesis and plant regeneration from immature zygotic embryos of *Cryptomeria japonica* D. Don. *Plant Cell Reports*, Vol. 22, No. 4, pp 239-243, ISSN 0721-7714.

Kaur-Sawhney, R.; Shekhawat, N.S. & Galston, A.W. (1985). Polyamine levels as related to growth, differentiation and senescence in protoplast-derivated cultures of *Vigna-aconitifolia* and *Avena-sativa*. *Plant Growth Regulation*, Vol. 3, No. 3-4, pp 329-337, ISSN 0167-6903.

Kaur-Sawhney, R.; Tiburcio, A.F.; Altabella, T. & Galston, A.W. (2003). Polyamines in plants: An overwiew. *Journal of Cell and Molecular Biology*, Vol. 2, No. 1, pp 1-12, ISSN 1303-3646.

Kevers, C.; Le Gal, N.; Monteiro, M.; Dommes, J. & Gaspar, T. (2000). Somatic embryogenesis of *Panax ginseng* in liquid cultures: a role polyamines and their metabolic pathways. *Plant Growth Regulation*, Vol. 31, No. 3, pp 209-214, ISSN 0167-6903.

Kong, L.; Attree, S.M. & Fowke, L.C. (1998). Effects of polyethylene glycol and methylglyoxal bis(guanylhydrazone) on endogenous polyamine levels and somatic embryo maturation in white spruce (*Picea glauca*). *Plant Science*, Vol. 133, No. 2, pp 211-220, ISSN 0168-9452.

Kusano, T.; Berberich, T.; Tateda, C. & Takahashi, Y. (2008). Polyamines: essential factors for growth and survival. *Planta*, Vol. 228, No. 3, pp 367-381, ISSN 0032-0935.

Kumar, A.; Altabella, T.; Taylor, M. & Tiburcio, A.F. (1997). Recent advances in polyamine research. *Trends in Plant Science*, Vol. 2, No. 4, pp. 124-130, ISSN 1360-1385.

Malá, J. (1991). Organogenesis and somatic embryogenesis in Norway spruce. *Communicationes Instituti Forestalis Cechoslovaca*, Vol. 17, pp 59–72, ISSN: 1211-2992.

Malá, J.; Dujíčková, M. & Kálal, J. (1995). The development of encapsulated somatic embryos of Norway Spruce (*Picea abies* /L./ Karst.). *Communicationes Instituti Forestalis Bohemicae*, Vol. 18, pp 59-73, ISSN: 1211-2992.

Mattoo, A.K.; Minocha, S.C.; Minocha, R. & Handa, A.K. (2010). Polyamines and cellular metabolism in plants: transgenic approaches reveal different responses to diamine putrescine versus higher polyamines spermidine and spermine. *Amino Acids*, Vol. 38, No. 2, pp 405-413, ISSN 0939-4451.

Mengoli, M.; Pistocchi, R. & Bagni, N. (1989). Effect of long-term treatment of carrot cell-cultures with millimolar concentrations of putrescin. *Plant Physiology and Biochemistry*, Vol. 27, No. 1, pp 1-8, ISSN 0981-9428.

Minocha, R.; Kvaalen, H.; Minocha, S.C. & Long, S. (1993). Polyamines in embryogenic cultures of Norway spruce (*Picea abies*) and red spruce (*Picea rubens*) Tree *Physiology*, Vol. 13, No. 4, pp. 365-377, ISSN 0829-318X.

Minocha, R.; Smith, D.R.; Reeves, C.; Steele, K.D. & Minocha, S.C. (1999). Polyamine levels during the development of zygotic and somatic embryos of *Pinus radiata*. *Physiologia Plantarum*, Vol. 105, No. 1, pp 155-164, ISSN 1399-3054.

Minocha, S.C & Long, S. (2004). Polyamines and their biosynthetic enzymes during somatic embryo development in red spruce (*Picea rubens* Sarg.) *In Vitro Cellular & Developmental Biology-Plant*, Vol. 40, No. 6, pp 572-580, ISSN 1054-5476.

Noceda, C.; Salaj, T.; Pérez, M.; Viejo, M.; Cañal, M.J.; Salaj, J. & Rodriguez R. (2009). DNA demethylation and decrease on free polyamines is associated with embryogenic capacity of *Pinus nigra* Arn. cell culture. *Trees – Structure and Function*, Vol. 23, No. 6, pp 1285-1293, ISSN 0931-1890.

Santanen, A. & Simola, L.K. (1992). Changes in polyamine metabolism during somatic embryogenesis in *Picea abies*. *Journal of Plant Physiology*. Vol.140, pp 475-480, ISSN 0176-1617.

Serapiglia, M.J.; Minocha, R. & Minocha, S.C. (2008). Changes in polyamines, inorganic ions and glutamine synthetase activity in response to nitrogen avaibility and form in red spruce (*Picea rubens*). *Tree Physiology*, Vol. 28, No. 12, pp 1793-1803, ISSN 0829-318X.

Silveira, V.; Floh, E.I.S.; Handro, W. & Guerra, M.P. (2004). Effect of plant growth regulators on the cellular growth and levels of intracellular proteins, starch and polyamines in embryogenic suspension cultures of *Pinus taeda*. *Plant Cell, Tissue and Organ Culture*, Vol. 76, No. 1, pp 53-60, ISSN 0167-6857.

Slocum, R.D.; Flores, H.E.; Galston, A.W. & Weinstein, L.H. (1989). Improved method for HPLC analysis of polyamines, agmatine and aromatic monoamines in plant tissue. *Plant Physiology*, Vol. 89, No. 2, pp 512–517, ISSN 0032-0889.

Tabor, C.W. & Tabor, H. (1984). Polyamines. *Annual Review of Biochemistry*, Vol. 53, pp 749-790, ISSN 0066-4154.

Tiburcio, A.F.; Altabella, T.; Borrell, A. & Masgrau, C. (1997). Polyamine metabolism and its regulation. *Physiologia. Plantarum*, Vol. 100, No. 3, pp 664-674, ISSN 1399-3054

Vágner, M.; Vondráková, Z.; Fischerová, L.; Vičánková, A. & Malbeck, J. (2005). Endogenous phytohormones during Norway spruce somatic embryogenesis. Procceding COST 843 and 851 Action, Stará Lesná (28.6.-3.7.), pp. 162-164.

Vera-Sirera, F.; Minquet, E.G., Singh, S.K.; Ljung, K., Tuominen, H., Blazquez, M.A. & Carbonell, J. (2010). Role of polyamines in plant vascular development. *Plant Physiology and Biochemistry*, Vol. 48, No. 7, pp 534-539, ISSN 0981-9428.

Vondráková, Z.; Cvikrová, M.; Eliášová, K., Martincová, O. & Vágner, M. (2010). Cryotolerance in Norway spruce and its association with growth rates, anatomical features and polyamines of embryogenic cultures. *Tree Physiology*. Vol. 30, No 10, pp 1335-1348, ISSN 0829-318X.

3

Somatic Embryogenesis in Forest Plants

Katarzyna Nawrot-Chorabik
Department of Forest Pathology, Faculty of Forestry,
University of Agriculture in Kraków, Kraków
Poland

1. Introduction

Somatic embryogenesis has become an increasingly applied *in vitro* method in plant breeding in many laboratories across the world. It provides potentially high micropropagation efficiency and other possibilities, such as: cryopreservation of plant material (Hargreaves & Menzies 2007, Misson *et al.* 2006), obtaining secondary metabolites (Mulabagal & Tsay 2004), obtaining valuable and selected planting stock in short time (Rodriguez *et al.* 2007), obtaining transformed plants (Walters *et al.* 2005) and conducting studies on pathogenicity on embryonic level (Hendry *et al.* 1993; Nawrot-Chorabik *et al.* 2011). Hence, there is considerable interest in application of this vegetative micropropagation method in many tree species.

When defining the method of somatic embryogenesis, it needs to be stressed, that this is an *in vitro* morphogenesis, in which adventive embryos, which are not the product of gametic fusion, are formed from plant somatic cells. This method is based on a theory of plant cell totipotency, according to which there is unlimited ability of living cells to divide and to reproduce the whole organism.

The first research on somatic embryogenesis was carried out in the 50s of the 20th century, when the first somatic embryos were obtained in carrot (*Daucus carota*) (Steward 1958). Seven years later embryos of a deciduous species – sandalwood (*Santalum album*) were obtained (Rao 1965). Pioneering studies on somatic embryogenesis in conifers were conducted in Canada in the period of 1968–1980 (Durzan & Steward 1968, Chalupa & Durzan 1973, Durzan & Chalupa 1976). In 1985 Hakman *et al.* (1985) and Chalupa (1995) initiated somatic embryogenesis in European spruce. In recent years *in vitro* studies of woody plants have become more important, since they created new perspectives for development of many industries, such as: pharmacy (e.g. obtaining Taxol from European yew) (Cusidó *et al.* 1999), cosmetology (e.g. obtaining Juglone from walnut and saponins from Conker tree) (Wilkinson and Brown 1999) or production of Christmas trees and decorations (establishment of plantation areas of Nordmann fir) (Misson *et al.* 2006).

Intensive research on improving and on the potential of somatic embryogenesis of economically important tree species are carried out on the following genera: *Abies* (Nawrot-Chorabik 2008; 2009; Salaj & Salaj 2003/4), *Picea* (Klimaszewska *et al.* 2010; Mihaljević & Jelaska 2005), *Pinus* (Lelu-Walter *et al.* 2008; Klimaszewska *et al.* 2001), *Taxus* (Nhut *et al.*

2007), *Acer* (Ďurkovič & Mišalová 2008), *Castanea* (Corredoira *et al.* 2003), *Quercus* (Toribo *et al.* 2005), *Salix* (Naujoks 2007) and *Ulmus* (Ďurkovič & Mišalová 2008; Mala *et al.* 2007).

Before entering the material from *in vitro* cultures into a commercial scale, it is required to conduct long-term observation of growth and development of large amount of somatic seedlings, that represent significant number of genotypes. This allows valuable species of trees to be produced *in vitro* in commercial tissue culture laboratories around the world, i.e. in United States of America (Plant Tissue Cultures Lab – West Lafayette) and in Canada, where Park et al. (2001) included somatic seedlings of white spruce (*Picea glauca*) – the most widespread species in this country - into the program of forest tree breeding and selection. Moreover, somatic seedlings are produced in Great Britain (Date Palm Developments), Israel (Ginosa Tissue Culture Nurseries Ltd.), France – planting stock of *Pinus pinaster* for establishment of forest cultivation (Cyr & Klimaszewska 2002) and in Poland (Tissue culture laboratory Vitroflora in Łochowo), in Italy (Department of Plant Production Di.Pro.Ve.) and in many other countries.

The purpose of this chapter is to present information on somatic embryogenesis of trees in a concise manner, which should introduce the reader into the most important aspects related to this topic. Various stages of the method will be explained indicating the difficulties encountered during *in vitro* culture of trees. Additionally, external factors affecting the breeding success will be discussed. Finally, a short history of research on somatic embryogenesis will be presented.

Concisely presented information on somatic embryogenesis of trees will introduce the reader to key concepts of this topic, briefly present the history of the research on this method, it will also explain each stage of the process indicating the difficulties encountered during *in vitro* cultures of trees with somatic embryogenesis and discuss external factors affecting the success of cultures.

All these aspects will help to identify the most appropriate future research directions in *in vitro* cultures of trees and to introduce the importance of somatic embryogenesis as an alternative method for vegetative reproduction and its contribution to the plant biotechnology development.

2. Material and experimental procedures

2.1 Primary explants used for initiation of *in vitro* cultures with somatic embryogenesis

The term "primary explant" refers to the initial plant material inoculated on a medium, i.e. a fragment of a plant from which the *in vitro* culture was initiated (other plant fragments are called secondary explants). The following types of primary explants may be used in the somatic embryogenesis method: mature zygotic embryos isolated from mature seeds of trees, megagametophytes - immature seeds collected from immature cones with embryo and endosperm, buds, needles or leaves of trees and progenitor cells. Next, the following aspects need to be considered when choosing the primary explant: age of tissues and organs of the parent plants. Most preferably, young organs should be collected, because they have grater potential for development. The minimum storage time, particularly for megagametophytes and seeds of coniferous trees, needs to be reduced. The highest frequency of embryogenesis is obtained from "fresh" seeds and megagametophytes. The location within the plant is also important – in the case of buds – initial, developed leaf buds should be collected. Moreover, the location of primary explant on the medium is significant too – zygotic embryos should be placed on solidified media in a horizontal position, since there is variation in the explants' development on media, which results from natural polarity of plant fragments (Tab. 2). The Author's own research showed, that mature zygotic embryos of silver fir, which did not adhere strictly to the medium, did not produce callus or the initiated callus tissue quickly decays. Another significant factor is the date of explant collection. Generally: in a temperate climate buds need to be collected in the early spring, megagametophytes should be collected from closed cones in June, while mature zygotic embryos should be isolated from non-stored seeds, acquired immediately after physiological maturity. This phenomenon is associated with the natural biological rhythm of parent plants, which affects the effect of embryogenesis. From a physiological point of view this is associated with the period of intensity of most metabolic and enzymatic activity of cells, which is the most intensive in spring. For example, the ability of the explants of white poplar (*Populus alba*) to form callus is maintained at high levels from spring to autumn, while it decreases in winter.

2.2 Disinfection of plant material

In the case of forest trees it is difficult to optimize the method of explants' disinfection due to large contamination of most plant organs with bacteria and endophytic fungi (Kowalski & Kehr 1992). The disinfection procedure should be optimized for a specific tree species, and even for the type of primary explant, for which the chemical agent will be effective against microorganisms. Disinfection of forest trees' explants needs to be conducted in several stages. Disinfection time is sometimes quite long (up to 2-day). Based on many experiments, the Author recommends that during disinfection the seeds of coniferous trees should be kept at 4°C for 24 hours in sterile water with the addition of ascorbic acid or PVP (Polyvinylpyrrolidone), which act as antioxidants. This promotes easier isolation of zygotic embryos, but also embryos inoculated onto media produce smaller amount of phenolic glycosides, which disrupt the process of callogenesis (Pict. 1d). Disinfecting solutions should be supplied with substances that reduce surface tension and facilitate the penetration of the surface of plant material, e.g. Tween 80. In specific cases, plant material is additionally disinfected with solutions of fungicides or antibiotics, which may sometimes have negative impact on reducing the frequency of callus initiation.

Explants' disinfection stages of forest trees can be presented in the following way:

-   initial disinfection: explants should be rinsed with running water in temperature about 18°C for 30 minutes (by the end with the addition of Tween 80) to get rid of the resin which in varying degrees covers the ligneous plant material. In some cases it is recommended to use brushes and sometimes even 2-second flaming is recommended to remove the epidermal products.
-   proper disinfection: it is performed under sterile conditions in a laminar air flow chamber Biohazard, in 70% solution of ethyl alcohol, followed by the selected disinfectant, e.g. sodium hypochlorite - NaOCl, calcium hypochlorite - Ca(OCl)$_2$, hydrogen peroxide – H$_2$O$_2$ or mercuric chloride (sublimate) – HgCl$_2$ (Tab. 2). Finally, the explants should be 3-5 times rinsed with sterile deionized water. Seeds should be tightly closed in a beaker with sterile deionized water with the addition of ascorbic acid

or PVP and placed at 4°C. After 24 hours in sterile conditions the primary explants are inoculated on the medium for initiation.

2.3 Chemical composition of media

Choosing the right type of culture medium, that contains the optimum concentrations of growth regulators for the species, is a key factor to achieve the desired effects in *in vitro* cultures of trees. In almost every step of somatic embryogenesis the composition of basic media needs to be modernized and plant hormone concentrations need to be adjusted. Culture media applied for woody species are rich in macro- and microelements, vitamins, carbon source and growth and development regulators, and sometimes a source of amino acids (enzymatic digest of casein) (Tab. 1).. Activated charcoal– AC is used in some stages of somatic embryogeniesis, usually during the change in the medium composition between the successive stages of embryogenesis. The consistency of media is usually solid and in bioreactors media are liquid. The pH of media is within the range of 5.6 – 5.8. Macro-and micronutrients and vitamins are prepared in a concentrated form of so-called stock solutions, that can be portioned in 10 or 100 dm^3 and stored at minus 20°C in plastic bags.

Macronutrients (N, K, P, Ca, Mg, S) are added in concentrations up to $3000 mg \times dm^{-3}$ in the form of inorganic salts. For induction of somatic embryogenesis it is necessary to maintain balance between cations – NH_4^+ and anions – NO_3^-. Macronutrients are necessary for synthesis of proteins, nucleic acids and for proper functioning of the water balance of plant cells. They ensure appropriate cytoplasmic membrane permeability and are involved in the synthesis of chlorophyll.

Micronutrients (Fe, Cu, Zn, Mn, B, Mo, I, Al) are added to media in concentrations from 0.03 to 100 $mg \times dm^{-3}$, in the form of inorganic salts. However, too low concentration of micronutrients in a medium may inhibit proliferation of embryogenic callus or its dieback. The Author's own research indicates that frequency and quality of embryogenic callus of trees may be increased by adding higher concentrations of zinc (Zn) in hydrated form into the media: $ZnSO_4 \times 7H_2O$. Micronutrients are essential for the synthesis of chlorophyll, they are involved in the functioning of chloroplasts, and assimilation of atmospheric nitrogen.

Vitamins such as thiamine (vitamin B_1), nicotinic acid (vitamin B_3), pyridoxine (vitamin B_6), folic acid (vitamin B_9) and myo-inositol (isomeric form of vitamin B_8, precursor of vitamins) and biotin (vitamin H) aim to improve the physiological condition of cells, they are also necessary for the proliferation of fir (*Abies*) and pine (*Pinus*) callus.

Disaccharides, mainly sucrose and sometimes maltose, are the carbon source in the media, necessary for synthesis of organic compounds. Carbohydrates act also as osmotic balance stabilizers of the media, which affect the absorbance of substances influencing the embryogenic cells' development (Tab. 1).

Substances that solidify media for forest trees are most frequently Phytagel and less frequently agar – natural extract from red algae. The concentration of these substances in the medium is important for the correct development of callus. Too high concentration hinders diffusion, and hence reduces the availability of nutrients for cells, while too low concentration favors the occurrence of "vitreous" explants i.e. callus is too hydrated, it is characterized by anatomical and physiological anomalies. Genus *Pinus* is an exception among coniferous trees, for which the medium should be less solidified. Based on the own research, the Author recommends application of Phytogel (Sigma-Aldrich) for the genus *Pinus* in the concentration of 3.8 g x dm^{-3}.

Growth regulators, so called phytohormones - auxins, cytokines, gibberellins and inhibitors affect callus growth and development through regulation of gene expression. An auxin to cytokine ratio is of particular importance in the early stages of embryogenesis. The presence of auxins: 2,4–D (2,4-dichlorophenoxy acetic acid), IBA (indolyl-3-butyric acid), picloram (4-amino-3,5,6-trichloro-2-pyridinecarboxylic acid) is essential for the induction of somatic embryogenesis and rooting of somatic embryos in the cotyledonary stage. The role of auxins is to stimulate the differentiation of primary explants, which leads to unleashing the embryogenic potential of cells. Such cells rapidly divide and form clusters of embryonic cells. Cytokines such as BA (benzylaminopurine), KIN (kinetin), TDZ (thidiazuron) promote the proliferation of callus and somatic embryo formation in the globular stage. Cytokines stimulate the biosynthesis of nucleic acids, structural proteins and enzymes, inhibit the activity of ribonuclease and protease, and accelerate cell division.

On the other hand, inhibitors such as ABA (abscisic acid) are used in the further stages of somatic embryogenesis. They cause the maturation of somatic embryos through globular, heart-shaped, torpedo and cotyledonary stages (Fig. 1), and in the final stage - their development into the seedling. Moreover, ABA increases the resistance of cells to stress conditions.

3. Stages of somatic embryogenesis of trees

The method of somatic embryogenesis is a multi-stage process. It consists of 5 basic phases (Fig. 1). For a planned outcome of the subsequent stages of embryogenesis, the media need to be selected and optimized for each phase separately. The selected medium, whose composition depends on the species of a tree, must be enriched with optimized concentration of growth regulators to enforce specific organogenetic changes in the plant material. Furthermore, one needs to determine what environmental conditions should be adjusted for the success of *in vitro* culture for each morphogenetic level.

3.1 Embryogenic callus initiation

Initiation, also called indution, is a process of unleashing the embryogenic potential of a single cell or a group of cells. Embryogenic callus, in trees having the form of floculent, usually transparent or white, often viscous, well-hydrated mass, originates on the initial explant (Pict. 1a). The first formed embryogenic cells are called proembryogenic masses (PEM). For about 2-3 weeks the initiated embryogenic callus grows on the initiation medium and then embryogenic cell mass is formed, otherwise known as embryogenic suspensor mass (ESM) (Pict. 1b). It creates a shapeless mass of rapidly dividing cells differentiated in size and shape (from izometric cells to loosely bound, large cells). Callus initiated on a single explant is called a line, which is a single genotype with single set of chromosomes. In this phase, it must be determined whether the callus is embryogenic, by staining with acetocarmine and microscopic observations of proembryos (Gupta & Durzan 1987), (Pict. 1c). In the callus of gymnosperm trees, zones of embryogenic masses may be distinguished – small cells whose nuclei stain red and long, colorless cells with small nuclei and large vacuoles. In some cases, one can find polyploid cells in the callus, originating from selective effect of cytokines, which favors the development of micro-seedlings with unfavorable characteristics.

Components:	Gupta & Durzan (1985) DCR	Litvay et al. (1985) LV	Bornman & Jansson (1981) MCM	Murashige & Skoog (1962) MS	Schenken & Hildebrandt (1972) SH
Macronutrients [mg/dm^3]					
NH_4NO_3	400	1601	-	1650	-
KNO_3	340	2022	2000	1900	2500
$CaCl_2 \times 2H_2O$	85	22	-	440	200
$Ca(NO_3)_2 \times 4H_2O$	556	-	500	-	-
$MgSO_4 \times 7H_2O$	370	1849	125	370	400
KH_2PO_4	170	408	135	170	-
$(NH_4)_2SO_4$	-	-	400	-	-
$NH_4H_2PO_4$	-	-	-	-	300
$Na\,H_2\,PO_4 \times H_2O$	-	-	-	-	-
$FeCL_3 \times 6H_2O$	-	-	-	-	-
$K_2\,SO_4$	-	-	-	-	-
KCl	-	-	75	-	-
Micronutrietns [mg/dm^3]					
KI	0.83	4.1	0.25	0.83	1.0
H_3BO_3	6.20	30.9	1.5	6.2	5.0
$MnSO_4 \times 4H_2O$	-	27.9	0.17	22.3	13.2
$MnSO_4 \times H_2O$	22.30	-	-	-	-
$ZnSO_4 \times 7H_2O$	86.5*	86.5*	3.0	8.6	86.5*
$Na_2MoO_4 \times 2H_2O$	0.25	1.21	0.25	0.25	0.1
$CuSO_4 \times 5H_2O$	0.25	0.5	0.025	0.025	0.2
$CoCl_2 \times 6H_2O$	0.03	0.12	0.025	0.025	0.1
$Na_2 \times EDTA$	37.30	37.3	37.25	37.3	20.0
$FeSO_4 \times 7H_2O$	27.80	27.8	27.85	27.8	15.0
Vitamins and other organic compounds [mg/dm^3]					
Meso-inositol	-	99.1	90	100	1000
Nicotinic acid	0.5	0.49	0.6	0.5	5.0
Pyridoxine	0.5	0.1	1.2	0.5	0.5
HCl	1.0	0.1	1.7	0.1	5.0
Thiamine HCl	2.0	-	2.0	2.0	-
Glycine Panthoten	-	-	0.5	-	-
Biotine	-	-	0.125	-	-
Folic acid	-	-	1.1	-	-
Glutamine	50.0	-	500	-	-
Enzymatic digest of casein	1000*	1500*	-	-	1500*
Sucrose	10 g ×l^{-1}	20 g×l^{-1}	20 g×l^{-1}	30 g×l^{-1}	20 g×l^{-1}
pH	5.8	5.7	5.6	5.8	5.6

Table 1. Examples of media used for embryogenic callus *in vitro* initiation (Author's own modification *)

Tree (Genus)	Primary explant	Disinfectant/ time of disinfection	Initiation medium
Coniferous trees:			
Abies	embryos isolated from mature seeds; megagametophytes	sodium hypochlorite - NaOCl 5 - 10% (5 - 30 min.)	SH[1], MCM[2],
Picea	embryos isolated from mature seeds; leaf buds; protoplasts	sodium hypochlorite 10% (15 min.) calcium hypochlorite 7% (20 min.)	MS[3], BM-3[4]
Pinus	embryos isolated from mature seeds;	hydrogen peroxide – H_2O_2 7 - 12% (5 15 min.)	DCR[5], LV[6]
Taxus	embryos isolated from mature seeds	sodium hypochlorite - NaOCl 5 - 10% (5 - 30 min.); calcium hypochlorite - $Ca(OCl)_2$ 5 - 10% (5 - 30 min.)	MS, WPM[7]
Deciduous trees:			
Betula	leaves, seeds	sodium hypochlorite 3 - 10%; calcium hypochlorite 7%	WPM, MS, N7[8]
Castanea	leaves	sodium hypochlorite 3 - 10%	GD[9], MS
Fagus	immature zygotic embryos isolated from seeds	commercial bleach (40 g x dm[-3])	WMP (1/2 concentration)
Quercus	mature embryos regenerating into shoot	mercuric chloride 0,1%	WPM

[1] - Schenken & Hildebrandt (1972); [2] – Bornman & Jansson (1981);
[3] - Murashige & Skoog (1972); [4]- Gupta & Durzan (1986); [5]- Gupta & Durzan (1985);
[6] - Litvay et al. (1985); [7] - Lyoyd & McCown's (1981); [8] – Simola (1985);
[9] - Gresshoff & Doy (1972)

Table 2. Most commonly used types of primary explants, disinfectants and media for the initiation of *in vitro* culture by somatic embryogenesis, depending on the generic name of a tree

3.2 Callus proliferation

Callus proliferation is necessary to obtain suitable quantities of embryogenic tissue. Cell proliferation is a result of callus passages (each 2–3 weeks) onto fresh media of identical or amended composition in relation to the composition of initial medium. Increasing the concentration of cytokines in the medium for propagation has a beneficial effect on callogenesis because these regulators stimulate cell division processes. Enzymatic digest of casein added to the medium in quantities of 1000 – 1500 mg x dm³, particularly in the case of gymnosperm trees, often has a beneficial effect on proliferation of callus (Nawrot-Chorabik 2008), (Fig. 1).

Not all callus lines are embryogenic and are capable of intensive proliferation. Only certain genotypes are characterized by high frequency of embryogenesis (Pict. 1e). The origin of plant material, particularly seeds, has a significant impact on embryogenic capacity of callus. First somatic embryos are formed in a globular stage. This is the induction of somatic embryos (Pict. 1f). Also the explant itself, e.g. a leaf, has a particular meaning in the process of embryo induction on a proliferated callus. Younger – the innermost leaf tissues (constituting the primary explant) produce callus with large quantities of somatic embryos, while further - more external - parts of leaves may produce smaller callus or the embryos are induced directly on this leaf fragment (Trigiano & Gray 2011).

3.3 Conversion of somatic embryos, gene expresion

Globally speaking, the phenomenon of conversion is understood in two ways – as the development of somatic embryos into plants of identical genotype as initial explant capable of *ex vitro* growth and development, morphologically developed - with a root, apical bud and first assimilation organs (Becwar et al. 1989) - and more rarely as survivability of seedlings regenerated after inoculation and adaptation to *ex vitro* conditions. For somatic embryogenesis of forest trees conversion also refers to the successive stages of somatic embryos' development. The following stages may be distinguished: globular, heart-shaped, torpedo, early-cotyledonary and cotyledonary (Fig. 1, Pict. 1 g, h). Subsequent stage of somatic embryos development (maturation) has gained importance due to its aftermath. With proper embryonic morphogenesis in the process of micropropagation one can obtain valuable plants with large capacities for uniform and rapid germination, with normal growth in the broadly understood range of environmental factors. Properly developed seedlings in terms of physiology are formed not only from embryos, that have the appropriate morphology, but also that gathered the necessary amount of reserve material. Otherwise, often only a rootless shoot is developed. In woody plants the embryo development starts from small clusters of embryogenic cells called proembryogenic masses (PEM I), composed of cells with dense cytoplasm, adjacent to a single vacuolated cell showing tendency to elongation. After about three days further elongated cells develop from a group of cells with dense cytoplasm and form PEM II. Then, after about two weeks large aggregates of cells are formed, classified as PEM III (von Arnold & Clapham 2008). Reducing the amount of auxins and cytokines sometimes stimulates the differentiation of somatic embryos, but the rule is that the addition of abscisic acid (ABA) to the medium is necessary to obtain a cotyledonary embryo. The first visible response of somatic embryos to abscisic acid is their change to colorless. From this point the embryo begins to elongate and form cotyledons.

During the conversion of somatic embryos **gene expression** occurs – which, if started at the right time, ensures proper construction and development of the embryo. So far the following types of genes have been identified: genes responsible for cell cycle and cell wall synthesis, genes responsive to hormones and transcription process associated with somatic embryogenesis. Cell division and growth requires a strict control in time and space.

Expression of genes responsible for cell-cycle process is therefore important for the further development of the embryo. Genes responsible for cell wall synthesis are also important, as somatic embryogenesis depends on proper formation of cell wall components. Among others, the following genes are included in the group responsible for cell-cycle and cell wall synthesis: cdc2M, CEM6, SERP, AGP. The expression of these genes at the right time ensures proper construction of the embryo (Yiang & Zhang 2011). Induction and growth of somatic cells can be stimulated by appropriate hormones that affect hormone-sensitive genes, among which one can distinguish genes responsive to abscisic acid (ABA) e.g. LEA, but also genes responsive to auxins (indolyl-3-acetic acid – IAA and picloram - PIC). These genes include GH3, PIN, ARF, SAUR. The proper course of somatic embryogenesis requires genes that regulate the individual stages and the entire process. These include transcription factors associated with somatic embryogenesis, such as LEC, BBM, WUC, AGL15. All these factors ensure the proper development of somatic embryo (Yiang & Zhang 2011).

3.4 *In vitro* seedling rhizogenesis

Only somatic embryos in cotyledonary stage with clearly developed cotyledons, hypocotyl and primordial root, with morphology similar or identical to zygotic embryo, will germinate and develop into somatic seedlings. Such embryos are transferred onto germination medium. The majority of these are media poor in macro- and micronutrients and sugar, often without growth regulators. These media should be supplemented with auxin, which acts as root inducer - *indole-3-butyric acid* (IBA). Some species, particularly gymnosperm trees, require additional treatments during the rhizogenesis stage, such as drying of embryos (so called desiccation) under conditions of high humidity (ca. 95%) or higher concentrations of solidifiers in the culture media. These treatments cause that the developed embryos in the cotyledonary stage have proper turgidity, which enhances their ability to germinate. According to the literature, the germination ability of somatic embryos obtained by somatic embryogenesis is relatively low and the average is around 15% (Cornu & Geoffrion 1990; Salajova et al. 1995).

3.5 Acclimation to environmental conditions

Adaptation of developed forest tree seedlings to *ex vitro* conditions is difficult due to physiological determinants of young tree seedlings. Their slenderness due to lack of woody tissue and the covering tissue – cuticle, poorly developed root system and assimilation apparatus and significant hydration of the tissues causes instability of seedlings in a new medium – cellulose-peat pots. The cultivated seedlings are transferred from media to pots with a volume of 0.065 liters distributed under different names, e.g. Fetlipots, Finnpots, Jiffypots. The substrate in pots should be watered with basic medium diluted in a 1:1 ratio. Due to the above-mentioned physiological determinants of seedlings, the following treatments should be used to facilitate acclimation: undercooling (important for conifer species) that prevent growth interruption, increasing the light intensity and the use of fogging and variable conditions of light (photoperiod) and temperature in computer-controlled greenhouses. One can also introduce LED (Light Emitting Diode) illumination. For better growth and development of seedlings of gymnosperm species, the Author recommends white LED light with color temperature of 5000 – 6500 K.

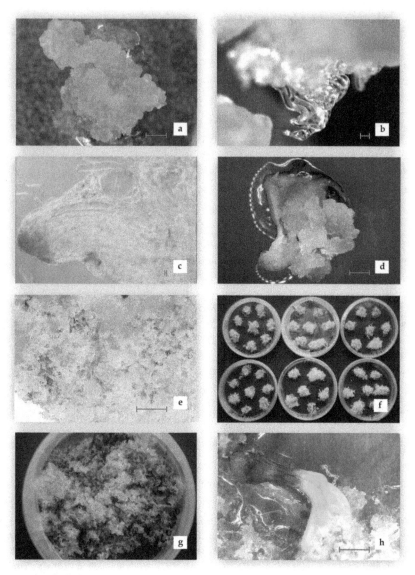

Picture 1. Initiated embryogenic callus on a zygotic embryo of *P. sylvestris* (a); characteristic, hydrated embryogenic callus of gymnosperm trees, the example of *P. sylvestris* (b); elongated embryogenic cells of *Abies alba* (c); callus of *Pinus sylvestris* with evident brown fragments resulting from the produced polyphenols (d); proliferated clones of *Abies nordmaniana* - genotype No. 19 (e); embryogenic callus of *Abies nordmaniana* with induced somatic embryos in globular stage (f); embryogenic callus of *Abies nordmaniana* with embryo in torpedo stage (g); embryogenic callus of *Abies nordmaniana* with embryos in cotyledonary stage (h); Bars a, d = 10 mm; Bar b = 5000 μm; Bar c = 1000 μm; Bar e, h = 15 mm; (Pictures: a – d, h T. Kowalski; e – g B. Chorabik)

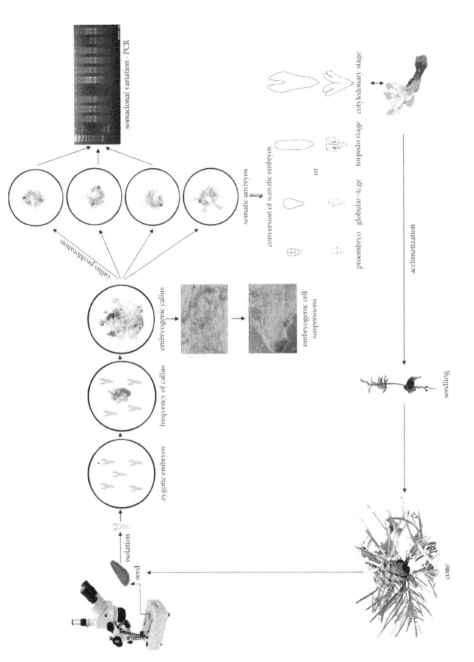

Fig. 1. Development phases of somatic embryogenesis methods

4. Physical conditions of *in vitro* cultures

The physical conditions affecting the state of *in vitro* culture during plant tissue morphogenesis are the abiotic factors such as temperature, light, relative humidity, pH, oxygen and carbon dioxide concentration. These factors must be closely coordinated, but also they must be coordinated with other chemical factors such as e.g. composition of media. For the somatic embryogenesis culture to be successful, usually a constant **temperature** is maintained *in vitro* both during the day and night. Only in rare cases it is necessary to apply proper temperature variation. Most frequently the temperature in a phytotron chamber varies from 23 to 25°C. The optimum temperature, however, should be determined experimentally, depending on the tested species and primary explant. Each stage of the culture may require different temperature. Furthermore, it should be noted that the temperature inside the room with cultures is a few degrees lower than inside the vessel with explants. Therefore, tree explants respond better to the temperature slightly decreased in relation to its optimum temperature than to increased temperature. The **light** impacts the morphogenetic changes, that often are induced by this important factor. The first two stages of somatic embryogenesis in most species of trees progress without light, due to the fact that development process of embryogenic callus does not require intensive photosynthesis. Biotechnological Laboratory of *in vitro* Cultures in the Department of Forest Pathology, University of Agriculture in Cracow, carries out research on the impact of light wavelength and light intensity on the morphogenetic changes in embryogenic callus with somatic embryos of basic, forest-forming gymnosperm tree species of Poland (fir, spruce, pine). It was found that white, diffused, low intensity LED light, which is a mixture of various wavelengths (380–780 nm), in 12-hours' photoperiod is needed only during the conversion of embryos. White light is the most favorable due to the similarity to the prevailing natural conditions. It matches the range of photosynthetically active light (Photosynthetically Active Radiation – PAR) with a wavelength of 400 – 700 nm. It affects the induction of chlorophyll synthesis, chloroplast development and formation of adventitious organs from callus cells. In the beginning of later stages, namely during rhizogenesis, darkness is required (similarly to initiation and proliferation of callus). Only after 10-14 days, the seedlings need to be transferred to white LED light (about 10 times lower intensity than in natural conditions). During the acclimation of somatic seedlings the plants should be placed in a higher intensity white LED light – of intensity similar to natural conditions, optimal for each species. Moreover, it was experimentally demonstrated that blue LED light in the wavelength of 440-490 nm, used for the 12-hour photoperiod has beneficial effect on callus with forming embryos. Embryogenic callus passaged onto activated charcoal medium without plant hormones for the period of 10 - 14 days, kept in blue light, is easier to purify from growth regulators obtained from proliferation medium. Thanks to blue light in the later stage, i.e. conversion, morphogenetic processes are launched more rapidly and the matured somatic embryos are correctly transformed into somatic seedlings in cotyledonary stage (Nawrot-Chorabik, unpublished data). The mechanism of biochemical processes occurring in embryogenic callus with developing embryos exposed to light of different wavelengths is not fully understood, therefore this issue should be carefully investigated. Other light wavelengths, i.e. green light (490-560 nm) and yellow light (560-590 nm) sometimes stimulates formation of adventitious buds from hypocotyl fragments. Conducting *in vitro* culture with different wavelengths argues for the introduction of light parameters (wavelength, light intensity and exposure time) control in phytotrons for better plant growth. **Relative humidity** - RH determines the content of water vapor in the gas phase of the vessel in which the culture is conducted (above the medium surface). RH depends among others on the temperature, chemical composition of a medium, size of explants and vessels. Once the medium and the interior of e.g. a Petri dish have the same temperature, and the vessel is sealed, then the relative humidity should theoretically be 98-99%. However, glass and plastic materials used *in vitro* are not sufficiently leakproof and water vapor gradually escapes on the outside. Therefore, phytotrons with humidity settings should be used, which in the case of micropropagation of trees should oscillate between 50 and 70%. A **pH** of media has significant impact on *in vitro* cultures of trees. Callus and somatic embryos of woody species are formed in acidic pH, i.e. within the range of pH 5.6 – 5.8. pH of a medium may change during the culture. Such changes may be observed particularly in liquid media, and in solid media they may result from too rarely conducted passages. **Oxygen** (product of photosynthesis) and **carbon dioxide** (product of respiration) are two gasses, components of air occurring at a concentration of 21% and 0.036%. However, in Petri dishes or in Erlenmeyer's flasks, in which the culture is carried out, the concentration of these gasses depends among others on: the size of explant, and thus the intensity of photosynthesis, respiration or transpiration and the composition of media (mainly carbohydrate content), light, temperature and the size and shape of the vessel. It is most preferred to maintain oxygen concentration in the vessel at a level higher than its concentration in the air. It was demonstrated that under such conditions, i.e. at the concentration of oxygen within the range of 60-70%, intensive cell divisions occur and the amount of callus, adventitious shoots and somatic embryos increases. Lower than in the air oxygen concentration generally inhibits the plant development. Even roots, which naturally grow in conditions of oxygen deprivation, develop more intensively on the medium in sufficient oxygen supply. Characteristic and frequently observed growth of roots over the level of the medium suggests a lack of oxygen. Sometimes, however, low level of oxygen (7.8%) induces formation of lower amounts of callus, but with the majority of embryogenesis-competent cells. Proper CO_2 concentration primarily determines the proper course of photosynthesis. It has been shown that concentration of this gas higher than in the air (approximately 1-5%) stimulates the explants. It accelerates the intensity of photosynthesis, the cell proliferation intensity in suspension and callus proliferation (Woźny & Przybył 2004).

5. Application of bioreactors

In large production *in vitro* laboratories somatic embryogenesis method is used for reproduction of selected tree species by using bioreactors. In the early stages of micropropagation, embryogenesis is induced on solid media, and after a certain time embryogenic tissue is transferred to liquid media, where it is propagated in the cell suspension in bioreactors. Bioreactors are constructed to enable conducting cell cultures under conditions appropriate to minimize or completely eliminate the possibility of infection. They allow not only the commercial tissue culture, but also the production of somatic embryos. They also allow to obtain embryogenic cells from non-embryogenic ones and to carry out microbiological and enzymatic processes. Bioreactors contain a number of control and measurement sensors, that measure and continuously maintain the following parameters: speed of mixing and aeration, concentration of dissolved oxygen, concentration of dissolved carbon dioxide, the amount of foam, overpressure in the tank, but also oxygen and carbon dioxide concentration in exhaust gasses. This is ensured by appropriate technological parameters and modern design solutions of bioreactors by applying specialized computer software. Bioreactors allow to precisely control the metabolism of plants and processes of proliferation and development of callus. The basic requirements for bioreactors designed for *in vitro* culture are: high efficiency of oxygen exchange and discharge of secreted heat. Air lift bioreactors are now the largest group, since they are very versatile. On the other hand, balloon-type bubble bioreactors are practical because of their shape which prevents media foaming. Bioreactors without forced mixing seem to be particularly useful in cell, tissue and plant organ cultures, as they do not generate the stress of mechanical or pneumatic agitation. The current review of bioreactors used in research and practice was presented in the papers by: Paek & Chakrabarty (2003), and Ziv (2005).

6. Cryopreservation as a method for long-term storage of material obtained by *in vitro* cultures

Cryopreservation is considered the best method for the long-term storage of plant tissues cultured *in vitro* at the temperature of liquid nitrogen (-196°C). When plant material is kept at such low temperature, cell divisions and metabolic processes are stopped for an indefinite time. Additional advantages of this method is low storage space and relatively low costs. Cryopreservation is a method of storage of callus, somatic embryos, pollen, buds and tree seeds in Dewar flasks. As a result of biotechnology development, genetic resources in gene banks have been supplemented with new, valuable genotypes of endangered, economically and ecologically important tree species. The success of cryopreservation depends on increasing tissue tolerance to dehydration stress and the stress caused by rehydration after thawing. Physical state that ensures cell survival in the process of dehydration and freezing is non-crystalline state i.e. vitrification. Under the stress of dehydration, the protein-lipid cytoplasmic membranes are the most vulnerable to damage, since polynusaturated fatty acids of membrane phospholipids are easily peroxidized during desiccation. Moreover, during freezing and thawing of plant material, spontaneous mutations as well as biochemical and structural changes at the cellular level may occur. Therefore, before and after freezing, the plant material should be analyzed using molecular biology techniques, e.g. by checking the somaclonal variation of embryogenic callus (Nawrot-Chorabik 2009).

Cryopreservation process may be carried out in different ways from placing the cryo-tubes in a Mr. Frosty vessel (NALGENE™ W USA), which ensure a slow temperature decrease by 1°C to the use of computerized cryobath equipment (CryolLogic) with freeze control system.

7. Short review of economically important coniferous and deciduous trees micropropagated with somatic embryogenesis including difficulties encountered during *in vitro* cultures

Pioneering studies on somatic embryogenesis of coniferous trees were carried out in 1968-1980, when the development and metabolism of callus and suspended cells was studied (Durzan & Steward 1968, Chalupa & Durzan 1973, Durzan & Chalupa 1976). Thorpe & Biondi (1984), Dunstan (1988) and Becwar et al. (1988) paid attention to the potential of a method, which could be used for vegetative proliferation of gymnosperm species. They conducted the *in vitro* culture of selected species of conifers. However, the first studies on deciduous trees originate from the forties, when reports on the *in vitro* regeneration of adventitious buds from the callus tissue of field elm *(Ulmus campestris)* were published (Gautheret 1940). Experiments on *in vitro* organogenesis of field elm were also conducted in 1949 by Jacquiot (1949). Results of his study were similar to the results obtained nine years earlier by Gautheret (1940) (Szczygieł 2005 after Gautheret 1940). It should be noted that field elm was cultivated in the forties as park and avenue tree. Currently this species is endangered due to its susceptibility to Dutch elm disease. Jacquiot (1949) simultaneously conducted research on organogenesis of silver birch (*Betula verrucosa*). Apart from adventitious buds Jacquiot (1949) obtained the beginnings of roots, however, he did not manage to grow fully developed plants (Szczygieł 2005). The first complete plant, which was obtained during this intensive research, was common aspen (*Populus tremula*) regenerated from a leaf in 1970 by Winton (1970). This achievement initiated greater interest in the method of vegetative *in vitro* propagation of trees.

Currently over 300 plant species, including 120 species of deciduous trees, are propagated by somatic embryogenesis (Bajaj 1995). Some of them are forest trees, such as: European beech (*Fagus sylvatica*), English oak (*Quercus robur*), ash (*Fraxinus* spp.), small-leaved lime (*Tilia cordata*), walnut tree (*Juglans* spp.), poplar (*Populus* spp.) and locust tree (*Robinia*). Among all *in vitro* propagated plants there are 50 coniferous trees, e.g.: European silver fir (*Abies alba*), European larch (*Larix decidua*), larch hybrids, Norway spruce (*Picea abies*), white spruce (*Picea glauca*), black spruce (*Picea mariana*), sugar cone pine (*Pinus lambertiana*), Caribbean pine (*Pinus caribea*), Loblolly pine (*Pinus taeda*) and others.

As indicated in Chapter 2.1, in order to initiate embryogenic tissue of coniferous trees, mature and immature zygotic embryos are mainly used as primary explants. The cotyledons of 7-day germinated embryos were used for initiation of embryogenic callus of *Picea abies* (Krogstrup 1986, Lelu et al. 1990), and 12-day cotyledons were used for initiation of *Picea glauca* and *Picea mariana* (Lelu & Bornman 1990). Other researchers applied also cotyledons of 12-30-day seedlings of *Picea glauca* and *Picea mariana* (Attree et al. 1990). Ruaud et al. (1992) used hypocotyls and cotyledons of 1-month somatic and zygotic seedlings and needles of 14-months somatic seedlings cultured in a greenhouse, as well as needles of 7-56 – day somatic and zygotic seedlings for initiation of embryogenic callus of *Picea abies* (Ruaud 1993). Harvengt et al. (2001) obtained embryogenic tissue on 3-year needles of somatic seedlings of *Picea abies*. These needles are the oldest spruce explant, from which embryogenic tissue was obtained (Szczygieł 2005). Nagmani & Bonga (1985) in *Larix decidua* and von Aderkas et al. (1990) in *L. decidua* and *L. leptolepis* used megagametophytes with removed immature embryos for initiation of haploid embryogenic tissue and megagametophytes with immature embryos to initiate diploid embryogenic callus. For initiation of somatic embryogenesis also protoplasts were used (Attree et al. 1987, Klimaszewska 1989, von Aderkas et al. 1990). Attree et al. (1987) regenerated somatic embryos from protoplasts isolated from embryogenic tissue of *Picea glauca* and Klimaszewska (1989) cultured seedlings of *Larix decidua* x *L. leptolepis* hybrid also using protoplasts from embryogenic tissue as explants in somatic embryogenesis. Similarly, von Aderkas et al. (1990) regenerated seedlings from protoplasts isolated from haploid callus of *Larix decidua*. Higher frequency of embryogenic callus initiation was obtained using immature emryos as explants. However, more practical is the use of mature zygotic embryos, isolated from seeds stored in cold rooms. Embryogenic tissue on mature zygotic embryos of European silver fir (*Abies alba*) was initiated by Hristoforoglu et al. (1995) in 40%, Szczygieł (2005) after Braumüller et al. (2001) in 52% and Nawrot – Chorabik (2008) in 6%. Currently, other sources of explants (needles, cotyledons, hypocotyls) and new, synthetic growth regulators that stimulate the process of somatic embryogenesis initiation are searched for (Szczygieł 2005). The first reports on somatic embryogenesis of Scots pine were focused predominantly on initiation from immature seeds and on studying the reactions of cut zygotic embryos at several developmental stages on various culture media (Lelu et al. 1999). The efforts during the regeneration of a small amount of somatic seedlings and young trees were not aimed at the development of mature somatic embryos, but the creation of a protocol for the efficient production of large amounts of plant clones. In another case, crossbreeding was conducted among selected parent trees in order to assess the impact of genotype of parents on somatic embryogenesis (Niskanen et al. 2004). During initiation, maternal effect was clearly visible, while paternal effect was predominantly invisible. A similar conclusion was reached during somatic embryogenesis studies in *Pinus taeda* (MacKay et al. 2006). In other conducted experiments the impact of several factors on the maturation of somatic embryos was investigated. These factors are: age of the culture, abscisic acid and sucrose concentration in the medium (Lelu-Walter et al. 2008). The most current research on somatic embryogenesis in genus *Pinus* has concentrated on examining the impact of parental genotypes and initiation that origin from controlled crossing between maternal trees, that had previously been tested for their response to initiation (Lelu et al. 1999).

The research on somatic embryogenesis in deciduous trees has shown that during *in vitro* culture many difficulties may be encountered. Based on the vine, it was found that although somatic and zygotic embryos are nearly identical in structural and functional characteristics, the ontogeny of somatic embryos tends to be more variable. Somatic embryos of most species, including grapevine (*Vitis* spp.), tend to exhibit several typical morphological abnormalities such as variation in shape, size, and number of cotyledons. In most species, somatic embryos are larger than zygotic embryos, and their regeneration rates are lower. For instance, in *Vitis rupestris* Scheele, only 3% of somatic embryos were capable of developing into complete plants (Jayasankar et al. 2003). This kind of abnormalities may also be found in forest tree species.

Penduculate oak (*Quercus robur*) is an endangered species due to the periodic dieback of oak stands in ecosystems of Troncais forest in France, but also in Austria, Hungary, Romania, Northern Germany, Russia and Ukraine. Oaks die also in Poland in The Krotoszyn Plateau and within the area of Mediterranean Basin. Oaks reproduce in nature by acorns, which are very sensitive to drying during storage, therefore their ability to germinate drops significantly. For this reason, this species should definitely be rescued by micropropagation in *in vitro* laboratories, although the oak is less susceptible to tissue cultures. Such attempt was undertaken in Slovakia, where the method of somatic embryogenesis was used for micropropagation of oak (*Querkus* spp.). Callus was initiated on mature oak seeds. Disinfection of seeds was slightly different than in the case of conifers, i.e.: the seeds were washed in 70% ethanol, extending the duration of the alcohol to 10 minutes and then they were treated for 15–20 with 0.1% solution of mercuric chloride. Embryogenic axes were aseptically isolated from the surrounding cotyledons with preservation of cotyledonary nodes and plumule. Isolated explants were treated with 100 mg x dm^{-3} solution of ascorbic acid to prevent oxidation for 30 min. For the cultivation of embryogenic axes in *Quercus* spp. WPM medium (Lyoyd & McCown 1980) with 20 g x dm^{-3} sucrose and 6 g x dm^{-3} Difco-Bacto Agar, supplemented with 1 mg x dm^{-3} BA and 0.01 mg x dm^{-3} NAA was used. In all experiments the medium pH was adjusted to 5.5 – 5.7. The culture required less frequent passages than in the case of conifers, i.e. in 4-5 week intervals. The number of shoots per explant formed during the subculture was recorded (Ostrolucká et al. 2007).

Silver birch (*Betula pendula*) is one of the most important birch species. It naturally occurs throughout Europe and in central parts of Asia. Silver birch is economically significant because of its medicinal properties. Birch leaves, that contain saponins, flavonoids and terpene compounds are used as medicinal substances. Chaga mushroom was used in folk medicine as an anticancer agent. Its therapeutic properties have been confirmed by research conducted in Poland and Russia. Forest yield can be enhanced significantly by large-scale multiplication of selected genotypes with improved growth rates, valuable quality of wood and high stress and disease tolerance. Vegetative propagation is an important tool of preserving unique characteristics of some of the selected trees of Silver birch (Chalupa 1995). In 1990 Kurtén et al. (1990) induced somatic embryogenesis of *B. pendula* using eight different families, obtained by crossing parent plants with different herbivore resistance. Embryogenic culture was initiated from seeds, seedlings and leaves from 1-year-old plants. Sodium hypochlorite was used for explant disinfection (leaves for 10 min., seeds for 30 min.), and N7 basal medium was used as the medium for initiation (Simola 1985). Nuutila et al. (1991) studied effect of different sugar and inorganic nitrogen concentrations in *in vitro* cultures of birch. Embryogenic callus was initiated on N7 medium, on which somatic embryos were obtained in a later stage (Simola 1985). The first stage of *in vitro* culture (callus initiation) and maturation of somatic embryos was carried out on solidified medium, while the second stage (proliferation) was carried out in liquid medium. Embryogenic tissue of birch in Chalupa's cultures (1987) was characterized by regions of meristematic cells, which were small, thin-walled and highly cytoplasmic. Globular somatic embryos developed at the periphery of the meristematic regions and consisted of densely cytoplasmic cells. At later stages, embryos were surrounded by epidermic cells. Late heart-shaped embryos developed only in small part of callus initiated from explants. Somatic embryos matured after 4 – 6 weeks (Chalupa 1987).

Sweet chestnut (*Castanea sativa*) is the tree valued mostly because of its fruit (chestnuts). This species of *Fagaceae* family occurs naturally in the Mediterranean Basin, Asia Minor and the Caucasus. A major limitation of the embryogenic systems used in chestnut is the maitenance of embryogenic competance and the low conversion rate of somatic embryos into plants. Experiments were performed to determine the influence of proliferation medium on the maintenance of embryogenic competence and on repetitive embryogenesis in *C. sativa* somatic embryos derived from leaf explants. Somatic embryo proliferation was carried out by both direct secondary embryogenesis and by the culture of nodular callus tissue originated from cotyledons of somatic embryos. Both systems led to the production of cotyledonary somatic embryos on MS proliferation medium supplemented with 0.1 mg x dm^{-3} BA and 0.1 mg x dm^{-3} NAA. A total of 39% of embryos eventually produced plants either through conversion to plantlets or indirectly through rooting of shoots. Shoots formed by somatic embryos could be excised, multiplied and rooted following the micropropagation procedures (Corredoira et al. 2003).

European beech (*Fagus sylvatica*) occurs almost in the whole Europe. Its wood is hard, compact, with no heartwood, which causes the beech to be the most commonly used in the art and practice of utility (high calorific value). Therefore, the trees selected for desirable genotypes can be cloned by *in vitro* cultures developing an efficient method of micropropagation. Until now, very few researchers dealt with this economically significant species. Among them were Vieitez et al. (1992) and Naujoks (2001). Vieitez et al. (1992) isolated immature embryos from seeds disinfected in commercial bleach (40 g x dm^{-3}). The culture was established on WPM medium with addition of 2,4-D (0.45; 2.36; 4.52 µM x dm^{-3}) and BA (2.2 µM x dm^{-3}). Embryogenic cell suspensions of 5 genotypes were successfully established in LM medium containing 2,4-D, on which the secondary explants were passaged. The earliest stage, identified as embryogenic, consisted of single cells, undergoing a first asymmetric division leading to the formation of two daughter cells of unequal size, a small cell with dense cytoplasm and strong affinity for acetocarmine and a larger cell with only slight affinity for acetocarmine. Subsequently, either the smaller cells underwent further polarized divisions to form pro-embryos or both daughter cells underwent a series of anarchical divisions leading to the formation of cell aggregates or proembryogenic masses, from which multiple embryos developed by cleavage polyembryony. Finally, 10% of embryos (63 plantlets from 638 embryos) obtained by somatic embryogenesis had developed roots, shoots and well-formed leaves (Vieitez et al. 1992). On the other hand, the research by Naujoks (2001) showed, that on WPM medium, the somatic embryos regenerated on 4 callus lines among 33 lines analyzed (7.9%). Somatic embryos of beech matured on a medium 50% WPM (with 2,4-D and BA) followed by rooting and acclimation in the ground. However, based on the experience and results gained, it was concluded that the long-term storage of embryogenic lines and somatic embryos should be carried out (Naujoks 2001).

The above examples show that the method of somatic embryogenesis is intensively tested in *in vitro* cultures. However, despite conducting numerous experiments on obtaining micro-seedlings of tree species, some of the issues related to this method remain still unresolved. More attention should be paid to the optimization methods, so that the plant material obtained *in vitro* (somatic embryos, callus tissue and properly developed seedlings) was useful for selection, breeding and molecular biology of forest trees and to be useful in other business sectors.

8. Acknowledgements

The author would like to express their special thanks to Prof. dr hab. Tadeusz Kowalski for the given valuable suggestions and doing microscopic photos.

9. References

Attree, S.M.; Bekkaoui, F.; Dunstan, D.L. & Fowke, L.C. (1987). Regeneration of somatic embryos from protoplasts isolated from an embryogenic suspension culture of white spruce (*Picea glauca*). *Plant Cell Reports*, Vol. 6, pp. 480-483.

Attree, S.M.; Budimir S. & Fowke, L.C. (1990). Somatic embryogenesis and plantlet regeneration from cultured shoots and cotyledons of seedlings from stored seeds of black and white spruce (*Picea mariana* and. *P. glauca*). *Canadian Journal of Botany*, Vol. 68, pp. 30-34.

Bajaj, Y.P.S. (1995). Somatic embryogenesis and its applications for crop improvement. In: Bajaj Y.P.S. (Ed.). Biotechnology in Agriculture and Forestry. Vol. 30, Somatic embryogenesis and synthetic seeds I, pp. 221-233. Springer-Verlag,

Becwar, M.R.; Noland T.J. & Wyckoff, J.L. (1989). Maturation, germination and conversion of Norway spruce (*Picea abies* L.) somatic embryos to plants. *In Vitro Cellular and Developmental Biology – Plant,* Vol. 25, pp. 575-580.

Becwar, M.R.; Wann, S.R.; Johnson, M.A.; Verhagen, S.A.; Feirer R. & Nagmani, R. (1988). Development and characterization of *in vitro* embryogenic systems in conifers. In: Ahuja M.R. (Ed.). Somatic cell genetics and woody plants. Kluwer Academic Publishers, Dordrecht. pp. 1-18.

Bornman, C.H. & Jansson, E. (1981). Regeneration of plants from conifer leaf, with special reference to *Picea abies* and *Pinus sylvestris. Colloque International sur la Culture In Vitro des Essances Forestieres.* Fontaineblau, AFOCEL. Nangis. pp. 41-53.

Diaumüller, S.; Ross, H.; Rahmad, A. & Zoglauer, K (2001). Somatic embryogenesis in silver fir (*Abies alba* Mill.). *International Conference on Wood, Breeding, Biotechnology and Industrial Expectations.* Abstract, Bordeaux, France, pp. 62.

Chalupa, V. & Durzan, D.J. (1973). Growth and development of resting buds of conifers *in vitro. Canadian Journal of Forest Research,* Vol. 3, pp. 196-208.

Chalupa, V. (1985). Somatic embryogenesis and plantlet regeneration from cultured immature and mature embryos of *Picea abies* (L.) Karst. *Communication Institute of Forest Czech Republik,* Vol. 14, pp. 57-63.

Chalupa, V. (1987). Somatic embryogenesis and plant regeneration in *Picea, Quercus, Betula, Tilia, Robinia, Fagus* and *Aesculus. Communication Institute of Forest Czech Republik,* Vol. 15, pp. 133-148.

Chalupa, V. (1995). Somatic embryogenesis in birch (*Betula pendula* Roth.). In: Jain, S.M., Gupta, P.K. & Newton, R.J. (Ed(s).). (1995). Vol. 2: 207-220. Somatic Embryogenesis in Woody Plants. Kluwer Academic Publishers, Dordrecht.

Chen, Z.; Yao, Y. & Zhang, L. (1988). Studies on embryogenesis of woody plants in China. In: Ahuja M.R. (Ed(s).). (1988). Somatic Cell Genetics of Woody Plants. Kluwer Academic Publishers, Dordrecht. pp. 19-25.

Cornu, D. & Geoffrion, C. (1990). Aspects de l'embryogenese somatique chez le meleze. *Bulletin de la Société Botanique de France,* Paris. Vol. 137, pp. 25-34.

Corredoira, E.; Balleste, A. & Vieitez, A.M. (2003). Proliferation, Maturation and Germination of *Castanea sativa* Mill. Somatic Embryos Orginated from Leaf Explants. *Annals of Botany,* Vol. 92, pp. 129-136.

Cusidó, R.M.; Palazón, J.; Navia-Osorio, A.; Mallol, A.; Bonfill, M.; Morales, C. & Piñol, M.T. (1999). Production of Taxol and baccatin III by a selected *Taxus baccata* callus line and its derived cell suspension culture. *Plant Science,* Vol. 146, pp. 101-107.

Cyr, D.R. & Klimaszewska K. (2002). Conifer somatic embryogenesis: II Applications. *Dendrobiology,* Vol. 48, pp. 41-49.

Dunstan, D.I. (1988). Prospects and progress in conifer biotechnology. *Canadian Journal of Forest Research,* Vol. 18, No. 12, pp. 1497-1506.

Ďurkovič, J. & Mišalová, A. (2008). Micropropagation of Temperate Noble Hardwoods: An Overview. *Functional Plant Science and Biotechnology,* Vol. 2, No. 1, pp. 1-9.

Durzan, D.J. & Chalupa, V. (1976). Growth and metabolism of cells and tissue of jack pine (*Pinus banksiana*). 3. Growth of cells in liquid suspension cultures in light and darkness. *Canadian Journal of Botany*, Vol. 54, pp. 456-467.

Durzan, D.J. & Steward, F.C. (1968). Cell and tissue culture of white spruce and jack pine. *Bimonthly research notes / Canadian Forestry Service*, Vol. 24, pp. 30.

Gautheret, R.J. (1940). Nouvelles recherches sur le bourgeonnernent du tissu cambial d'*Ulmus campestris* cultive *in vitro*. *C. R. Academy of Science Paris*, 210: 744-746.

Gresshoff, P.M. & Doy, C.H. (1972). Development and differentiation of haploid *Lycopersicon esculentum*. *Planta*, Vol. 107, pp. 161-170.

Gupta, P.K. & Durzan, D.J. (1985). Shoot multiplication from mature trees of Douglas – fir (*Pseudotsuga menziesii*) and sugar pine (*Pinus lambertiana*). *Plant Cell Reports*, Vol. 4, pp. 177-179.

Gupta, P.K. & Durzan, D.J. (1987). Biotechnology of somatic polyembryogenesis and plantlet regeneration in loblolly pine. *Bio/Technology*, Vol. 5, pp. 147-151.

Gupta, P.K., Durzan, D.J. (1986). Plantlet regeneration via somatic embryogenesis from subcultured callus of mature embryos of *Picea abies* (Norway spruce). *In Vitro Cellular and Developmental Biology Plant*, Vol. 22, pp. 685-688.

Hakman, I.; Fowke, L.C; von Arnold, S. & Eriksson, T. (1985). The development of somatic embryos in tissue cultures initiated from immature embryos of *Picea abies* (Norway spruce). *Plant Science*, Vol. 38, pp. 53-59.

Hargreaves, C. & Menzies, M. (2007). Organogenesis and cryopreservation of juvenile radiata pine. Vol. 6, pp. 51-66. In: Jain, S.M. & Häggman H. (Ed(s).). (2007). *Protocols for Micropropagation of Woody Trees and Fruits*, Springer-Verlag.

Harvengt, L.; Trontin, J.F.; Reymond, L.; Canlet, F. & Paques, M. (2001). Molecular evidence of true-to-type propagation of 3-year old Norway spruce through somatic embryogenesis. *Planta*, Vol. 213, No. 5, pp. 828-832.

Hendry, S.J.; Boddy, L. & Lonsdale, D. (1993). Interactions between callus cultures of European beech, indigenous ascomycetes and derived fungal extracts. *New Phytologist*, Vol. 123, pp. 421-428.

Hristoforoglu, K.; Schmidt, J. & Bolhar-Nordenkampf, H. (1995). Development and germination *of Abies alba* somatic embryos. *Plant Cell Tissue and Organ Culture*, Vol. 40, pp. 277-284.

Jacquiot, C. (1949). Observations sur la neoformation de burgeons chez le tissu cambial d'*Ulmus campestris* cultive *in vitro*. *C. R. Academy of Science Paris*, Vol. 229, pp. 529-530.

Jayasankar, S.; Bondada, B.R., Li Z. & Gray, D. J. (2003). Comparative anatomy and morphology of *Vitis vinifera* (*Vitaceae*) somatic embryos from solid- and liquid-culture-derived proembryogenic masses. *American Journal of Botany*, Vol. 90, pp. 973-979.

Klimaszewska, K. (1989). Recovery of somatic embryos and plantlets from protoplast cultures of *Larix* x *eurolepsis*. *Plant Cell Reports*, Vol. 8, pp. 440-444.

Klimaszewska, K.; Overton C.; Stewart D. & Rutledge R.G. (2010). Initiation of somatic embryos and regeneration of plants from primordial shoots of 10-year-old somatic white spruce and expression profiles of 11 genes followed during the tissue culture process. *Planta*, Vol. 233, No. 3, pp. 635-647.

Klimaszewska, K.; Park, Y.-S.; Overton, C.; Maceacheron, I. & Bonga, J.M. (2001). Optimized somatic embryogenesis in *Pinus strobus* L. *In Vitro Cellular and Developmental Biology Plantarum*, Vol. 37, pp. 392-399.

Kowalski, T. & Kehr, R.D. (1992). Endophytic fungal colonization of branch bases in several forest tree species. *Sydowia*, Vol. 44, pp. 137-168.

Krogstrup, P. (1986). Embryo-like structures from cotyledons and ripe embryos of Norway spruce *(Picea abies)*. *Canadian Journal of Forest Research*, 16: 664-668.

Kúrten, U.; Nuutila, A.-M.; Kauppinen, V. & Rousi, M. (1990). Somatic embryogenesis in cell cultures of birch *(Betula pendula* Roth). *Plant Cell Tissue and Organ Cultures*, Vol. 23, pp. 101-105.

Lelu, M.A. & Bornman, C.H. (1990). Induction of somatic embryogenesis in excised cotyledons of *Picea glauca* and *Picea mariana*. *Plant Physiology and Biochemistry*, Vol. 28, No. 6, pp. 785-791.

Lelu, M.A.; Bastein, K.; Drugeault, A. & Klimaszewska, K. (1999). Somatic embryogenesis and plantles development in *Pinus sylvestris* and *Pinus pinaster* on medium with and without growth regulators. *Physiologia Plantarum*, Vol. 105, pp. 719-728.

Lelu, M.A.; Boulay, M.P. & Bornman, C.H. (1990). Somatic embryogenesis in cotyledons of *Picea abies* is enhanced by an adventitious bud-inducing treatment. *New Forests*, Vol. 4, pp. 125-135.

Lelu-Walter, M.A.; Bernier-Cardou, M. & Klimaszewska, K. (2008). Clonal plant production from self- and cross-pollinated seed families of *Pinus sylvestris* (L.) through somatic embryogenesis. *Plant Cell Tissue and Organ Cultures*, Vol. 92, pp. 31-45.

Litvay, J.D.; Verma, D.C. & Johnson, M.A. (1985). Influence of loblolly pine *(Pinus taeda* L.) culture medium and its components on growth and somatic embryogenesis of wild carrot *(Daucus carota* L.). *Plant Cell Reports*, Vol. 4, pp. 325-328.

Lloyd, G. & McCown, B. (1981). Commercially-feasible micropropagation of Mountain laurel, *Kalmia latifolia* , by use of shoot tip culture. *International Plant Propagators Society*, Vol. 30, pp. 421-427.

MacKay, J.J.; Becwar, M.R. & Park, Y.S. (2006). Genetic control of somatic embryogenesis initiation in loblolly pine and implications for breeding. *Tree Genetics and Genomes*, Vol. 2, pp. 1-9.

Malá, J.; Cvikrová, M. & Chalupa, V. (2007). Micropropagation of mature trees of *Ulmus glabra*, *Ulmus minor* and *Ulmus laevis*. In: Jain S.M. & Häggman H. (Ed(s).). *Protocols for Micropropagation of Woody Trees and Fruits*, Vol. 22, pp. 237-248. Springer-Verlag.

Mihaljevic, S. & Jelaska, S. (2005). Omorica spruce *(Picea omorica)*. In: Jain, S.M. & Gupta P. (Ed(s).). *Protocol for somatic embryogenesis in woody plants. Forestry Sciences*, 77. Vol. 4, pp. 35-46. Springer-Verlag.

Misson, J.-P.; Druart, P.; Panis, B. & Watillon, B. (2006). Contribution to the study of the maintenance of somatic embryos of *Abies nardmaniana* Lk: culture media and cryopreservation method. *Propagation of Ornamental Plants*, Vol. 6, No. 1, pp. 17-23.

Mulabagal, V. & Tsay, H.-S. (2004). Plant Cell Cultures – An Alternative and efficient source for the production of biologically important secondary metabolites. *International Journal of Engineering Science*, Vol. 2, No. 1, pp. 29-48.

Murashige, T. & Skoog, F. (1962). A revised medium for rapid growth and bioassays with tobacco tissue cultures. *Physiologia Plantarum*, Vol. 15, pp. 473-494.

Nagmani, R. & Bonga, J.M. (1985). Embryogenesis in subcultured callus of *Larix decidua*. *Canadian Forest Research*, Vol. 15, pp. 108-1091.

Naujoks, G. (2001). Investigations on somatic embryogenesis of beech (*Fagus sylvatica* L.). *Poster at the Xth International Conference on Plant Embryology -"From Gametes to Embryos"*, 5. – 8. September 2001, Nitra, Slovak Republik.

Naujoks, G. (2007). Micropropagation of *Salix caprea* L. In: Jain S.M. & Häggman, H. (Ed(s).). *Protocols for Micropropagation of Woody Trees and Fruits*, Vol. 20, pp. 213-220. Springer-Verlag.

Nawrot – Chorabik, K. (2008). Embryogenic callus induction and differentiation in silver fir (*Abies alba* Mill.) tissue culture. *Dendrobiology*, Vol. 59, pp. 31-40.

Nawrot – Chorabik, K. (2009). Somaclonal variation in embryogenic cultures of silver fir (*Abies alba* Mill.). *Plant Biosystems*, Vol. 143, pp. 377-385.

Nawrot – Chorabik, K.; Jankowiak, R. & Grad, B. (2011). Growth of two blue-stain fungi associated with *Tetropium* beetles in the presence of callus cultures of *Picea abies*. *Dendrobiology*, Vol. 66, pp. 41-47.

Nhut, D.T.; Hien, N.T.T.; Don, N.T. & Khiem, D.V. (2007). *In vitro* shoot development of *Taxus wallichiana* Zucc., a valuable medicinal plant. In: Jain, S.M. & Häggman, H. (Ed(s).). *Protocols for Micropropagation of Woody Trees and Fruits*, Vol. 10, pp. 107-116. Springer-Verlag.

Niskanen, A.M.; Lu, J.; Seitz, S.; Keinonen, K.; Weissenberg, K. & Pappinen, A. (2004). Effect of parent genotype on somatic embryogenesis in Scots pine (*Pinus sylvestris*). *Tree Physiology*, 24, pp. 1259-1265.

Nuutila, A.M.; Kúrten, U. & Kauppinen, V. (1991). Optimization of sucrose and inorganic nitrogen concentration for somatic embryogenesis of birch (*Betula pandula* Roth.) callus cultures: A statistical approach. *Plant Cell, Tissue and Organ Culture*, Vol. 24, pp. 73-77.

Ostrolucká, M.G.; Gajdošová, A. & Libiaková, G. (2007). Protocol for micropropagation of *Quercus* spp. In: Jain S.M. & Häggman H. (Ed(s).). *Protocols for Micropropagation of Woody Trees and Fruits*, Vol. 8, pp. 85–92. Springer-Verlag.

Paek, K.Y. & Chakrabarty, D. (2003). Micropropagation of woody plants using bioreactor. In: Jain, S.M. & Ishii, K. (Ed(s).). (2003) Micropropagation of woody trees and fruits. Dordrecht, The Netherlands: Kluwer Academic Publishers, pp. 735-756.

Park, Y.S. (2001). Implementation of somatic embryogenesis in clonal forestry: technical requirements and deployment strategies. *International Conference on: Wood, Breeding, Biotechnology and Industrial expectations*, Abstract, Bordeaux, France, pp. 106.

Rao, P.S. (1965). *In vitro* induction of embryonal proliferation in *Santalum album* L. *Phytomorphology*, Vol. 15, pp. 175-179.

Rodríguez, R.; Valledor, L.; Sánchez, P.; Fraga, M.F.; Berdasco, M.; Hasbún, R.; Rodríguez, J.L.; Pacheco, J.C.; García, I.; Uribe, M.M.; Ríos, D.; Sánchez-, M.; Materán, M.E.; Walter, C. & Cañal, M.J. (2007). Propagation of selected *Pinus* genotypes regardless of age. In: Jain, S.M. & Häggman, H. (Ed(s).). *Protocols for Micropropagation of Woody Trees and Fruits*, Vol. 13, pp. 137-146. Springer-Verlag.

Ruaud, J.N. (1993). Maturation and conversion into plantlets of somatic embryos derived from needles and cotyledons of 7-56-day old *Picea abies*. *Plant Sciences*, Vol. 92, pp. 213-220.

Ruaud, J.N.; Bercetche, J. & Paques, M. (1992). First evidence of somatic embryogenesis from needles of 1-yearold *Picea abies* plants. *Plant Cell Reports*, Vol. 11, pp. 563-566.

Salaj, T. & Salaj, J. (2003/4). Somatic embryo formation on mature *Abies alba* × *Abies cephalonica* zygotic embryo explants. *Biologia Plantarum*, Vol. 47, pp. 7– 11.

Salajova, T.; Salaj, J., Jasik, J. & Kormutak, A. (1995). Somatic embryogenesis in *Pinus nigra* Arn. In: Jain, S.M.; Gupta, P.K. & Newton, R.J. (Ed(s).). Somatic Embryogenesis in Woody Plants. Kluwer Academic Publishers, Dordrecht. pp. 207-220.

Schenck, R.U. & Hildebrandt, A.C. (1972). Medium and techniques for induction and growth of monocotyledonous and dicotyledonous plant cell cultures. *Canadian Journal of Botany*, Vol. 50, pp. 199-204.

Simola, L.K. (1985). Propagation of plantlets from leaf callus *Betula pendula* f. *purpurea*. *Scientia Horticulare*, Vol. 26, pp. 77-85.

Steward, F.C. (1958). Growth and development of cultivated cells. III. Interpretations of the growth from free cell of carrot. *American Journal of Botany*, Vol. 45, pp. 709-713.

Szczygieł, K. (2005). Somatyczna embriogeneza – alternatywny sposób uzyskiwania wyslekcjonowanego materiału sadzeniowego gatunków drzew iglastych. *Leśne Prace Badawcze*, 3: 71-92. [Somatic embryogenesis – alternative way of obtaining selected planting stock of coniferous tree species]. *Forest Research Papers*, Vol. 3, pp. 71-92.

Thorpe, T.A. & Biondi, S. (1984). Conifers. Handbook of plant cell culture. In: Sharp, W.R.; Evans, A.; Ammirato, P.V. & Yamada, Y. (Ed(s).). MacMillian Publishing Company of New York, pp. 435-470.

Toribo, M.; Celestino, C. & Molinas, M. (2005). Cork oak, *Quercus suber* L. In: Jain, SM. & Gupta, P. (Ed(s).). *Protocol for somatic embryogenesis in woody plants. Forestry Sciences 77*, Vol. 35: 445-458. Springer-Verlag.

Trigiano, R.N. & Gray, D.J. (Ed(s).). (2011). Plant Development and Biotechnology. Section IV Propagation and development concepts. CRC PRESS.

Troch, V.; Werbrouck, S.; Geelen D. & Van Labeke, M-Ch. (2009). Optimization of horse chestnut (*Aesculus hippocastanum* L.) somatic embryo conversion. *Plant Cell, Tissue and Organ Culture*, Vol. 98, No. 1, pp. 115-123.

Vieitez, J.F.; Ballester, A. & Vieitez, A.M. (1992). Somatic embryogenesis and plantlet regeneration from cell suspension cultures of *Fagus sylvatica* L. *Plant Cell Reports*, Vol. 11, pp. 609-613.

von Aderkas, P.; Klimaszewska, K. & Bonga, J. (1990). Diploid and haploid embryogenesis in *Larix letlolepis*, *L. decidua*, and their reciprocal hybrids. *Canadian Forest Research*, Vol. 20, pp. 9-14.

von Arnold, S. & Clapham, D. (2008). Spruce embryogenesis. In: Suárez M.F., & Bozhkov P.V.) (Ed(s).). Methods in molecular biology, *Plant embryogenesis*. Vol. 427, pp. 31-45. *Humana Press*, Inc. in *Totowa, NJ*, US (United States).

Walters, C.; Find, J.I. & Grace, L.J. (2005). Somatic embryogenesis and genetic transformation in *Pinus radiata*. In: Jain S.M. & Gupta P. (Ed(s).). *Protocol for somatic embryogenesis in woody plants. Forestry Sciences 77*, Vol. 2, pp. 11-24. Springer-Verlag.

Wilkinson, J.A. & Brown, A.M.G. (1999). Horse Chestnut - *Aesculus hippocastanum*: Potential applications in cosmetic Skin-care products. *International Journal of Cosmetic Science*, Vol. 21, No. 6, pp. 437-447.

Winton, L.L. (1970). Shoot and tree production from aspen tissue cultures. *American Journal of Botany*, Vol. 57, pp. 904-909.

Woźny, A. & Przybył, K. (Ed(s).). (2004). Komórki roślinne w warunkach stresu. Tom II Komórki *in vitro*. [Plant cells under stress conditions]. Vol. 2. *In vitro* cells. Adam Mickiewicz University. pp. 72-91.

Yang, X. & Zhang, X. (2011). Developmental and molecular aspects of nonzygotic (somatic) embryogenesis In: Trigiano R.N. & Gray D.J. (Ed(s).). (2011). *Plant tissue culture, development, and biotechnology*, CRC Press. pp. 307-326.

Ziv, M. (2005). Simple bioreactors for mass propagation of plants. *Plant Cell, Tissue and Organ Culture*, Vol. 81, pp. 277-285.

4

Somatic Embryogenesis and Efficient Plant Regeneration in Japanese Cypresses

Tsuyoshi E. Maruyama and Yoshihisa Hosoi
Forestry and Forest Products Research Institute,
Department of Molecular and Cell Biology, Tsukuba,
Japan

1. Introduction

There are six species in the genus *Chamaecyparis* worldwide, of which two, namely the Hinoki cypress (*Chamaecyparis obtusa* Sieb. *et* Zucc.) and Sawara cypress (*Chamaecyparis pisifera* Sieb. *et* Zucc.) are distributed in Japan (Maruyama *et al.*, 2002). The Hinoki cypress is one of the most important commercial timber trees in Japan, representing about 25% of the plantation area in the country. However, the plantation areas are subject to various pests and diseases. In addition, Hinoki cypress pollinosis is reportedly one of the most serious allergic diseases in Japan (Maruyama *et al.*, 2005). The wood quality of Sawara cypress is considered inferior to Hinoki cypress, but grows faster (Fukuhara, 1978), is highly adaptable to humid and poor soils, and is considered more resistant to termite injury (Maeta, 1982) and far more tolerant to cold than the Hinoki cypress (Fukuhara, 1978). Fukuhara (1989) and Yamamoto and Fukuhara (1980) reported the possibility of obtaining natural hybrids between *C. obtusa* and *C. pisifera*.

We are interested in the development of a transgenic Japanese cypress with disease resistance and allergen-free pollen grains. Genetic engineering offers a significant tool to improve forest trees within a relatively short period. However, unfortunately, the major limitation to transformation is the difficulty in regenerating whole plants from target cells, making it vital to develop an efficient and stable plant regeneration system for genetic engineering and somatic hybridization breeding in order to develop disease-resistant hybrids. Somatic embryogenesis is an ideal procedure for effective propagation; not only of plus trees but also target tissue for genetic transformation. Since somatic embryogenesis and the plantlet regeneration of gymnosperm woody species was first reported in Norway spruce (*Picea abies* L. Karst.) (Hakman *et al.*, 1985; Chalupa, 1985; Hakman and von Arnold, 1985), studies in many other conifers have been reported (Tautorus *et al.*, 1991; Attree and Fowke, 1993; Gupta and Grob, 1995; Jain *et al.*, 1995; Hay and Charest, 1999). However, except for the *Larix* or *Picea* species and *Pinus radiata* (Lelu *et al.*, 1994a; Lelu *et al.*, 1994b; Klimaszewska *et al.*, 1997; Kong and Yeung, 1992; Kong and Yeung, 1995; Walter *et al.*, 1998), the regeneration of plants for most species is sometimes difficult or poor and effective utilization remains problematic.

In this chapter we describe a stable and efficient plant regeneration system for the Hinoki and Sawara cypress via somatic embryogenesis. The initiation of embryogenic cultures (EC), their maintenance and proliferation, maturation of somatic embryos, germination and plant conversion, and *ex vitro* acclimatization and field transfer will be discussed in subsequent sections.

2. Embryogenic culture initiation (ECI)

Immature open-pollinated cones of the Hinoki and Sawara cypress (Fig. 1A and Fig. 2A) were collected in June and July from plus mother trees. The collected cones were subsequently disinfected by 1 min immersion in 99.5% ethanol and dried in the laminar flow cabinet before dissection. The excised seeds were disinfected with 1% (w/v available chlorine) sodium hypochlorite solution for 15 min and then rinsed five times with sterile distilled water. After the seed coats had been removed, the megagametophytes containing immature zygotic embryos were used as explants for ECI initiation.

The explants were cultured in 24-well tissue culture plates (one per well) containing 1/2 MS medium (Murashige and Skoog medium)(Murashige and Skoog, 1962)(MS medium with basal salts reduced to half the standard concentration but replacing all NH_4NO_3 with 1000 mgL^{-1} glutamine) or 1/2 EM medium (Embryo Maturation medium) (Maruyama *et al.*, 2000) (EM medium with basal salts, vitamins, and myo-inositol reduced to half the standard concentration and with KCl concentration reduced to 40 mgL^{-1}), supplemented with 0.5 gL^{-1} casein hydrolysate, 1.0 gL^{-1} glutamine, 10 μM 2,4-dichlorophenoxyacetic acid (2,4-D), 5 μM 6-benzylaminopurine (BAP), 10 gL^{-1} sucrose, and 3 gL^{-1} gelrite. The pH of the media was adjusted to 5.8 prior to autoclaving for 15 min at 121°C. The cultures were kept in darkness at 25±1°C. The presence or absence of the distinct early stages of embryos characterized by an embryonal head with a suspensor system (Fig. 1C and Fig. 2C) from the explants was observed under an inverted microscope weekly for up to 3 months.

Embryogenic tissues (ET) extruding from the micropylar ends of explants appeared mostly after 2-4 weeks of culture, while the mean initiation frequency of ET from immature seeds of the Sawara cypress (Fig. 1B) varied from 12.5 to 33.3%. Initiation of ET was also possible in the absence of exogenous plant growth regulators (PGR) as reported for pine species (Smith, 1996; Lelu *et al.* 1999). In the Hinoki cypress, a medium without PGR containing 2 gL^{-1} activated charcoal (AC) was also effective for the induction of EC. The mean initiation frequencies of ET on a medium with and without PGR were 14.5 and 17.2%, respectively, which indicated that when explants are cultured at the appropriate developmental stage, the absence of exogenous PGR did not impede ECI.

The results of experiments for somatic embryogenesis initiation in both cypresses, where relatively small variations were achieved, suggested that the medium was not the most critical factor for ECI when explants were collected from late June to early July. The induction response and the beginning of germination were observed in some explants collected in mid-July. Since the physiological maturation of a seed is determined by its ability to germinate, this result indicates that the zygotic embryos from immature seeds collected in mid-July was the critical limit for ECI on a medium with no PGR. At this time, no germination was observed on PGR-supplemented medium. Among the factors influencing the somatic embryogenesis initiation, the appropriate developmental stage of zygotic embryos seemed the most critical. The optimal developmental stage for many conifer species has been reported in terms of seed collection date or time after fertilization (Becwar *et al.*, 1990; Tautorus *et al.*, 1991; Lelu *et al.*, 1994b; Jain *et al.*, 1995; Zoglauer *et al.*, 1995; Klimaszewska *et al.*, 1997; Lelu *et al.*, 1999; Kim *et al.*, 1999). However, due to the

Fig. 1. Somatic embryogenesis in Sawara cypress.
A: Open-pollinated cones. B: Excised immature seeds. C: Embryogenic cells. D-F: Development of embryogenic cells. G-I: Embryo maturation. J: Different maturated embryo sizes in function to PEG concentration in maturation media. K: Germination. L: Plant conversion. M: Acclimatized plant derived from a somatic embryo. N-O: Somatic plants growing out in the field. *Bars* 1mm (B-H), 1cm (A, I-M), 1m (N-O)

difficulty in determining the precise time of fertilization in open-pollinated cones and the fact that the variation in the zygotic embryo development depends on weather and location, the criteria for explant collection for ECI cannot be easily generalized. In addition, variation in the developmental stage of embryos may be observed among trees and even the same tree when individual cones are compared. In the case of the Hinoki and Sawara cypress, most of the immature embryos collected from late June to early July seemed at the late embryogeny stage. Observation of the developmental stage of individual embryos is thus likely to be the most appropriate method to determine the optimal time for embryo selection.

In the present study, relatively high initiation frequencies were achieved for both species and almost all the initiated lines continued to proliferate, even after several years of culture, resulting in stable embryogenic lines. Sometimes however, the initiation of somatic embryogenesis may not result in the establishment of an embryogenic line because the ensuing ET ceases to proliferate, making it important to distinguish between the initial extrusion from an explants and continuous growth, when assessing the success rate (Klimaszewka et al., 2007). Kim et al. (1999) reported that from 294 lines initiated in *Larix leptolepis*, only one embryogenic cell line could be proliferated. These results suggest that the capture of stable cell lines should be the optimal criterion by which to compare the ability of somatic embryogenesis initiation among species and families.

3. Maintenance and proliferation of embryogenic cultures

The maintenance and proliferation of EC was possible in several media containing a combination of 2,4-D plus BAP. The principal characteristics of these media were the reduction in the concentration of inorganic components from the standard and the addition of glutamine as an organic nitrogen source. The growth and proliferation of EC on media with a high concentration of inorganic components as a nitrogen source was suboptimal. These media supported growth only for short culturing whereupon the cell condition deteriorated over time. A similar response was also observed for the Japanese cedar EC (Maruyama et al., 2000).

The positive effect of organic nitrogen sources in the medium on the maintenance and proliferation of EC have been reported for many species (Boulay et al., 1988; Finer et al., 1989; Becwar et al., 1990; Gupta and Pullman, 1991; Tremblay and Tremblay, 1991; Smith, 1996; Klimaszewska and Smith, 1997). In our study, filter-sterilized glutamine in a medium combined with a reduction of the nitrate content increased the proliferation rate and the number of mature cotyledonary embryos of the Sawara cypress. In contrast, Zoglauer et al. (1995) reported that the continuous subculture of embryogenic suspensions of *Larix decidua* on organic nitrogen-supplemented medium resulted in a dramatic decline in the number of mature embryos obtained. Jalonen and von Arnold (1991) demonstrated the dependence of embryo maturation on the type of embryo morphology during proliferation.

In our culture routines, EC were maintained and proliferated by 2 to 3-week interval subcultures on 90-mm diameter Petri dishes containing 1/2 EM medium or 1/2 LP medium (Aitken-Christie and Thorpe, 1984) supplemented with 30 gL^{-1} sucrose, 3 μM 2,4-D, 1 μM BAP, and 1.5 gL^{-1} glutamine. These media supported the growth of the embryogenic cell lines captured. ET proliferated readily and retained their original translucent and mucilaginous appearance. The fresh weight of ET on the maintenance-proliferation medium increased about 5- to 12-fold after a 2- to 3-wk culture period. In general, solid media were used for the maintenance routine and liquid media for rapid proliferation of the cultures. The low-density subculture helped maintain suitable conditions for EC (densely embryonal head with a distinct suspensor system) in the suspension culture. Before the maturation step, about 10-20 mg FW of ET from the solid medium were transferred to a 100 mL flask containing 30-40 mL of medium (of a composition equivalent to that used for the maintenance and proliferation but without gelrite) and cultured for about 2 weeks on a rotary shaker at 50-70 rpm, in darkness at 25±1°C.

Although the initiation of ET was also possible without any additional auxin and cytokinin supplements required, exogenous PGR were found to be essential for the continuous maintenance and proliferation of ET (Fig. 2B). Conversely, the maintenance and proliferation of EC on media with no PGR was reported for *Pinus radiata* (Smith, 1996). He indicated that the use of PGR is not necessary, and that some cell lines maintained on a medium with 2,4-D and BAP lose their plant-forming potential much sooner than others, which have been maintained on a medium without PGR. However, in our experiments, the EC of Japanese cypresses maintained in the absence of PGR showed a tendency to embryo development and a decline in proliferation over time. This result was consistent with the results reported for other Japanese conifers (Maruyama *et al.*, 2000; Maruyama *et al.*, 2005; Maruyama *et al.*, 2007).

4. Maturation of somatic embryos

About 100-200 mg FW of ET suspended in 2-3 mL of medium were plated on 70-mm diameter filter paper disks over 90-mm diameter Petri dishes containing 30-40 mL of maturation medium that contained sugar, abscisic acid (ABA), AC, polyethylene glycol 4,000 (PEG), and EMM amino acids (Smith, 1996) (gL^{-1}: glutamine 7.3, asparagine 2.1, arginine 0.7, citrulline 0.079, ornithine 0.076, lysine 0.055, alanine 0.04, and proline 0.053). The petri dishes were sealed with Novix-II film (Iwaki Glass Co., Ltd., Chiba, Japan) and kept in darkness at 25±1°C for 6-12 weeks.

4.1 Effect of kind and concentration of sugar

Table 1 shows the effect of different kinds of sugar on the maturation of Sawara cypress somatic embryos. At the tested sugar concentrations, optimal results were achieved by using maltose as a carbohydrate source. Although 30 and 50 gL^{-1} did not result in any statistical difference in terms of cotyledonary embryos per Petri dish, the peak embryo maturation frequency resulted from the medium containing 50 gL^{-1} maltose with an average of 372 mature embryos. In contrast, when sucrose was used, 50 gL^{-1} resulted in a decrease of maturation frequency. Maltose has been considered a better carbohydrate and/or osmoticant source than sucrose or glucose for embryo maturation (Uddin *et al.*, 1990; Uddin, 1993). Similarly, Nørgaard and Krogstrup (1995) reported the beneficial effect of maltose for embryo maturation of *Abies* spp. A medium containing maltose as a carbohydrate source and PEG as osmoticum was reported as an effective combination to enhance somatic embryo maturation in the Loblolly pine (Li *et al.*, 1998). These authors inferred that about a 10-fold enhancement was achieved by using maltose to replace sucrose, and that the morphology of cotyledonary embryos was improved. In our results, the morphologies of cotyledonary embryos induced on the medium with sucrose or maltose were relatively similar. The main difference came in terms of the embryo maturation efficiency.

Kind of sugar	Concentration of sugar	
	30 gL^{-1}	50 gL^{-1}
Sucrose	108 B	158 B
Maltose	316 A	387 A

[1] Cotyledonary embryos per Petri dish. Means followed by same letter are not significantly different at p < 0.05. Three dishes for each treatment were used.

Table 1. Effect of kind and concentration of sugar on maturation of Sawara cypress somatic embryos[1]

4.2 Effects of ABA and AC

Table 2 showed the beneficial effect of increased ABA content in media supplemented with AC on the maturation of Sawara cypress somatic embryos. The best result was achieved with 100 μM ABA in the presence of AC, obtaining an average of 348 cotyledonary embryos per petri dish. The higher the ABA concentration, the greater the number of mature embryos. A similar result was reported in *Pinus strobus* (Klimaszewska and Smith, 1997), *Picea glauca-engelmannii* complex (Roberts *et al.*, 1990a), and *P. glauca* (Dunstan *et al.*, 1991). The addition of AC into the media notably enhanced the maturation efficiency, with around a 4-fold enhancement achieved by using 33.3 to 100 μM in combination with 2 gL^{-1} AC. Pullman and Gupta (1991) reported further improved Loblolly pine embryo development using a combination of ABA and AC, while Gupta *et al.* (1995) reported further improved quality of cotyledonary embryos of Douglas-fir (*Pseudotsuga menziesii*) by a combination of ABA, AC, and PEG. Similarly, Lelu-Walter el al. (2006) indicated that coating the cells with AC reduced ET proliferation and significantly enhanced the maturation of maritime pine somatic embryos. AC is widely used in tissue culture media, where it is believed to function as an adsorbent for toxic metabolic products and residual hormones (von Aderkas *et al.*, 2002; Pullman and Gupta, 1991).

ABA-free media or those supplemented with a low concentration (10 μM) failed to stimulate appropriate embryo maturation, producing only a few mature cotyledonary embryos (Table 2). Embryogenic cells on media without ABA did not develop beyond the embryo stage 1, as described elsewhere (von Arnold and Hakman, 1988). Most of the proembryos arrested development, with the proliferation of EC evident. Lelu *et al.* (1999) reported that mature embryos of *Pinus sylvestris* and *P. pinaster* were produced in far higher numbers and that the development of cotyledonary somatic embryos versus abnormal, shooty ones was enhanced with the addition of 60 μM ABA in comparison with media without ABA. Somatic embryos of the hybrid larch (*Larix x leptoeuropaea*) developed normally on a medium supplemented with 60 μM ABA, but abnormally on a medium with no ABA (Gutmann *et al.*, 1996). Most of the studies on somatic embryogenesis in conifers have reported ABA as a key hormone in embryo development and that the number and quality of embryo produced was vastly reduced in its absence (Durzan and Gupta, 1987; von Arnold and Hakman, 1988; Hakman and von Arnold, 1988; Attree and Fowke, 1993; Dunstan *et al.*, 1998).

ABA (µM)	AC (0 gL^{-1})	AC (2 gL^{-1})
0	1 D	3 D
10	7 D	16 D
33.3	48 CD	178 B
100	84 C	348 A

[1] Cotyledonary embryos per Petri dish. Means followed by same letter are not significantly different at $p < 0.05$. Three dishes for each treatment were used.

Table 2. Effect of ABA and AC on maturation of Sawara cypress somatic embryos [1]

Several authors have suggested that the role of ABA in somatic embryogenesis is to inhibit cleavage polyembryony with the consequent development of individual somatic embryos (Durzan and Gupta, 1987; Boulay et al., 1988; Krogstrup et al., 1988; Gupta et al., 1991), to stimulate the accumulation of nutrients, lipids, proteins, and carbohydrates (Hakman and von Arnold, 1988), and suppress precocious germination (Roberts et al., 1990a). In addition, Gupta et al. (1993) reported improved desiccation tolerance to less than 10% water content with 80 to 90% germination rates in Norway spruce embryos produced with a combination of ABA and AC. The use of ABA for somatic embryo maturation in gymnosperms is extensively reported in the compilation of Jain et al. (1995).

4.3 Effect of PEG

As shown in Table 3, the addition of PEG stimulated the mature embryo production of Sawara cypress (Fig.1D-I), with a higher concentration of PEG in the medium resulting in a higher maturation frequency. The best result was obtained at a concentration of 150 gL^{-1} with an average number of 1,043 cotyledonary embryos collected per Petri dish, in comparison with 382, 215, and 13 embryos per dish at concentrations of 75, 50, and 0 gL^{-1}, respectively. In the absence of PEG, most of the proembryos did not develop into cotyledonary embryos. Embryogenic cell proliferation was evident and most of them developed into structures consisting of small embryonal heads from which elongated suspensors extended (stage 1 somatic embryos).

Although in recent years, several studies have reported promotion of the maturation of somatic embryos by the addition of ABA into media solidified with a high concentration of gellan gum (gelrite) in the absence of PEG (Klimaszewska and Smith, 1997; Lelu et al., 1999), the use of PEG in combination with ABA has become routine for stimulating somatic embryo maturation in many gymnosperms. In our study, a high concentration of gellan gum in the absence of PEG was not effective in promoting the somatic embryo maturation of Hinoki and Sawara cypress as described above (data not shown). In contrast, some authors have reported that PEG promotes maturation but inhibits the further development of Picea glauca (Kong and Yeung, 1995) and P. abies somatic embryos (Bozhkov and von Arnold, 1998). The results of our experiments indicated that the positive effect of PEG on maturation did not inhibit the further development of somatic embryos in Japanese cypresses. Almost all mature cotyledonary embryos germinated (Fig. 1K) and developed normal plants (Fig. 1L). The beneficial effect of PEG on embryo maturation may be related to a water stress induction similar to that generated by desiccation (Attree and Fowke, 1993), to an increase in the accumulation of storage reserves, such as storage proteins, lipids, and polypeptides (Roberts *et al.*, 1990a; Attree *et al.*, 1992; Misra *et al.*, 1993), and to a tolerance to water loss (Attree *et al.*, 1991).

Morphological differences among somatic embryos of Sawara cypress obtained on media supplemented with different concentrations of PEG was restricted to size (Fig. 1J). The higher the PEG concentration, the smaller the resulting embryos (Table 3). However, the embryo size was not found to be influential in germination and subsequently plant conversion. Cotyledonary embryos germinated and converted in plants at high frequencies independent of their size (Table 3). More compact PEG-treated embryos were also reported for *Larix laricina* (Klimaszewska *et al.*, 1997) and *Picea abies* (Find, 1997). Iraki *et al.* (1989) reported that small cell size was a typical symptom of PEG-induced osmotic stress. Low external osmotic potential may have led to alterations in the cell wall composition, decreasing the ratio of cellulose to hemicellulose. This results in decreased cell wall tensile strength and the reduced ability of cells to expand (Iraki *et al.*, 1989). Therefore, the presence of PEG in the maturation medium may have influenced the subsequent growth of somatic embryos. Bozhkov and von Arnold (1998) determined that the morphology of mature somatic embryos of *Picea abies* had changed after PEG-treatment (smaller, irregularly shaped embryos, smaller cell size, larger root caps with intercellular spaces in pericolumn, degraded quiescent center), which could further restrict the embryo growth. However, in our study the subsequent development of PEG-treated embryos was no different to untreated ones. Germination frequencies and plant conversion rates of Sawara cypress were similar in somatic embryos derived from different PEG-treated media (Table 3).

PEG (gL⁻¹)	Somatic embryos	Size range (mm)	Germination (%)	Conversion (%)
0	13 C	3-10	97 A	92 A
50	215 BC	2- 8	98 A	93 A
75	382 B	2- 6	97 A	93
150	1,043 A	1- 3	97 A	92 A

[1] Cotyledonary embryos per Petri dish. Means followed by same letter are not significantly different at *p* < 0.05. Three dishes for each PEG concentration were used.

Table 3. Effect of PEG concentration in maturation media on production, size, germination and plant conversion of Sawara cypress somatic embryos [1]

The development of a proembryo mass of Hinoki cypress was encouraged by the transfer of EC onto a PGR free-medium. Cells developed gradually to form an individual and compact mass showing the early stages of somatic embryos going to a mature stage (Fig. 2D-E). Embryo maturation was induced by the transfer of cultures onto a medium containing maltose, PEG, AC, ABA and a higher concentration of amino acids. The embryos continue to develop and after 3-4 weeks of culture the initial formation of the cotyledonary embryo was observed (Fig. 2F). For most cell lines, the development of somatic embryos to the cotyledonary stage was observed after about 6-8 weeks of culture (Fig. 2G).

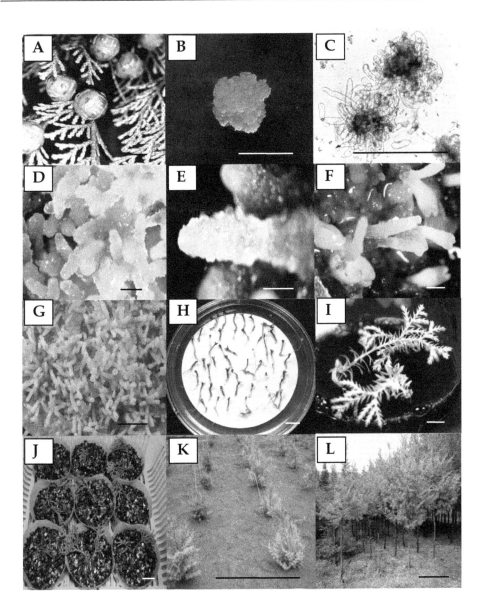

Fig. 2. Somatic embryogenesis in Hinoki cypress. A: Collected open-pollinated cones. B: Proliferation of induced embryogenic tissue on medium containing auxin and cytokinin. C: Embryogenic cells. D-F: Different developmental maturation stages of somatic embryos. G: Production of somatic embryos. H: Germination of somatic embryos. I: Plantlets growing *in vitro*. J: Acclimatized plants derived from somatic embryos. K-L: Somatic plants growing out in the field. *Bars* 1mm (C-F), 1cm (A-B, G-J), 1m (K-L)

Cell line	Total number of somatic embryos	Number of somatic embryos per Petri dish (Mean ± SE [1])	Germination frequency (%) (germinants/total tested)	Conversion frequency (%) (plants/total tested)
HNO7-1	1403	467.7 ± 21.3	94 (659/700)	91 (637/700)
HNO7-2	32	10.7 ± 1.2	50 (5/10)	40 (4/10)
HNO7-3	450	150.0 ± 15.0	82 (41/50)	78 (39/50)
HNO7-4	312	104.0 ± 18.6	83 (33/40)	80 (32/40)
HH2-1	47	15.7 ± 4.3	80 (8/10)	80 (8/10)
HH2-2	30	10.0 ± 1.7	70 (7/10)	70 (7/10)
HN1-1	54	18.0 ± 5.7	50 (5/10)	40 (4/10)
HN1-2	57	19.0 ± 3.8	76 (38/50)	72 (36/50)
HHA2-1	1536	512.0 ± 34.8	93 (219/236)	91 (215/236)
HHA2-2	188	62.7 ± 22.6	86 (43/50)	80 (40/50)
HHA2-5	14	4.7 ± 1.2	70 (7/10)	60 (6/10)
HHA2-6	1724	574.7 ± 78.3	94 (317/336)	92 (308/336)
HF4-1	565	188.3 ± 34.6	95 (123/130)	94 (122/130)
HF4-11	12	4.0 ± 2.1	NT[2]	NT
HF4-15	170	56.7 ± 9.0	98 (47/48)	96 (46/48)
HF4-19	181	60.3 ± 11.8	90 (18/20)	90 (18/20)
HF4-21	7	2.3 ± 1.9	NT	NT
HK7-17	4	1.3 ± 0.9	NT	NT
HK7-25	209	69.7 ± 11.0	93 (28/30)	93 (28/30)
HK7-29	8	2.7 ± 1.5	NT	NT
HK7-30	1511	503.7 ± 86.0	100 (130/130)	100 (130/130)
HK7-33	1052	350.7 ± 47.7	99 (286/290)	97 (280/290)
HK7-39	33	11.0 ± 1.0	70 (7/10)	60 (6/10)
HK7-45	19	6.2 ± 1.9	50 (5/10)	50 (5/10)
HK7-46	3	1.0 ± 0.6	NT	NT
HK7-57	280	93.3 ± 19.3	98 (41/42)	95 (40/42)
HK7-58	10	3.3 ± 0.3	NT	NT
HK7-60	18	6.0 ± 2.6	NT	NT
HK7-72	10	3.3 ± 1.9	NT	NT
HK7-75	1428	476.0 ± 56.0	100 (50/50)	100 (50/50)
HK7-83	7	2.3 ± 1.5	NT	NT
HK7-88	30	10.0 ± 5.1	NT	NT
HK7-105	66	22.0 ± 6.2	NT	NT
HK7-107	66	22.0 ± 1.5	60 (6/10)	60 (6/10)
Total	11536	113.1±18.1	93 (2123/2282)	91 (2067/2282)

[1]SE: standard errors of means of 3 replicates for each cell line
[2]NT: non-tested

Table 4. Somatic embryo production, germination and plant conversion for 34 cell lines of Hinoki cypress

Mature cotyledonary embryos were produced in 34 of 50 embryogenic cells lines tested (68%), and the mean number of somatic embryos per Petri dish produced varied from 1 to 575 (Table 4). This result indicates that the potential to develop cotyledonary somatic embryos varied among the cell lines. Similar results were reported for the Japanese cedar (Igasaki et al., 2003), maritime pine (Ramarosandratana et al., 2001; Miguel et al., 2004; Lelu-Walter et al., 2006), and Japanese pines (Maruyama et al., 2005; Maruyama et al., 2007).

5. Germination and plant conversion

Mature cotyledonary somatic embryos were collected from the maturation medium and transferred to the germination medium (1/2 LP or a 1/2 EM PGR free-medium with 2 gL^{-1} AC and 10 gL^{-1} agar). Cultures were kept at 25±1°C under a photon flux density of about 65 μmol m^{-2}s^{-1} with cooling and fluorescent lamps for 16 h daily.

The start of germination (Fig. 2H) was observed as early as 3-5 days after transfer to the germination medium, and after 2-4 weeks of culture, most of the somatic embryos germinated and were converted into plantlets. The mature cotyledonary somatic embryos from 23 embryogenic cell lines of the Hinoki cypress were tested, with mean germination and plantlet conversion frequencies of 93 and 91%, respectively (Table 4). This result was similar to that achieved for the Sawara cypress (Table 3). No morphological difference among the germinants and plantlets was observed among the genotypes.

Regenerated emblings of Hinoki (Fig. 2I) and Sawara cypress (Fig. 1L) were transferred to 300 ml flasks containing 100 mL of fresh medium (same composition used for the germination and conversion but with 30 gL^{-1} sucrose and 5 gL^{-1} AC) and kept under the same conditions described above for 8-12 weeks before ex vitro acclimatization.

6. *Ex vitro* acclimatization and field transfer

The developed emblings of the Hinoki (Fig.2J) and Sawara cypress (Fig.1M) were transplanted into plastic pots filled with vermiculite and acclimatized in plastic boxes inside a growth cabinet. During the first 2 weeks, emblings were kept under high relative humidity by covering the plastic boxes with transparent plastic covers and irrigating with tap water. Subsequently, the cover was gradually opened and the pots were fertilized with a nutrient solution modified from Nagao (1983) containing in mgL^{-1}: NH$_4$NO$_3$ 143, NaH$_2$PO$_4$ · 2H$_2$O 55.1, KCl 47.1, CaCl$_2$ · 2H$_2$O 52.5, MgSO$_4$ · 7H$_2$O 61, Fe-III EDTA 25, Cu EDTA 0.1, Mn EDTA 0.1, Zinc EDTA 0.1, H$_3$BO$_3$ 1.5, KI 0.01, CoCl$_2$ · 6H$_2$O 0.005, and MoO$_3$ 0.005. The covers were completely removed about 4 weeks after transplanting. Survival rates ranging from 90 to 100% were achieved after acclimatization. Subsequently, the acclimatized plants were transferred to a greenhouse and grown under controlled conditions for 6-8 months before transplanting to the field. No indication of any morphological abnormality was reported, and the growth of established plants is currently being monitored in the field (Fig. 1N-O and Fig. 2K-L).

7. Concluding remarks

An effective plant regeneration system has been achieved for Japanese cypresses via the specified procedure. In addition to high somatic embryo maturation efficiency, the subsequent high germination and plant conversion frequencies attained demonstrated the high quality of the somatic embryos produced. These somatic embryos have a zygotic embryo-like morphology, are generally longer than they are wide, with radial symmetry, and have the ability to produce normal plants like the zygotic one. The maturation frequency and the quality of embryo produced are the key criteria for the optimization of an efficient plant regeneration system via somatic embryogenesis. The cotyledonary somatic embryos of the Hinoki and Sawara cypress readily germinated after transfer to a PGR-free medium without any kind of post-maturation treatment, as was previously reported as necessary to promote the germination of somatic embryos of some other species (Roberts *et al.*, 1990b; Roberts *et al.*, 1991; Kong and Yeung, 1992; Kong and Yeung, 1995; Jones and van Staden, 2001). Thus, most of the germinants developed epycotyl and grew into normal plants. The present system should permit, in the near future, the large-scale clonal propagation of selected trees and the genetic engineering of Japanese cypresses.

8. References

Aitken-Christie, J. and Thorpe, T, A. (1984) Clonal Propagation: Gymnosperms. *In* Cell Culture and Somatic Cell Genetics of Plants, Vol. 1., Vasil, I.K. (Ed.), 480pp, Academic Press Inc., San Diego, 82-95.

Attree, S.M. and Fowke, L.C. (1993) Somatic embryogenesis and synthetic seeds of conifers. Plant Cell Tiss. Org. Cult., 35: 1-35.

Attree, S.M., Moore, D., Sawhney, V.K., and Fowke, L.C. (1991) Enhanced maturation and desiccation tolerance of white spruce [*Picea glauca* (Moench) Voss] somatic embryos: effects of a non-plasmolysing water stress and abscisic acid. Ann. Bot., 68: 519-525.

Attree, S.M., Pomeroy, M.K., and Fowke, L.C. (1992) Manipulation of conditions for the culture of somatic embryos of white spruce for improved triacylglycerol biosynthesis and desiccation tolerance. Planta, 187: 395-404.

Becwar, M.R., Nagmani, R., and Wann, S.R. (1990) Initiation of embryogenic cultures and somatic embryo development in loblolly pine (*Pinus taeda*). Can. J. For. Res., 20: 810-817.

Boulay, M.P., Gupta, P.K., Krogstrup, P., and Durzan, D.J. (1988) Development of somatic embryos from cell suspension cultures of Norway spruce (*Picea abies* Karst.). Plant Cell Rep., 7: 134-137.

Bozhkov, P.V. and von Arnold, S. (1998) Polyethylene glycol promotes maturation but inhibits further development of *Picea abies* somatic embryos. Physiol. Plant., 104: 211-224.

Chalupa, V. (1985) Somatic embryogenesis and plantlet regeneration from cultured immature and mature embryos of *Picea abies* (L.). Karst. Communi. Inst. For. Cech., 14: 57-63.

Dunstan, D.I., Bekkaoui, F., Pilon, M., Fowke, L.C., and Abrams, S.R. (1998) Effects of abscisic acid and analogues on the maturation of white spruce (*Picea glauca*) somatic embryos. Plant Sci., 58: 77-84.

Dunstan, D.I., Bethune, T.D., and Abrams, S.R. (1991) Racemic abscisic acid and abscisyl alcohol promote maturation of white spruce (*Picea glauca*) somatic embryos. Plant Sci., 76: 219-228.

Durzan, D.J. and Gupta, P.K. (1987) Somatic embryogenesis and polyembryogenesis in Douglas-fir cell suspension cultures. Plant Sci., 52: 229-235.

Find, J.I. (1997) Changes in endogenous ABA levels in developing somatic embryos of Norway spruce (*Picea abies* [L.] Karts.) in relation to maturation medium, desiccation and germination. Plant Sci., 128: 75-83.

Finer, J.J., Kriebel, H.B., and Becwar, M.R. (1989) Initiation of embryogenic callus and suspension cultures of easter white pine (*Pinus strobus* L.). Plant Cell Rep., 8: 203-206.

Fukuhara, N. (1978) Meiotic observation in the pollen mother cell of interspecific hybrid between *Chamaecyparis obtusa* and *C. pisifera*. J. Jpn. For. Soc., 60: 437-441.

Fukuhara, N. (1989) Fertility in interspecific-crossing between hinoki (*Chamaecyparis obtusa* Endl.) and Sawara (*C. pisifera* Endl.) and identification of the hybrids. Bulletin of the Forestry and Forest Products Research Institute, 354: 1-38.

Gupta, P.K. and Grob, J.A. (1995) Somatic embryogenesis in conifers. *In* Somatic embryogenesis in woody plants, Vol. 1., Jain, S.M. *et al.* (Eds.), 460pp, Kluwer Academic Publishers, Netherlands, 81-89.

Gupta, P.K. and Pullman, G.S. (1991) Method for reproducing coniferous plants by somatic embryogenesis using abscisic acid and osmotic potential variation. United States Patent # 5,036,007.

Gupta, P.K., Pullman, G., Timmis, R., Kreitinger, M., Carlson, W., Grob, J., and Welty, E. (1993) Forestry in the 21st century: The biotechnology of somatic embryogenesis. Bio/Technology, 11: 454-459.

Gupta, P.K., Timmis, R., Timmis, K.A., Carlson, W.C., and Welty, E.D.E. (1995) Somatic embryogenesis in Douglas-fir (*Pseudotsuga menziessi*). *In* Somatic Embryogenesis in Woody Plants, Vol. 3, Gymnosperms. Jain, S.M., Gupta, P.K., and Newton, R. (Eds.), 388pp, Kluwer Academic Publishers, Netherlands, 303-313.

Gutmann, M., von Aderkas, P., Label, P., and Lelu, M.-A. (1996) Effects of abscisic acid on somatic embryo maturation of hybrid larch. J. Exp. Bot., 35: 1905-1917.

Hakman, I., Fowke, L.C., von Arnold, S., and Eriksson, T. (1985) The development of somatic embryos in tissue cultures initiated from immature embryos of *Picea abies* (Norway spruce). Plant Sci., 38: 53-59.

Hakman, I. and von Arnold, S. (1985) Plantlet regeneration through somatic embryos in *Picea abies* (Norway spruce). J. Plant Physiol., 121: 149-158.

Hakman, I. and von Arnold, S. (1988) Somatic embryogenesis and plant regeneration from suspension cultures of *Picea glauca* (white spruce). Physiol. Plant., 72: 579-587.

Hay, E.I. and Charest, P.J. (1999) Somatic embryo germination and desiccation tolerance in conifers. *In* Somatic Embryogenesis in Woody Plants, Vol. 4. Jain, S.M. *et al.* (ed), 547 pp, Kluwer Academic Publishers, Dordrecht, 61-96.

Igasaki, T., Sato, T., Akashi, N., Mohri, T., Maruyama, E., Kinoshita, I., Walter, C., Shinohara, K. (2003) Somatic embryogenesis and plant regeneration from immature zygotic embryos of *Cryptomeria japonica* D. Don. Plant Cell Rep., 22: 239-243.

Iraki, N.M., Bressan, R.A., Hasegawa, P.M., and Carpita, N.C. (1989) Alteration of the physical and chemical structure of the primary cell wall of growth-limited plant cells adapted to osmotic stress. Plant Physiol., 91: 39-47.

Jain, S.M., Gupta, P.K., and Newton, R. (1995) Somatic embryogenesis in woody plants, Vol. 3, Gymnosperms, 388 pp. Kluwer Academic Publishers, Netherlands.

Jalonen, P. and von Arnold, S. (1991) Characterization of embryogenic cell lines of *Picea abies* in relation to their competence for maturation. Plan Cell Rep., 10: 384-387.

Jones, N.B. and van Staden, J. (2001) Improved somatic embryo production from embryogenic tissue of *Pinus patula*. In Vitro Cell. Dev. Biol.-Plant, 37:543-549

Kim, Y.W., Youn, Y., Noh, E.R., and Kim, J.C. (1999) Somatic embryogenesis and plant regeneration from immature zygotic embryos of Japanese larch (*Larix leptolepis*). Plant Cell Tiss. Org. Cult., 55: 95-101.

Klimaszewska, K., Devantier, Y., Lachance, D., Lelu, M.-A., and Charest, P.J. (1997) *Larix laricina* (tamarack): somatic embryogenesis and genetic transformation. Can. J. For. Res., 27: 538-550.

Klimaszewska, K. and Smith, D.R. (1997) Maturation of somatic embryos of *Pinus strobus* is promoted by a high concentration of gellam gum. Physiol. Plant., 100: 949-957.

Klimaszewska, K., Trontin, J-F., Becwar M.R., Devillard, C., Park, Y-S., Lelu-Walter, M-A. (2007) Recent progress in somatic embryogenesis of four *Pinus* spp. Tree and Forestry Science and Biotechnology, 1:11-25.

Kong, L. and Yeung, E. (1992) Development of white spruce somatic embryos: II. Continual shoot meristem development during germination. *In vitro* Cell. Dev. Biol., 28P: 125-131.

Kong, L. and Yeung, E. (1995) Effects of silver nitrate and polyethylene glycol on white spruce (*Picea glauca*) somatic embryo development: enhancing cotyledonary embryo formation and endogenous ABA content. Physiol. Plant., 93: 298-304.

Krogstrup, P., Eriksen, E.N., Moller, J.D., and Roulund, H. (1988) Somatic embryogenesis in Sitka spruce (*Picea sitchensis* (Bong.) Carr.). Plant Cell Rep., 7: 594-597.

Lelu, M.-A., Bastien, C., Drugeault, A., Gouez, M.-L., and Klimaszewska, K. (1999) Somatic embryogenesis and plantlet development in *Pinus sylvestris* and *Pinus pinaster* on medium with and without growth regulators. Physiol. Plant., 105: 719-728.

Lelu, M.-A., Bastien, C., Klimaszewska, K., and Charest, P.J. (1994a) An improved method for somatic plantlet production in hybrid larch (*Larix* x *leptoeuropaea*). Part 2. Control of germination and plantlet development. Plant Cell Tiss. Org. Cult., 36:117-127.

Lelu, M.-A., Bastien, C., Klimaszewska, K., Ward, C., and Charest, P.J. (1994b) An improved method for somatic plantlet production in hybrid larch (*Larix* x *leptoeuropaea*). Part 1. Somatic embryo maturation. Plant Cell Tiss. Org. Cult., 36:107-115.

Lelu-Walter, M.A., Bernier-Cardou, M., and Klimaszewska, K. (2006) Simplified and improved somatic embryogenesis for clonal propagation of *Pinus pinaster*. Plant Cell Rep., 25:767-776.

Li, X.Y., Huang, F.H., Murphy, J.B., and Gbur, E.E.JR. (1998) Polyethylene glycol and maltose enhance somatic embryo maturation in loblolly pine (*Pinus taeda* L.). *In vitro* Cell. Dev. Bio.-Plant, 34: 22-26.

Maeta, T. (1982) Effects of gamma-rays irradiation on interspecific hybridization between *Chamaecyparis obtusa* S. et Z. and *C. pisifera* S. et Z. Hoshasen Ikusyujo Kenkyu Hokoku 5: 1-87 (in Japanese).

Maruyama, E., Hosoi, Y., and Ishii, K. (2002) Somatic embryogenesis in Sawara cypress (*Chamaecyparis pisifera* Sieb. et Zucc.) for stable and efficient plant regeneration, propagation and protoplast culture. J. For. Res., 7: 23-34.

Maruyama, E., Hosoi, Y., and Ishii, K. (2003) Somatic embryo culture for propagation, artificial seed production, and conservation of Sawara cypress (*Chamaecyparis pisifera* Sieb. et Zucc.). J. For. Res., 8: 1-8.

Maruyama, E., Hosoi, Y., and Ishii, K. (2005) Somatic embryo production and plant regeneration of Japanese black pine (*Pinus thunbergii*). J. For. Res., 10: 403-407.

Maruyama, E., Hosoi, Y., and Ishii, K. (2007) Somatic embryogenesis and plant regeration in yakutanegoyou, *Pinus armandii* Franch. var. *amamiana* (Koidz.) Hatusima, an endemic and endangered species in Japan. In Vitro Cell. Dev. Biol.-Plant, 43:28-34.

Maruyama, E., Ishii, K., and Hosoi, Y. (2005) Efficient plant regeneration of Hinoki cypress (*Chamaecyparis obtusa*) via somatic embryogencsis. J. For. Res., 10:73-77.

Maruyama, E., Tanaka, T., Hosoi, Y., Ishii, K., and Morohoshi, N. (2000) Embryogenic cell culture, protoplast regeneration, cryopreservation, biolistic gene transfer and plant regeneration in Japanese cedar (*Cryptomeria japonica* D. Don). Plant Biotechnology, 17: 281-296.

Miguel, C., Gonçalves, S., Tereso, S., Marum, L., Maroco, J., and Olivera, M.M. (2004) Somatic embryogenesis from 20 open-pollinated families of Portuguese plus trees of maritime pine. Plant Cell Tiss. Orga.Cult., 76:121-130.

Misra, S., Attree, S.M., Leal, I., and Fowke, L.C. (1993) Effects of abscisic acid, osmoticum and desiccation on synthesis of proteins during the development of white spruce somatic embryos. Ann. Bot., 71: 11-22.

Murashige, T. and Skoog, F. (1962) A revised medium for rapid growth and bioassays with tobacco cultures. Physiol. Plant., 15: 473-497.

Nagao, A. (1983) Differences of flower initiation of *Cryptomeria japonica* under various alternating temperatures. J. Jap. For. Soc., 65: 335-338 (in Japanese).

Nørgaard, J.V. and Krogstrup, P. (1995) Somatic embryogenesis in *Abies* spp. *In* Somatic Embryogenesis in Woody Plants, Vol. 3, Gymnosperms. Jain, S.M., Gupta, P.K., and Newton, R. (Eds.), 388pp, Kluwer Academic Publishers, Netherlands, 341-355.

Pullman, G.S. and Gupta, P.K. (1991) Method for reproducing coniferous plants by somatic embryogenesis using absorbent materials in the development stage media. U.S. Patent No. 5,034,326.

Ramarosandratana, A., Harvengt, L., Bouvet, A., Calvayrac, R., and Pâques, M. (2001) Effect of carbohydrate source, polyethylene glycol and gellum gum concentration on embryonal-suspensor mass (ESM) proliferation and maturation of maritime pine somatic embryos. In Vitro Cell. Dev. Biol.-Plant, 37:29-34.

Roberts, D.R., Flinn, B.S., Webb, D.T., Webster, F.B., and Sutton, B.C.S. (1990a) Abscisic acid and indole-3-butyric acid regulation of maturation and accumulation of storage proteins in somatic embryos of interior spruce. Physiol. Plant., 78: 355-360.

Roberts, D.R., Lazaroff, W.R., and Webster, F.B. (1991) Interaction between maturation and high relative humidity treatments and their effects on germination of Sitka spruce somatic embryos. J. Plant Physiol., 138: 1-6.

Roberts, D.R., Sutton, B.C.S., and Flinn, B.S. (1990b) Synchronous and high frequency germination of interior spruce somatic embryos following partial drying at high relative humidity. Can. J. Bot., 68: 1086-1090.

Smith, D.R. (1996) Growth medium. U. S. Patent No. 5,565,455.

Tautorus, T.E., Fowke, L.C., and Dunstan, D.I. (1991) Somatic embryogenesis in conifers. Can. J. Bot., 69: 1873-1899.

Tremblay, I. and Tremblay, F.M. (1991) Carbohydrate requirements for the development of black spruce (*Picea mariana*) and red spruce (*P. rubens*) somatic embryos, Plant Cell Tiss. Org. Cult., 27: 95-103.

Uddin, M. (1993) Somatic embryogenesis in Gymnosperms. U.S. Patent No. 5,187,092.

Uddin, M.R., Dinus, R.J., and Webb, D.T. (1990) Effects of different carbohydrates on maturation of *Pinus taeda* somatic embryos. Abstracts VII International Congress on Plant Tissue and Cell Culture, Amsterdam, Netherlands, p. 272.

von Aderkas, P., Label P., and Lelu, M.A. (2002) Charcoal effects on early developmental and hormonal levels of somatic embryos of hybrid larch. Tree Physiology, 22:431-434.

von Arnold, S., and Hakman, I. (1988) Regulation of somatic developments in *Picea abies* by abscisic acid (ABA). J. Plant Physiol., 132:164-169.

Walter, C., Grace, L.J., Wagner, A., White, D.W.R., Walden, A.R., Donaldson, S.S., Hinton, H., Gardner, R.C., Smith, D.R. (1998) Stable transformation and regeneration of transgenic plants of *Pinus radiata* D. Don. Plant Cell Rep., 17: 460-468.

Yamamoto, Ch. and Fukuhara, N. (1980) Cone and seed yields after open-, self-, intraspecific-, and interspecific-pollinations in *Chamaecyparis obtusa* (Sieb. et Zucc.) Endl. and *C. pisifera* (Sieb. et Zucc.) Endl. Bulletin of the Forestry and Forest Products Research Institute, 311: 65-92.

Zoglauer, K., Dembny, H., Behrendt, U., and Korlach, J. (1995) Developmental patterns and regulating factors in direct somatic embryogenesis of European larch (*Larix decidua* Mill.). Med. Fac. Landbouww. Univ. Gent, 60: 1627-1636.

5

Mechanisms of Lumen Development in *Drosophila* Tubular Organs

Na Xu[1,2,*], Carolyn Pirraglia[1,*], Unisha Patel[1] and Monn Monn Myat[1]

[1]Department of Cell and Developmental Biology, Weill Medical College of Cornell University, New York
[2]Department of Cell Biology, Albert Einstein College of Medicine, New York
USA

1. Introduction

Tubular organs of both invertebrate and vertebrate animals serve many important physiological functions, such as the delivery of gases, nutrients and hormones and removal of waste. All tubular organs contain a central lumen that is formed through a variety of mechanisms and whose size and shape is essential for organ function. While some lumens form from pre-polarized cells, others form *de novo* from single cells or solid cords of cells (Andrew and Ewald, 2010). Studies of lumen formation in tubular organs in the *Drosophila* embryo have benefited from the genetic analysis available in *Drosophila* and the advent of sophisticated microscopic techniques that allow lumen formation to be visualized *in vivo* in real time in a developing embryo. In this chapter we will review recent advances on the cellular and molecular mechanisms by which lumens form and their size is controlled in the salivary gland, trachea and dorsal vessel of the *Drosophila melanogaster* embryo.

2. Dorsal vessel

The *Drosophila* cardiac tube, or dorsal vessel, is a hemolymph pumping organ that constitutes the entire cardiovascular system of the *Drosophila* open circulatory system. The dorsal vessel is established during embryogenesis and is composed of two rows of 52 contractile myoendothelial cells (cardioblasts [CBs]) enclosing a central lumen surrounded by loosely attached non-muscular pericardial cells (Figure 1A and B) (Tao and Schulz, 2007). The dorsal vessel is derived from mesodermal cells that acquire certain epithelial characteristics to form two bilateral rows of CBs that migrate dorsally and meet at the dorsal midline to create a lumen exclusively formed by the membrane walls of the CBs (Figure 1B). At the end of cardiac morphogenesis, the posterior portion of the dorsal vessel becomes enlarged and constitutes the definitive heart, whereas the anterior portion has a narrow diameter and is equivalent to the aorta (Figure 1A). The heart is the only region of the dorsal vessel that exhibits automatic and synchronized beating to act as a myogenic pump and promote circulation of the hemolymph throughout the cardiovascular system.

---

* These authors contributed equally

In this section, we discuss the genetic networks that control lumen formation of the *Drosophila* dorsal vessel. In particular, we discuss the necessary changes in cell shape and cell-cell adhesion that occur during lumen formation, and the requirement of G-protein signaling for maintenance of the cardiac tube.

2.1 Dorsal vessel lumen formation

The cardiac myoendothelium originates from mesodermal cells that form two bilateral rows of CBs. During dorsal closure, when the dorsal epidermis from opposing sides of the embryo migrates as a sheet to seal the opening at the dorsal surface, the two rows of aligned CBs, together with adjacent pericardial cells, migrate as a sheet of cells, in association and in coordination, with the overlying ectoderm towards the dorsal midline. Lateral alignment and dorsal migration of CBs are critical for the proper formation of the mature dorsal vessel, as mutations in genes that regulate these processes result in structural and lumenal defects (Reim and Frasch, 2010; Tao and Schulz, 2007). As the lateral rows of CBs approach the dorsal midline, the CBs adopt a pear-like shape through constriction of their cellular surfaces facing the dorsal midline (Figure 1C) (Medioni et al., 2008; Santiago-Martinez et al., 2008). Actin-rich protrusions extend from this membrane domain, which constitutes the leading edge of the dorsally migrating CBs (Medioni et al., 2008). CBs from each of the two lateral rows initiates contact with its contralateral counterpart at their dorsal-most leading edge and join at the dorsal midline. Subsequently, the CBs adopt a crescent-like shape, thereby allowing contralateral CBs to join ventrally to close the tube and form a central lumen (Figure 1C) (Medioni et al., 2008; Santiago-Martinez et al., 2008).

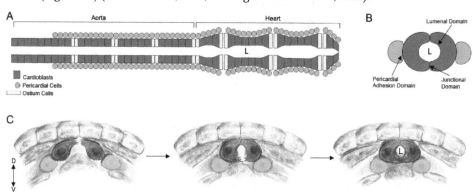

Fig. 1. Lumen formation in the dorsal vessel.

(A) The *Drosophila* embryonic dorsal vessel consists of the aorta and the heart proper where the lumen is lined by cardioblasts (red) and ostium cells (yellow) which in turn are surrounded by the pericardial cells (green). (B) Cross section of the dorsal vessel with a central lumen (L) showing the lumenal and junctional domains of the cardioblasts (red) and the pericardial adhesion domain between the cardioblasts and pericardial cells (green). (C) Lumen formation in the dorsal vessel is preceded by the dorsal migration of a row of cardioblasts (red) in contact with pericardial cells (green) on each side of the embryo, followed by cell-cell contact at the dorsal and then the ventral sides of the cardioblasts to form a central lumen. D: dorsal; V: ventral. Panel C was kindly provided by F. Macabenta and S. Kramer.

Concomitant with these cell shape changes, the membrane domains of CBs undergo significant remodeling to alter their cellular polarity. As CBs approach the dorsal midline, proteins required for cell adhesion, junctional domain formation and attractive/repulsive signals become sub-localized within distinct CB membrane domains. The lumenal domain, which encloses the lumen itself, is characterized by the presence of basal membrane matrix proteins, including Dystroglycan and Perlecan, and the attractant/repellant proteins, Slit and Roundabout (Robo) (Figure 1B). The junctional domain, located at the ventral and dorsal membrane regions, is characterized by the accumulation of adherens junction (AJ) proteins, such as E-Cadherin, Discs-Large and β-catenin. The pericardial adhesion domain, which exists at the contact points between the CBs and the pericardial cells (PCs), is distinguished by the presence of extracellular matrix (ECM) proteins, such as pericardin (Chartier et al., 2002).

The specification and maintenance of these distinct membrane domains and the dynamic changes in cell shape require specific genetic regulators, the loss of which disrupts lumen formation. *Shotgun* (*shg*), which encodes the *Drosophila* homolog of E-cadherin, is specifically required for adhesion of opposing rows of CBs to form the junctional domain (Haag et al., 1999; Santiago-Martinez et al., 2008). The loss of *shg* function results in a loss of adhesion between contralateral CBs, whereas the overexpression of E-cadherin in CBs results in an expansion of the junctional domain and the inability to form a lumen (Haag et al., 1999; Santiago-Martinez et al., 2008). One key regulator of E-cadherin-mediated adhesion between contralateral CBs is the Slit/Robo signaling pathway. Slit is an EGF- and LRR-containing secreted extracellular matrix protein that functions as the ligand for the Robo family of transmembrane receptors and has been shown previously to regulate repulsive axonal guidance in the *Drosophila* nervous system (Kidd et al., 1999; Qian et al., 2005; Rothberg et al., 1990). During migration of the bilateral rows of CBs, Slit and Robo accumulate at the presumptive lumenal domain as the CBs align at the dorsal midline. This polarization of Slit/Robo signaling is critical for its function in regulating lumen formation. In *robo* and *slit* mutants, E-cadherin mediated adhesion between the two opposing CBs is expanded preventing critical cell shape changes and blocking lumen formation (Medioni et al., 2008; Santiago-Martinez et al., 2008). In contrast, when Slit is ectopically expressed on all CB surfaces, a loss of cell adhesion was observed, resulting in the formation of multiple lumens (Santiago-Martinez et al., 2008). These studies indicate that polarized Slit/Robo repulsion is required for inhibition of E-cadherin mediated adhesion at the presumptive lumenal domain to form a central lumen.

Restriction of Slit localization to the lumenal domain is regulated by the transmembrane heparin sulfate proteoglycan, Syndecan (Knox et al., 2011). Syndecans are known to interact with a diversity of extracellular ligands, often in conjunction with other cell surface receptors, and are thought to play a dual role in adhesion and as regulators of signaling from the extracellular matrix (ECM). In *Drosophila*, the single Syndecan homolog, Sdc, regulates axon guidance by acting as a co-receptor with Robo to mediate Slit signaling (Chanana et al., 2009). Embryos that lack Sdc function fail to localize Slit and Robo to the luminal domain and fail to properly form a lumen, indicating that Sdc may also act as a co-receptor for Slit to regulate lumen formation in the dorsal vessel (Knox et al., 2011).

Formation of the dorsal vessel lumen also depends on the transmembrane receptor, Uncoordinated 5 (Unc5). Unc5 represents the single *Drosophila* homolog of a conserved receptor family that binds to the secreted ligand, Netrin (Net) (Keleman and Dickson, 2001). Unc5/Net signaling, like Slit/Robo signaling, plays a role in repulsive axonal guidance and has a localization pattern in the dorsal vessel similar to that of Slit and Robo, where Unc5, and its ligand, NetB, accumulate at the lumenal domain of CBs (Albrecht et al., 2011; Keleman and Dickson, 2001; von Hilchen et al., 2010). In embryos mutant for *unc5* or *netB*, CBs migrate and initiate contact with their contralateral counterparts normally but fail to form a central lumen (Albrecht et al., 2011). Thus, Unc5/Netrin acts as a repulsive force to inhibit contralateral CBs from attaching to one another at their presumptive lumenal domains.

2.2 Maintenance of the dorsal vessel

Genetic analysis has identified the mechanisms by which the lumen of the dorsal vessel is maintained. In particular, at the end of dorsal vessel development, the pericardial cells and CBs must adhere tightly to maintain the structure and integrity of the dorsal vessel. The loss of pericardial and CB adhesion results in the disruption of the dorsal vessel lumen and loss of cardiac function (Yi et al., 2006). The mevalonate pathway, which is important for the synthesis of isoprene derivatives that modify the C termini of proteins containing a CAAX motif (C, cysteine; A, aliphatic amino acids; X, any amino acid), is required for proper pericardial and CB adhesion and dorsal vessel maintenance. In mutants for HMGCR (hydroxymethylglutaryl (HMG)-coenyme A (CoA) reductase), an important regulator of the mevalonate pathway, CBs and pericardial cells properly align at the dorsal midline to form a central lumen; however, at the end of embryogenesis, pericardial cells dissociate from the CBs resulting in CB misalignment and loss of lumen integrity (Yi et al., 2006).

Dorsal vessel defects of *HMGCR* mutant embryos result from the failure of G protein γ subunit 1 (Gγ1) to be post-translationally modified with a geranylgeranyl moiety (Yi et al., 2006; Yi et al., 2008). G proteins form heterotrimers with subunits designated α, β and γ and act as intracellular effectors of G protein coupled receptors (GPCRs) (Malbon, 2005). Gγ1 functions with the β and α subunits, Gβ13F and G-oα47A, respectively, to regulate dorsal vessel maintenance, where loss of Gβ13F or G-oα47A results in pericardial cell-CB dissociation (Yi et al., 2008). Genetic analysis indicates that regulation between the Gα and Gβγ subunits, in coordination with Loco, a member of the regulators of G-protein signaling (RGS) protein family, ensure proper maintenance of the dorsal vessel (Yi et al., 2008). One mechanism by which heterotrimeric G proteins regulate CB-pericardial cell adhesion is by regulating septate junction (SJ) components (Yi et al., 2008). In *Drosophila*, SJs are spoke and ladder septa that connect adjacent plasma membranes and are functionally similar to tight junctions in mammalian systems (Banerjee et al., 2006). Although SJs are absent in the embryonic dorsal vessel, SJ proteins are present, suggesting that SJ proteins perform non-canonical functions during dorsal vessel morphogenesis. Gγ1 regulates the cellular localization of the SJ proteins, Coracle (Cora), Sinuous (Sinu), Neurexin-IV (Nrx-IV) and Nervana2 (Nrv2), and mutants for these SJ proteins have defects in pericardial cell-CB adhesion (Yi et al., 2008). In embryos mutant for SJ proteins, CBs properly align and adhere their ventral and lateral membrane domains at the dorsal midline to form a central lumen; however, the lumen is not maintained and becomes twisted and flattened (Yi et al., 2008). This is in contrast to embryos mutant for AJ proteins, where CBs fail to initialize adhesion with contralateral CBs at the dorsal midline.

These studies suggest a novel pathway in which heterotrimeric G-protein signaling controls proper localization and function of SJ proteins at the pericardial adhesion domain of CBs, which leads to the establishment of stable "SJ-like" adhesive contacts with pericardial cells to maintain the mature dorsal vessel lumen.

3. Salivary gland

The *Drosophila* salivary gland is a secretory organ and consists of a pair of elongated secretory tubes (hereafter referred to the salivary gland) that are connected to the larval mouth through the finer set of duct tubes. The glands are formed during embryogenesis and become functional in the larval stage when they synthesize and secrete proteins necessary for lubrication, digestion and taste. The salivary gland consists of a layer of polarized epithelial cells surrounding a central lumen that is formed from two placodes of epithelial cells, approximately 100 cells each. Salivary glands invaginate through constriction of apical domains and basal migration of nuclei to form a tube that is initially oriented dorsally. After all salivary gland cells have invaginated, the gland turns and migrates posteriorly until it reaches its final position in the embryo (Pirraglia and Myat, 2010). In this section, we will focus on our current understanding of how the salivary gland lumen achieves and maintains its size and shape.

3.1 Growth and remodeling of the apical membrane

The salivary gland lumen forms concomitantly with invagination of gland cells from the embryo surface. During the early migratory step of salivary gland development when the internalized gland turns and migrates posteriorly, gland lumen length doubles and lumen width in the proximal region (the region closest to the ventral surface) is reduced by half (Figure 2A-C) (Pirraglia et al., 2010). Salivary gland lumen size is controlled, at least in part, by the dynamic growth and remodeling of the apical membrane. Transmission electron micrographs (TEMs) revealed that after all salivary gland cells have invaginated from the embryo surface, the gland lumen is characterized by abundant apical protrusions into the luminal space (Myat and Andrew, 2002). Measurements of the length of the apical surface membrane per individual salivary gland cell showed an increase in apical surface membrane, suggesting dramatic growth of the apical membrane (Myat and Andrew, 2002). This rapid phase of membrane growth is followed by elongation of the apical domain of individual gland cells in the proximal-distal (Pr-Di) direction, the direction in which the salivary gland lumen elongates concomitant with posterior migration of the gland (Figure 2D) (Myat and Andrew, 2002; Pirraglia et al., 2010). The Sp1/egr-like transcription factor, Huckebein (Hkb), regulates the size and shape of the salivary gland lumen through control of apical membrane growth (Myat and Andrew, 2000a; Myat and Andrew, 2002). In *hkb* mutant salivary gland cells, the apical surface membrane fails to grow and the apical domain fails to elongate resulting in spherical lumens (Myat and Andrew, 2002).

In the salivary gland placode, the pattern of *hkb* RNA precedes the order in which salivary gland cells invaginate (Myat and Andrew, 2000a). This pattern of *hkb* RNA expression is controlled by Hairy, a basic helix-loop-helix (bHLH) transcription factor (Carroll et al., 1988; Hooper et al., 1989). In *hairy* mutant embryos, *hkb* RNA is expressed in all gland cells and

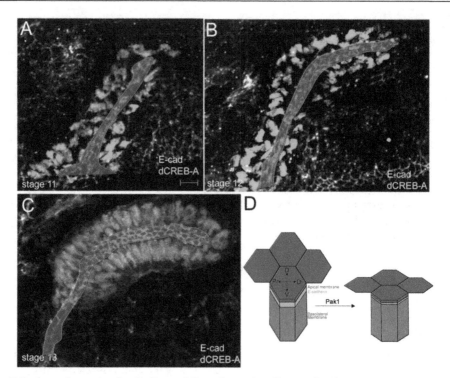

Fig. 2. Lumen elongation in the *Drosophila* embryonic salivary gland.

(A) The salivary gland lumen is formed as salivary gland cells invaginate from the embryo surface at
stage 11. (B and C) As the gland migrates posteriorly, lumen length increases. (D) Apical domain
elongation is controlled by differential localization of E-cadherin in a Pak1-dependent manner. Panels in
A-C are projected confocal images of wild-type embryos stained for dCREB-A (green) to mark the gland
nuclei and E-cadherin (E-cad; white) to mark the gland lumen that is outlined in red. Scale bar in A
represents 5 μm.

the lumens that form are expanded or branched (Myat and Andrew, 2002). Similar to loss of
hairy function, overexpression of *hkb* in salivary gland cells leads to expanded and rounded
lumens, instead of elongated lumens, but not branched lumens (Myat and Andrew, 2002).
Hkb controls salivary gland lumen size through two downstream target genes, *klarsicht*
(klar), which encodes a *Drosophila* KASH (Klar, Anc-1, Syne-1 homology) domain protein
(Mosley-Bishop et al., 1999), and *crumbs* *(crb)*, which encodes an apical transmembrane
protein that confers apical identity (Myat and Andrew, 2002; Wodarz et al., 1995). In
Drosophila ovaries and eye, Klar is present on the nuclear envelope and is required for
nuclear migration whereas in the early embryo, Klar localizes to lipid droplets and is
required for lipid droplet transport (Guo et al., 2005; Kracklauer et al., 2007; Welte et al.,
1998). Similar to Klar, the mammalian KASH domain proteins, such as Nesprins, localize to
nuclear membranes (Zhang et al., 2001) and regulate nuclear positioning (Zhang et al., 2010).
Crumbs (Crb) is important for the establishment and maintenance of apical polarity in both
Drosophila and mammalian epithelia and photoreceptor cells (Bulgakova and Knust, 2009;
Izaddoost et al., 2002; Pellikka et al., 2002; Tepass et al., 1996; Tepass and Knust, 1990;

Wodarz et al., 1995). Hkb regulates not only the levels of *crb* RNA in salivary gland cells but also Crb protein level and/or localization together with Klar (Myat and Andrew, 2002). Considering Klar likely mediates dynein-dependent cargo transport along microtubules (Mosley-Bishop et al., 1999; Welte et al., 1998), it is thought that Hkb mediates the apical delivery of vesicles, such as those containing Crb, through Klar, to promote apical membrane growth and polarized elongation of apical domains during salivary gland lumen elongation.

3.2 Ribbon function in salivary gland lumen elongation

crb RNA expression in salivary gland cells is also controlled by Ribbon (Rib), a BTB (bric-a-brac, tramtrack, broad-complex)/POZ (poxvirus and zinc finger) domain transcription factor required for the proper morphology of multiple tubular organs in *Drosophila*, such as the salivary gland, trachea, Malpighian tubules and the hindgut (Blake et al., 1998; Bradley and Andrew, 2001; Jack and Myette, 1997; Kerman et al., 2008; Shim et al., 2001). In salivary gland cells, Rib controls lumen elongation by simultaneously promoting *crb* RNA expression and limiting apical localization of active phosphorylated Moesin (Moe), a *Drosophila* Ezrin-Radixin-Moesin (ERM) family protein that links the actin cytoskeleton to the plasma membrane (Kerman et al., 2008). The salivary gland phenotype of *rib* mutants is phenocopied by gland specific expression of Moe^{T559D}, a phosphomimetic mutation in Moe where threonine (T) 559 is replaced by aspartic acid (D) and functions as a constitutively-active form of Moe in *Drosophila* developing eyes (Karagiosis and Ready, 2004; Kerman et al., 2008). Since Moe normally links the actin cytoskeleton to the plasma membrane, it is thought that inhibition of Moe activity by Rib decreases the linkage of the apical membrane to the actin cytoskeleton, which in turn, reduces apical membrane stiffness to allow lumen elongation. Furthermore, Rib may control lumen elongation by promoting Rab11-dependent delivery of apically targeted vesicles since *rib* mutant gland cells have a reduced number of apical Rab-11 positive vesicles (Kerman et al., 2008). Rab11 is a small GTPase that mediates apical trafficking of cargo proteins through recycling endosomes or directly from the Golgi (Satoh et al., 2005). Based on these observations, a model for Rib regulation of salivary gland lumen elongation is proposed where Rib promotes *crb* RNA expression and Rab11-dependent apical vesicle delivery to facilitate apical membrane growth, and limits apical Moe activity to reduce apical membrane stiffness which allows the salivary gland lumen to elongate (Kerman et al., 2008). This model is supported by computational models based on live imaging, which suggest that *rib* mutant salivary glands have increased apical stiffness and apical viscosity compared to wild-type salivary glands (Cheshire et al., 2008).

3.3 Pak1 is required for correct salivary gland lumen width

Recent studies from our laboratory demonstrate an essential role for the p21 activated kinase (Pak) 1 in control of salivary gland lumen size through the cell-cell adhesion protein, E-cadherin. Pak proteins are serine-threonine kinases that control vascular integrity in zebrafish blood vessels (Buchner et al., 2007; Liu et al., 2007) and lumen formation by human endothelial cells cultured in three-dimensional collagen matrices (Koh et al., 2008; Koh et al., 2009). In the *Drosophila* embryonic salivary gland, Pak1 functions downstream of the small

GTPase, Cdc42, to regulate gland lumen size (Pirraglia et al., 2010). Loss of *pak1* results in expansion of lumen diameter in the medial and distal regions of the gland without affecting lumen length. The widened lumen of *pak1* mutant salivary glands is not due to increased cell proliferation, and instead, is due to failure to limit apical domain size and to elongate the apical domain in the direction of lumen elongation. These changes in apical domain size and elongation in *pak1* mutant gland cells is accompanied by increased localization of E-cadherin, at the adherens junctions (AJs) and reduced localization at the basolateral membrane (Figure 2D). Pak1 controls this differential localization of E-cadherin in salivary gland cells through Rab5- and Dynamin-dependent endocytosis; not only does inhibition of either *Rab5* or Dynamin in salivary gland cells phenocopy the *pak1* mutant lumen defects, but expression of constitutively-active *Rab5* in *pak1* mutant gland cells restores normal distribution of E-cadherin and restores normal apical domain size and elongation (Pirraglia et al., 2010). Pak1 may regulate E-cadherin endocytosis indirectly through its downstream effector Merlin, the *Drosophila* homologue of the human neurofibromatosis 2 gene (McClatchey and Fehon, 2009), since expression of dominant-negative Merlin phenocopies the salivary gland lumen defects of *pak1* and *Rab5* mutant embryos. Thus, Pak1-dependent localization of E-cadherin at the AJs and at the basolateral membrane is important for apical domain elongation and control of salivary gland lumen size (Figure 2D). A role for Pak1 in lumen size control through membrane transport of E-cadherin is further supported by the demonstration that expression of an activated membrane-bound form of Pak1 in the salivary gland forms multiple intercellular lumens instead of a single central lumen. Induction of multiple intercellular lumens by activated Pak1 is due to the internalization of E-cadherin and apical membrane proteins into early endosomes (Pirraglia et al., 2010).

3.4 Control of salivary gland lumen size through secretory activity

While dynamic changes at the apical membrane and differential localization of E-cadherin control salivary gland lumen size early in gland development, directed secretion into the lumenal space expands lumen width and allows formation of a patent lumen in the mature gland. Secretory products are detected as electron dense material by TEM within apical vesicular structures and in the luminal space. As embryogenesis proceeds, the salivary gland lumen continues to fill with electron-dense secreted products and lumen width increases uniformly throughout the length of the lumen (Myat and Andrew, 2002; Seshaiah et al., 2001). Secretory function of salivary gland cells is controlled by *pasilla* (*ps*) which encodes a *Drosophila* homologue of the human Nova family RNA-binding proteins that function in RNA splicing (Jensen et al., 2000; Seshaiah et al., 2001), and by *PH4αSG1* and *PH4αSG2*, which encode homologues of the α-subunit of resident endoplasmic reticulum enzymes that hydroxylate proline in select secreted proteins (Abrams et al., 2006; Kivirikko and Pihlajaniemi, 1998). In *ps* mutant salivary glands, secretory contents within the lumen and apical vesicles is reduced, and the lumen fails to expand uniformly (Seshaiah et al., 2001). Similar to *ps* mutant embryos, *PH4αSG1* and *PH4αSG2* mutant embryos have reduced secretory products in the salivary gland lumens and are characterized by abnormally shaped lumens with regions of expansion, constriction and closure (Abrams et al., 2006). Together these studies show that *ps*, *PH4αSG1* and *PH4αSG2* control salivary gland lumen size at later stages of embryogenesis by affecting secretion into the gland lumen.

The expression of *PH4αSG1* and *PH4αSG2* is regulated by the single *Drosophila* FoxA family transcription factor Fork head (Fkh), that affects 59% of gene expression in the salivary gland (Maruyama et al., 2011) and is required for cell survival and cell shape change during salivary gland invagination (Myat and Andrew, 2000b). Fkh regulates the expression of *sage*, encoding a salivary gland specific basic helix-loop-helix (bHLH) protein, and functions with Sage to directly regulate the expression of *PH4αSG2* and to indirectly regulate the expression of *PH4αSG1* (Abrams et al., 2006). In addition to Fkh, secretory activity in the *Drosophila* salivary glands is controlled by CrebA which belongs to the CrebA/Creb3-like family of bZip transcription factors (Abrams and Andrew, 2005; Andrew et al., 1997; Fox et al., 2010). CrebA can bind directly to the enhancers of genes encoding components in secretory pathways and upregulate the expression of genes encoding both the general protein machinery required for secretion and of cell type-specific secreted proteins (Fox et al., 2010). Consistent with the role of CrebA in salivary gland secretion, lumens of CrebA mutant embryos are smaller and have reduced secretory material (Fox et al., 2010).

In summary, salivary gland lumen size in early stages of gland development is controlled by apical membrane growth and apical domain elongation in individual gland cells through processes regulated by transcription factors, Hairy, Hkb and Rib, and their downstream targets, Klar, Crb and Moe as well as by Cdc42 and its effector Pak1 through differential localization of E-cadherin. During late embryogenesis, uniform expansion of the gland lumen is controlled by directed secretion into the lumen through the activities of *ps*, *PH4αSG1*, *PH4αSG2* and *CrebA*.

4. Trachea

The *Drosophila* trachea serves as the respiratory organ of the animal, and like the vertebrate lung, salivary gland and vasculature it is a branched network of tubes. The pattern of the larval trachea is established during embryogenesis when cells from ten tracheal placodes or plates of approximately 90 ectodermal epithelial cells on each side of the embryo, invaginate into the underlying mesoderm to form elongated sacs (Figure 3A). In response to Fibroblast Growth Factor (FGF) or Branchless (Bnl), which is expressed in surrounding ectodermal and mesodermal cells (Ohshiro et al., 2002; Sutherland et al., 1996; Zhan et al., 2010), the invaginated tracheal cells which express the FGF receptor, Breathless (Btl), migrate towards the Bnl source to form the six primary branches (Figure 3B and C). Some of the primary branches, such as the visceral branch (VB) and the anterior and posterior dorsal trunk (DT), grow along the anterior-posterior axis, whereas other branches, such as the dorsal branch (DB), lateral trunk (LT) and ganglionic branch (GB), grow along the dorsal-ventral axis (Figure 3D). Tracheal cell migration is followed by fusion between the contralateral DBs, DT and LT branches of adjacent segmentally arranged metameres on each side of the embryo to form an interconnected tracheal network with a single central lumen (Figure 3F).

Similar to the *Drosophila* embryonic salivary gland, the lumen of the trachea is formed during the invagination step when cells of the placode become internalized and form elongated sacs (Casanova, 2007). As the internalized tracheal cells migrate out to form the six primary branches, the lumen extends simultaneously with the elongating branches. In this section, we focus on how lumen size is controlled in the trachea and how lumens form *de novo* at two distinct stages of tracheal development, first, during anastomosis of the tracheal DT, and second, during intracellular lumen formation in the specialized terminal cells.

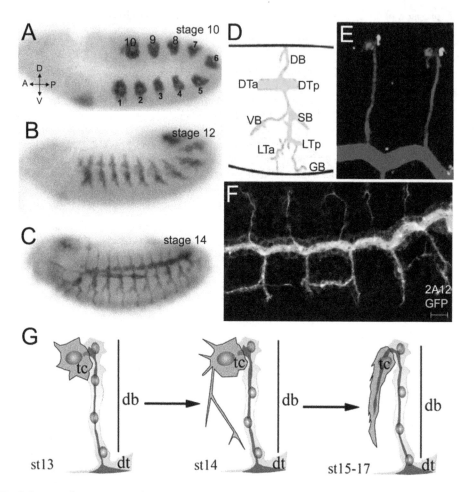

Fig. 3. Lumen formation in the tracheal branches.

(A) The embryonic trachea is formed from 10 placodes of ectodermal cells that invaginate into the interior of the embryo. D: dorsal; V: ventral; A: anterior and P: posterior. (B) Internalized tracheal cells migrate out to form the six primary branches, some of which fuse with branches from adjacent metameres to form an interconnected network (C). (D) Schematic diagram of one tracheal metamere showing the dorsal trunk (anterior and posterior; Dta and Dtp), lateral trunk (anterior and posterior; Lta and Ltp), dorsal branch (DB), visceral branch (VB), spiracular branch (SB) and ganglionic branch (GB). (E) Two DBs, each with a terminal cell (TC, red) and a fusion cell (FC, green). (F) The interconnected tracheal network of a stage 15 embryo has a single central lumen. (G) Terminal cell (TC) forms a lumen *de novo* as the cell elongates; db: dorsal branch, dt: dorsal trunk. Embryos in A-C were processed for RNA *in situ* hybridization to *tracheless* to label tracheal cells. Embryo in E was stained for DSRF (red) to label TC, Dysfusion to label FC (green) and 2A12 (blue) to label the lumen of the DB. Embryo in F was stained for 2A12 (white) to label the lumen and GFP to detect actin-GFP expressed specifically in the trachea with *breathless*-GAL4. Diagram in D is not drawn to scale. Panel G was kindly provided by J. Casanova with permission from the Nature Publishing Group.

4.1 Regulation of tracheal tube/lumen size and shape

Morphometric and genetic analyses in the *Drosophila* embryonic trachea were among the first to show that tube and lumen size are under genetic control. Tracheal tube length increases gradually, whereas tube diameter increases abruptly at distinct times during development. By the larval stage, tracheal tube diameter can be 40x times its original size (Beitel and Krasnow, 2000). These morphometric studies by Beitel and Krasnow (2000) were the first to show that tracheal tube size is not controlled by the number, size or shape of the cells that comprise the tube, and instead, is controlled at the apical surface of the tracheal cells and by the overall identity of each branch. The role of the apical surface membrane in control of tracheal tube size is supported by studies of the *grainy head* (*grh*) mutant (Hemphala et al., 2002). In *grh* mutant embryos, tracheal DT length is increased by 40% and is characterized by the dramatic growth of the apical surface membrane. Grh encodes a transcription factor that is expressed in a number of epithelial tissues (Bray and Kafatos, 1991; Ostrowski et al., 2002), including the epidermis, where Grh controls re-epithelialization during wound healing through the tyrosine kinase Stitcher (Wang et al., 2009). In the trachea, Grh acts downstream of Bnl/FGF signaling to limit lumen elongation and thus, ensure that branches with lumens of the correct size are formed (Hemphala et al., 2002).

A second mechanism by which tracheal lumen size is controlled is through the luminal secretion and modification of chitin, a fibrous substance composed of polysaccharides. Secretion of chitin occurs prior to expansion of the DT lumen and continues throughout growth of the DT lumen. Transient accumulation of chitin is thought to coordinate and stabilize expansion of the lumen (Araujo et al., 2005; Devine et al., 2005; Tonning et al., 2005). While genes encoding proteins that synthesize and secrete chitin control the uniform expansion of tracheal lumen diameter and elongation of lumen length, genes encoding proteins that modify chitin, specifically control lumen length. In mutants where chitin fibers do not form, the DT lumen at points of anastomosis between branches of neighboring hemisegments fail to expand, whereas the lumen throughout the rest of the DT is excessively dilated (Moussian et al., 2006; Tonning et al., 2005). By contrast, in mutants for Vermiform (Verm) and Serpentine (Serp), which encode chitin deacetylating proteins, the DT lumen is excessively elongated (Luschnig et al., 2006; Wang et al., 2006). Restriction of tracheal tube length also depends on genes encoding components of the SJs, a structure located basal to the adherens junctions, which like the vertebrate tight junctions functions as a paracellular diffusion barrier and is comprised in part by the claudin family proteins (Behr et al., 2003; Nelson et al., 2010; Tepass et al., 2001; Wu et al., 2004). Mutations in several SJ proteins affect both tracheal tube length and diameter but not early aspects of tracheal development (Behr et al., 2003; Beitel and Kransnow, 2000; Llimargas et al., 2004; Paul et al., 2003; Wu et al., 2004). One mechanism by which SJ proteins control tracheal tube size is through apical secretion of Verm and Serp (Luschnig et al., 2006; Wang et al., 2006). SJ-associated polarity proteins, such as Discs Large (Dlg) and Scribble (Scrib), also control tracheal tube length independent of chitin and without affecting the paracellular diffusion barrier function (Laprise et al., 2010). The FERM domain protein, Yurt, which belongs to the Yurt/Coracle group of basolateral polarity proteins controls tracheal lumen length by antagonizing the apical determinant protein, Crumbs. However, unlike Yurt, Scrib controls lumen size independent of Crb. The SJ-associated protein, Coracle (Cora), regulates tracheal lumen length by limiting Crb activity independent of Yrt as well as by promoting Verm secretion (Laprise et al., 2010). Thus, independent of Verm/Serp-dependent chitin modification, tracheal lumen length is controlled by a Yurt/Cora pathway dependent on Crb activity, and by a Scrib pathway independent of Crb. SJs likely control tracheal tube size by other mechanisms, such as cell shape since mutations in the SJ proteins, encoded by *megatrachea* and *lachesin*, cause tracheal cells to adopt an irregularly stretched morphology (Behr et al., 2003; Llimargas et al., 2004).

Apical secretion of luminal contents precedes expansion of tube diameter and occurs in a sudden burst through COPI-, COPII- and Sec24-dependent membrane transport (Forster et al., 2010; Grieder et al., 2008; Jayaram et al., 2008; Tsarouhas et al., 2007) and is dependent on Rho-Diaphanous-Myosin V transport (Massarwa et al., 2009). In addition to secretion of chitin, apical secretion may also play a role in tube length by contributing to apical membrane growth and/or targeting other as of yet unidentified, regulators of tube size to the luminal surface. At the end of embryogenesis, the chitin scaffold that has served to control lumen size in the trachea is removed in time for larval hatching when the trachea gets filled with oxygen and other gases. This is achieved through an endocytic pulse that allows the tracheal cells to internalize and clear away the luminal contents (Tsarouhas et al., 2007). The small GTPases, Rab5 and Dynamin are required for the endocytic pulse and luminal protein clearance (Tsarouhas et al., 2007). These studies highlight the important role that chitin plays in tracheal lumen size control; however, it is not known how chitin fibers allow uniform diametric growth of the tracheal tube and restrict tube length. Independent of chitin and SJs, tracheal lumen size is likely to be controlled by additional mechanisms, such as that demonstrated by Convoluted/dALS (Swanson et al., 2009) and by *serrano* mutants which implicate the planar cell polarity pathway in control of tracheal tube length (Chung et al., 2009).

Lumen shape in the tracheal tubes is controlled by receptor tyrosine phosphatases. In embryos double mutant for two receptor-linked protein-tyrosine phosphatases (RPTPs), *Ptp4E* and *Ptp10D*, tracheal branches, such as the ganglionic branches and terminal branches, form large bubble-like cysts with dilated lumens that stain positive for apical marker proteins (Jeon and Zinn, 2009). Cyst size and number are increased upon expression of activated Egfr (epidermal growth factor receptor) and decreased with reduction of Egfr. Thus, proper lumen shape in the trachea is achieved through downregulation of Egfr signaling by the Ptp4E and Ptp10D RPTPs.

4.2 Dorsal trunk anastomosis

During primary branch outgrowth, the tracheal lumen is initially closed at the branch tips. Later in development, a continuous tubular network is formed during anastomosis, when specialized cells, known as fusion cells, which are found at the tips of migrating branches such as the DT and DB, recognize each other's partner in the adjacent metamere and connect to form a continuous lumen (Figure 3E) (Baer et al., 2009). Although these specialized cells are called fusion cells, they, in fact, do not fuse themselves and instead mediate the fusion of two separate tubular structures. The tracheal fusion process occurs in four distinct steps. In the first step, tracheal cells at the tip of adjacent branches contact each other through filopodial extensions. In the second step, fusion cells form a cytoskeletal track at the site of contact consisting of F-actin, microtubules, the plakin Short Stop (Shot) and E-cadherin-based adhesion complexes that are assembled *de novo* at the contact site (Lee and Kolodziej, 2002). Structure function studies showed that distinct sites within the cytoplasmic domain of E-cadherin control the initial assembly of the F-actin track, recruitment of Shot and subsequent maturation of the track in a microtubule-dependent manner (Lee et al., 2003). In the third step, the cytoskeletal track expands to span the fusion cells and bridge the apical surfaces of the DT lumens with apical membrane formed *de novo* at the contact site. In the fourth and final step, the cytoskeletal track disassembles, the apical surfaces become continuous and the narrow lumen that is initially formed expands to its final size (Baer et al., 2009). Connection of the pre-existing DT lumens to the new lumen is dependent on targeted exocytosis and remodeling of the plasma membrane by the Arf-like 3 small GTPase (Arl3) which is known to associate with microtubules and vesicles (Jiang et al., 2007; Kakihara et al., 2008), and the COPI coatomer complex that mediates membrane transport of small vesicles (Grieder et al., 2008). Therefore, tracheal branch fusion is a complex and highly regulated process involving precise coordination of cytoskeletal proteins, adhesion proteins and components of the vesicular trafficking machinery.

4.3 Terminal cell lumen formation

Terminal cells (TCs) at the tips of some tracheal branches form intracellular lumens *de novo* (Figure 3E and G). Although *de novo* lumen formation in TCs was initially thought to occur by the "cell hollowing" mechanism (Lubarsky and Krasnow, 2003), recent studies by Gervais and Casanova (2010) show that the intracellular lumen forms by the inward growth of new apical membrane from the surface that is in contact with the adjacent tracheal cell and not through a cell-hollowing mechanism, shedding significant insight into this process. The TC elongates as its lumen is formed intracellularly and both these processes are accompanied by the asymmetric accumulation of the actin and microtubule cytoskeletal systems (Figure 3G). Genetic perturbation of the microtubule network results in defects in TC lumen elongation suggesting a critical role for microtubules in TC lumen formation. The Bnl/FGF signaling pathway, known to regulate multiple aspects of tracheal morphogenesis as described above, also regulates TC lumen elongation; in embryos with reduced gene dosage of *bnl*, TC lumen length is shortened (Zhan et al., 2010). Bnl/FGF signaling controls TC elongation and intracellular lumen formation by regulating actin and microtubules through Drosophila Serum Response Factor (DSRF) and Enabled, a VASP protein (Gervais and Casanova, 2010); however, DSRF is not required for Bnl-dependent initiation of TC elongation and lumen formation, and instead allows these processes to progress under normal conditions (Gervais and Casanova, 2011).

In addition to the requirement of Bnl signaling for intiation and progression of TC elongation and lumen formation, integrin-mediated adhesion between the terminal branches and the surrounding extracellular matrix is necessary for maintaining these tubes and for proper organization of the intracellular lumen (Levi et al., 2006). The amenability of *Drosophila* to large-scale genetic screens has allowed the generation of many new mutants affecting TC lumen formation (Ghabrial et al., 2011). Analysis of these new mutants is bound to bring novel insights to lumen formation in the tracheal TCs in the years to come.

5. Conclusion

In this chapter, we have reviewed our current understanding of how lumens form and are maintained in the dorsal vessel, salivary gland and trachea of the *Drosophila* embryo. Lumen formation in the *Drosophila* embryonic salivary gland and primary branches of the trachea occurs concomitantly with invagination of the salivary gland and tracheal cells from the embryo surface. Thus, it is not entirely surprising that lumen size control in these two epithelial-based organs share similar cellular and molecular mechanisms, such as the roles of apical membrane growth and luminal secretion in defining tube and lumen size. Although a role for chitin fibers in uniform diametric growth and restriction of tube length has not been documented for the salivary gland as in the trachea, evidence does exist for luminal secretion in forming a patent lumen in the salivary gland.

The dorsal vessel forms by entirely distinct mechanisms from that of the salivary gland and trachea; however, in terms of lumen formation and size control, there are conserved mechanisms between the dorsal vessel and salivary gland and trachea. For example, E-cadherin-mediated cell-cell adhesion is important for correct lumen size in the salivary gland and for forming a luminal space between CBs in the dorsal vessel. Although endocytic trafficking of E-cadherin is important for lumen size control in the salivary gland, it is not known whether Slit/Robo inhibition of E-cadherin occurs by a similar or distinct mechanism. In addition to E-cadherin, SJ proteins, such as Coracle, are required for correct lumen size in the trachea and the dorsal vessel. In the trachea, it is well established that synthesis, secretion and modification of chitin affects tube and lumen size. Although dorsal vessel shape is affected in SJ mutants, it is not known whether vessel diameter and/or length are affected as well. It was recently reported that the lumen of the mouse dorsal aorta forms by a "cord-hollowing" mechanism where lumen formation between two cells is initiated extracellularly through repositioning of cadherin-based adherens junctions and through repulsion of apposed lateral membranes (Lubarsky and Krasnow, 2003; Strillic et al., 2009). Due to the similarities between lumen formation in the *Drosophila* dorsal vessel and the mouse aorta, studies of lumen formation in the *Drosophila* dorsal vessel will continue to yield insight into lumen formation in the vertebrate vasculature.

Studies in the *Drosophila* embryonic salivary gland showed that growth of the apical membrane and modulation of E-cadherin localization at the adherens junctions and the basolateral membrane can influence lumen size and number. In the *Drosophila* embryonic salivary gland, the regulated process of invagination ensures that only a single central lumen is formed; however, the single central lumen can be replaced by multiple intercellular lumens, such as by expression of activated Pak1, as described above. By contrast, the formation of multiple lumens is a normal intermediate step in the formation of a single central lumen during zebrafish gut tube morphogenesis (Bagnat et al., 2007) and in pathological conditions, such as pre-invasive breast cancer, where multiple lumens characterize cribriform ductal carcinoma *in situ* (DCIS) (Jaffar and Bleiweiss, 2002). Thus, understanding the mechanisms by which tubular organs can transition between single and multiple lumens will increase our understanding of more complex processes, such as DCIS.

In the *Drosophila* embryonic trachea, *de novo* lumen formation occurs during anastomosis of specific branches, such as the DT, between adjacent hemisegments. A similar process of *de novo* lumen formation occurs during anastomosis of vascular sprouts during angiogenesis. Like the tracheal tip cells of the *Drosophila* trachea, vascular tip cells extend filopodia to explore the surrounding enrivonment. Moreover, the presence of vascular E-cadherin (VE-cadherin) at the tips of filopodia in cultured human endothelial cells (Almagro et al., 2010) and at tip-tip contact sites between neighboring sprouts during formation of the zebrafish intersegmental vessel (Blum et al., 2008) suggest a role for VE-cadherin in vascular anastomosis that parallels the role played by E-cadherin in *Drosophila* tracheal anastomosis.

Although the structure and function of the *Drosophila* embryonic dorsal vessel, salivary gland and trachea may differ from more complex organs of other organisms, it is clear that there are conserved mechanisms for lumen formation. Thus, the study of tube and lumen formation in *Drosophila* tubular organs will continue to yield novel mechanisms and shed significant insight into how lumens form in more complex organisms.

6. Acknowledgements

We are grateful to F. Macabenta and S. Kramer for providing Figure 1C and for critical reading of the manuscript. We also thank J. Casanova for providing Figure 3G.

7. References

Abrams, E. & Andrew, D. (2005). CrebA regulates secretory activity in the *Drosophila* salivary gland and epidermis. *Development* 132, 2743-2758.

Abrams, E., Milhoulides, W. & Andrew, D. (2006). Fork head and Sage maintain a uniform and patent salivary galnd lumen through regulation of two downstream target genes, PH4alphaSG1 and PH4alphaSG2. *Development* 133, 3517-3527.

Albrecht, S., Altenhein, B. & Paululat, A. (2011). The transmembrane receptor Uncoordinated5 (Unc5) is essential for heart lumen formation in *Drosophila melanogaster*. *Developmental Biology* 350 89-100.

Almagro, S., Durmort, C., Chervin-Petinot, A., Heyraud, S., Dubois, M., Lambert, O., Maillefaud, C., Hewat, E., Schaal, J., Huber, P. et al. (2010). The motor protein myosin-X transports VE-cadherin along filopodia to allow the formation of early endothelial cell-cell contacts. *Mol Cell Biol* 30, 1703-1717.

Andrew, D., Baig, A., Bhanot, P., Smolik, S. & Henderson, K. (1997). The Drosophila dCreb-A gene is required for dorsal/ventral patterning of the larval cuticle. *Development*, 181-193.

Andrew, D. & Ewald, A. (2010). Morphogenesis of epithelial tubes: Insights into tube formation, elongation and elaboration. *Dev Biol* 341, 34-55.

Araujo, S., Aslam, H., Tear, G. & Casanova, J. (2005). mummy/cystic encodes an enzyme required for chitin and glycan synthesis, involved in trachea, embryonic cuticle and CNS development - Analysis of its role in Drosophila tracheal morphogenesis. *Developmental Biology* 288, 179-193.

Baer, M. M., Chanut-Delalande, H. & Affolter, M. (2009). Cellular and Molecular Mechanisms Underlying the Formation of Biological Tubes. *Current Topics in Developmental Biology* 89, 137-162.

Bagnat, M., Cheung, I., Mostov, K. & Stainier, D. (2007). Genetic control of single lumen formation in the zebrafish gut. *Nat Cell Biol* 954-960.

Banerjee, S., Sousa, A. & Bhat, M. (2006). Organization and function of septate junctions: an evolutionary perspective. *Cell Biochem Biophys* 46, 65-77.

Behr, M., Riedel, D. & Schuh, R. (2003). The Claudin-like Megatrachea is essential in septate junctions for the epithelial barrier function in *Drosophila*. *Developmental Cell* 5, 611-620.

Beitel, G. & Krasnow, M. A. (2000). Genetic control of epithelial tube size in the *Drosophila* tracheal system. *Development* 127, 3271-3282.

Blake, K., Myette, G. & Jack, J. (1998). The products of ribbon and raw are necessary for proper cell shape and cellular localization of nonmuscle myosin in *Drosophila*. *Dev Biol* 147, 177-88.

Blum, Y., Belting, H., Ellertsdottir, E., Herwig, L., Luders, F. & Affolter, M. (2008). Complex cell rearrangements during intersegmental vessel sprouting and vessel fusion in the zebrafish embryo. *Dev Biol* 316, 312-322.

Bradley, P. B. & Andrew, D. J. (2001). Ribbon encodes a novel BTB/POZ protein required for directed cell migration in *Drosophila melanogaster*. *Development* 128, 3001-3015.

Bray, S. & Kafatos, F. (1991). Developmental function of Elf-1: an essential transcription factor during embryogenesis in Drosophila. *Genes Dev* 5, 1672-1683.

Buchner, D., Su, F., Yamaoka, J., Kamei, M., Shavit, J., Barthel, L., McGee, B., Amigo, J., Kim, S., Hanosh, A. et al. (2007). pak2a mutations cause cerebral hemorrhage in *redhead* zebrafish. *PNAS* 104, 13996-14001.

Bulgakova, N. & Knust, E. (2009). The Crumbs complex: from epithelial cell-polarity to retinal degeneration. *J Cell Sci* 122, 2587-2596.

Carroll, S., Laughon, A. & Thalley, B. (1988). Expression, function and regulation of the hairy segmentation protein in the *Drosophila* embryo. *Genes and Development* 2, 883-890.

Casanova, J. (2007). The emergence of shape: notions from the study of the Drosophila tracheal system. *EMBO Rep.* 8, 335-339.

Chanana, B., Steigemann, P., Jackle, H. & Vorbruggen, G. (2009). Reception of Slit requires only the chondroitin-sulphate-modified extracellular domain of Syndecan at the target cell surface. *PNAS* 106, 11984-11988.

Chartier, A., Zaffran, S., Astier, M., Semeriva, M. & Gratecos, D. (2002). Pericardin, a *Drosophila* type IV collagen-like protein is involved in the morphogenesis and maintenance of the heart   epithelium during dorsal ectoderm closure. *Development* 129, 3241-3253.

Cheshire, A., Kerman, B., Zipfel, W., Spector, A. & Andrew, D. (2008). Kinetic and mechanical analysis of live tube morphogenesis. *Developmental Dynamics*, 2874-2888.

Chung, S., Vining, M., Bradley, P., Chan, C., Wharton, K. J. & Andrew, D. (2009). Serrano (sano) functions with the planar cell polarity genes to control tracheal tube length. *PLoS Genet* 5, e1000746.

Devine, P., Lubarsky, B., Shaw, K., Luschnig, S., Messina, L. & Krasnow, M. A. (2005). Requirement for chitin biosynthesis in epithelial tube morphogenesis. *PNAS* 102, 17014-17019.

Forster, D., Armbruster, K. & Luschnig, S. (2010). Sec24-dependent secretion drives cell-autonomous expansion of tracheal tubes in Drosophila. *Current Biology* 20, 62-68.

Fox, R., Hanlon, C. & Andrew, D. (2010). The CrebA/Creb-like transcription factors are major and direct regulators of secretory capacity. *J Cell Biol* 191, 479-92.

Gervais, L. & Casanova, J. (2010). In vivo coupling of cell elongation and lumen formation in a single cell. *Current Biology* 20, 359-366.

Gervais, L. & Casanova, J. (2011). The Drosophila homologue of SRF acts as a boosting mechanism to sustain FGF-induced terminal branching in the tracheal system. *Development* 138, 1269-1274.

Ghabrial, A., Levi, B. & Krasnow, M. (2011). A systematic screen for tube morphogenesis and branching genes in the Drosophila tracheal system. *PLoS Genet* 7, e1002087.

Grieder, N. C., Caussinus, E., Parker, D. S., Cadigan, K. & Affolter, M. (2008). γCOP is required for apical secretion and epithelial morphogenesis in *Drosophila melanogaster*. *PLoS One* 3, e3241.

Guo, Y., Jangi, S. & Welte, M. (2005). Organelle-specific control of intracellular transport: distinctly targeted isoforms of the regulator Klar. *Mol Biol Cell.* 16, 1406-1416.

Haag, T., Haag, N., Lekven, A. & Hartenstein, V. (1999). The Role of Cell Adhesion Molecules in *Drosophila* Heart Morphogenesis: *Faint Sausage, Shotgun*/DE-Cadherin, and *Laminin A* Are Required for Discrete Stages in Heart Development. *Developmental Biology* 208, 56-69.

Hemphala, J., Uv, A., Cantera, R., Bray, S. & Samakovlis, C. (2002). Grainy head controls apical membrane growth and tube elongation in response to Branchless/FGF signaling. *Development* 130, 249-258.

Hooper, K. L., Parkhurst, S. M. & Ish-Horowicz, D. (1989). Spatial control of *hairy* protein expression during embryogenesis. *Development* 107, 489-504.

Izaddoost, S., Nam, S. C., Bhat, M. A., Bellen, H. J. & Choi, K. W. (2002). *Drosophila* crumbs is a positional cue in photoreceptor adherens junctions and rhabdomeres. *Nature* 416, 178-183.

Jack, J. & Myette, G. (1997). The genes raw and ribbon are required for proper shape of tubular epithelial tissues in *Drosophila*. *Genetics* 147, 243-253.

Jaffar, S. & Bleiweiss, I. (2002). Histologic classification of ductal carcinoma in situ. *Microscopy research and technique* 59, 92-101.

Jayaram, S., Senti, K., Tiklova, K., Tsahouras, V., Hemphala, J. & Samakovlis, C. (2008). COPI vesicle transport is a common requirement for tube expansion in Drosophila. *PLoS One* 3, e1964.

Jensen, K., Dredge, B., Stefani, G., Zhong, R., Buckanovich, R., Okano, H., Yang, Y. & Darnell, R. (2000). Nova-1 regulates neuron-specific alternative splicing and is essential for neuronal specificity. *Neuron* 25, 359-371.

Jeon, M. & Zinn, K. (2009). Receptor tyrosine phosphatases control tracheal tube geometries through negative regulation of Egfr signaling. *Development* 136, 3121-3129.

Jiang, L., Rogers, S. L. & Crews, S. T. (2007). The *Drosophila* dead end Arf-like 3 GTPase controls vesicle trafficking during tracheal fusion cell morphogenesis. *Dev. Biol* 311, 487-499.

Kakihara, K., Shinmyozu, K., Kato, K., Wada, H. & Hayashi, S. (2008). Conversion of plasma membrane topology during epithelial tube connection requires Arf-like 3 GTPase in *Drosophila*. *Mechanisms of Development* 125, 325-336.

Karagiosis, S. & Ready, D. (2004). Moesin contributes an essential structural role in *Drosophila* photoreceptor morphogenesis. *Development* 131, 725-32.

Keleman, K. & Dickson, B. (2001). Short- and Long-Range Repulsion by the *Drosophila* Unc5 Nectrin Receptor. *Neuron* 32, 605-617.

Kerman, B., Chesire, A., Myat, M. & Andrew, D. (2008). Ribbon modulates apical membrane during tube elongation through Crumbs and Moesin. *Developmental Biology* 320, 278-288.

Kidd, T., Bland, K. & Goodman, C. (1999). Slit is the midline repellent for the robo receptor in *Drosophila*. *Cell* 96, 785-94.

Kivirikko, K. & Pihlajaniemi, T. (1998). Collagen hydroxylases and the protein disulfide isomerase subunit of prolyl 4-hydroxylases. *Adv Enzymol Relat Areas Mol Biol* 72, 325-98.

Knox, J., Moyer, K., Yacoub, N., Soldaat, C., Komosa, M., Vassilieva, K., Wilk, R., Hu, J., de Lourdes Vasquez Pas, L., Syed, Q. et al. (2011). Syndecan contributes to heart cell specification and lumen formation during *Drosophila* cardiogenesis. *Developmental Biology* 356, 279-290.

Koh, W., Mahan, R. & Davis, G. (2008). Cdc42- and Rac1-mediated endothelial lumen formation requires Pak2, Pak4 and Par3, and PKC-dependent signaling. *J Cell Sci* 121, 989-1001.

Koh, W., Sachidanandam, K., Stratman, A., Sacharidou, A., Mayo, A., Murphy, E., Cheresh, D. &Davis, G. (2009). Formation of endothelial lumens requires a coordinated PKC_{E-}, Src-, Pak- and Raf- kinase-dependent signaling cascade downstream of Cdc42 activation. *J Cell Sci* 122, 1812-1822.

Kracklauer, M., Banks, S., Xie, X., Wu, Y. & Fischer, J. (2007). Drosophila klaroid encodes a SUN domain protein required for Klarsicht localization to the nuclear envelope and nuclear migration in the eye. *Fly (Austin)* 1, 75-85.

Laprise, P., Paul, S., Boulanger, J., Robbins, R., Beitel, G. J. & Tepass, U. (2010). Epithelial polarity proteins regulate *Drosophila* tracheal tube size in parallel to the luminal matrix pathway. *Current Biology* 20, 55-61.

Lee, M., Lee, S., Zadesh, A. D. & Kolodziej, P. A. (2003). Distinct sites in E-cadherin regulate different steps in *Drosophila* tracheal tube fusion. *Development* 130, 5989-5999.

Lee, S. & Kolodziej, P. A. (2002). The plakin Short Stop and the RhoA GTPase are requiered for E- cadherin-dependent apical surface remodeling during tracheal tube fusion. *Development* 129, 1509-1520.

Levi, B., Ghabrial, A. & Krasnow, M. A. (2006). Drosophila talin and integrin genes are required for maintenance of tracheal terminal branches and luminal organization. *Development* 133, 2383-93.

Liu, J., Fraser, S., Faloon, P., Rollins, E., Vom Berg, J., Starovic-Subota, O., Laliberte, A., Chen, J., Serluca, F. & Childs, S. (2007). A betaPix pak2a signaling pathway regulates cerebral vascular stability in zebrafish. *Proc Natl Acad Sci USA* 104, 13990-13995.

Llimargas, M., Strigini, M., Katidou, M., Karagogeos, D. & Casanova, J. (2004). Lachesin is a component of a septate junction-based mechanism that controls tube size and epithelial integrity in the *Drosophila* tracheal system. *Development* 131, 181-190.

Lubarsky, B. & Krasnow, M. A. (2003). Tube morphogenesis: making and shaping biological tubes. *Cell* 112, 19-28.

Luschnig, S., Batz, T., Armbruster, K. & Krasnow, M. A. (2006). Serpentine and Vermiform encode matrix proteins with chitin binding and deacetylation domains that limit tracheal tube length in *Drosophila*. *Current Biology* 16, 186-194.

Malbon, C. (2005). G proteins in development *Nat Rev Mol Cell Biol* 6, 689-701.

Maruyama, R., Grevengoed, E., Stempniewicz, P. & Andrew, D. (2011). Genome-wide analysis reveals a major role in cell fate maintenance and an unexpected role in endoreduplication for the Drosophila FoxA gene Fork head. *PLoS One* 6, e20901.

Massarwa, R., Schejter, E. & Shilo, B. (2009). Apical secretion in epithelial tubes of the Drosophila embryo is directed by the Formin-family protein Diaphanous. *Dev Cell.* 16, 877-888.

McClatchey, A. & Fehon, R. (2009). Merlin and the ERM proteins--regulators of receptor distribution and signaling at the cell cortex. *Trends Cell Biol* 19, 198-206.

Medioni, C., Astier, M., Zmojdzian, M., Jagla, K. & Semeriva, M. (2008). Genetic control of cell morphogenesis during *Drosophila melanogaster* cardiac tube formation. *J Cell Biol* 182, 249-261.

Mosley-Bishop, K. L., Li, Q., Patterson, L. & Fischer, J. A. (1999). Molecular analysis of the *klarsicht* gene and its role in nuclear migration within differentiating cells of the *Drosophila* eye. *Current Biology* 9, 1211-1220.

Moussian, B., Tang, E., Tonning, A., Helms, S., Schwarz, H., Nusslein-Volhard, C. & Uv, A. (2006). Drosophila Knickkopf and Retroactive are needed for epithelial tube growth and cuticle differentiation through their specific requirement for chitin filament organization. *Development* 133, 163-171.

Myat, M. M. & Andrew, D. J. (2000a). Organ shape in the *Drosophila* salivary gland is controlled by regulated, sequential internalization of the primordia. *Development* 127, 679-691.

Myat, M. M. & Andrew, D. J. (2000b). Fork head prevents apoptosis and promotes cell shape change during formation of the *Drosophila* salivary glands. *Development* 127, 4217-4226.

Myat, M. M. & Andrew, D. J. (2002). Epithelial tube morphology is determined by the polarized growth and delivery of apical membrane. *Cell* 111, 879-891.

Nelson, K., Furuse, M. & Beitel, G. (2010). The Drosophila Claudin Kune-kune is required for septate junction organization and tracheal tube size control. *Genetics* 185, 831-839.

Ohshiro, T., Emori, Y. & Saigo, K. (2002). Ligand-dependent activation of breathless FGF receptor gene in *Drosophila* developing trachea.. *Mechanisms of Development* 114, 3-11.

Ostrowski, S., Dierick, H. & Bejsovec, A. (2002). Genetic control of cuticle formation during embryonic development of Drosophila melanogaster. *Genetics* 161, 171-182.

Paul, S. M., Ternet, M., Salvaterra, P. M. & Beitel, G. J. (2003). The Na+/K+ ATPase is required for septate junction function and epithelial tube-size control in the *Drosophila*. *Development* 130, 4963-4974.

Pellikka, M., Tanentzapf, G., Pinto, M., Smith, C., McGlade, C. J., Ready, D. F. & Tepass, U. (2002). Crumbs, the *Drosophila* homologue of human CRB1/RP12, is essential for photoreceptor morphogenesis. *Nature* 416, 143-149.

Pirraglia, C. & Myat, M. M. (2010). Genetic Regulation of Salivary Gland Development in *Drosophila melanogaster*. In *Salivary Glands: Development, Adaptations and Disease.*, vol. 14 (ed. A. Tucker and I. Miletich), pp. 32-47: S. Karger.

Pirraglia, C., Walters, J. & Myat, M. M. (2010). Pak1 control of E-cadherin endocytosis regulates salivary gland lumen size and shape. *Development* 137, 4177-4189.

Qian, L., Liu, J. & Bodmer, R. (2005). Slit and Robo Control Cardiac Cell Polarity and Morphogenesis. *Current Biology* 15, 2271-2278.

Reim, I. & Frasch, M. (2010). Genetic and genomic dissection of cardiogenesis in the Drosophila model. *Pediatr Cardiol* 31, 325-334.

Rothberg, J., Jacobs, J., Goodman, C. & Artavanis-Tsakonas, S. (1990). Slit: an extracellular protein necessary for development of the midline glia and commissural axon pathway contains both EGF and LRR domains. *Genes and Development* 4, 2169-2187.

Santiago-Martinez, E., Soplop, N., Patel, R. & Kramer, S. (2008). Repulsion by Slit and Roundabout prevents Shotgun/E-cadherin-mediated cell adhesion during *Drosophila* heart tube lumen formation. *J Cell Biol* 182, 241-248.

Satoh, A., O'Tousa, J., Ozaki, K. & Ready, D. (2005). Rab11 mediates post-Golgi trafficking of rhodopsin to the photosensitive apical membrane of Drosophila photoreceptors. *Development* 132, 1487-97.

Seshaiah, P., Miller, B., Myat, M. M. & Andrew, D. J. (2001). Pasilla, the *Drosophila* homologue of the human Nova-1 and Nova-2 proteins, is required for normal secretion in the salivary gland. *Developmental Biology* 239, 309-22.

Shim, K., Blake, K. J., Jack, J. & Krasnow, M. A. (2001). The *Drosophila* ribbon gene encodes a nuclear BTB domain protein that promotes epithelial migration and morphogenesis. *Development* 128, 4923-4933.

Strillic, B., Kucera, T., Eglinger, J., Huges, M., McNagny, K., Tsukita, S., Dejana, E., Ferrara, N. & Lamert, E. (2009). The molecular basis of vascular lumen formation in the developing mouse aorta. *Developmental Cell* 17, 505-515.

Sutherland, D., Samakovlis, C. & Kransnow, M. A. (1996). Branchless encodes a *Drosophila* FGF homolog that controls tracheal cell migration and the pattern of branching. *Cell* 87, 1091-1101.

Swanson, L., Yu, M., Nelson, K., Laprise, P., Tepass, U. & Beitel, G. (2009). Drosophila convoluted/dALS is an essential gene required for tracheal tube morphogenesis and apical matrix organization. *Genetics* 181, 1281-1290.

Tao, Y. & Schulz, R. (2007). Heart development in *Drosophila*. *Seminars in Cell & Developmental Biology* 18, 3-15.

Tepass, U., Gruszynski-DeFeo, E., Haag, T. A., Omatyar, L., Torok, T. & Hartenstein, V. (1996). Shotgun encodes *Drosophila* E-cadherin and is preferentially required during cell rearrangement in the neurectoderm and other morphogenetically active epithelia. *Genes & Development* 10, 672-685.

Tepass, U. & Knust, E. (1990). Phenotypic and developmental analysis of mutations at the crumbs locus, a gene required for the development of epithelia in *Drosophila* melanogaster. *Roux's Arch Dev   Biol* 199, 189-206.

Tepass, U., Tanentzapf, G., Ward, R. & Fehon, R. (2001). Epithelial cell polarity and cell junctions in *Drosophila. Annu Rev Genet* 35, 747-784.

Tonning, A., Hemphala, J., Tang, E., Nannmark, U., Samakovlis, C. & Uv, A. (2005). A transcient luminal chitinous matrix is required to model epithelial tube diameter in the *Drosophila* Trachea. *Developmental Cell* 9, 423-430.

Tsarouhas, V., Senti, K., Jayaram, S., Tiklova, K., Hemphala, J., Adler, J. & Samakovlis, C. (2007). Sequential pulses of apical epithelial secretion and endocytosis drive airway maturation in drosophila. *Dev Cell* 13.,214-225.

von Hilchen, C., Hein, I., Technau, G. & Altenhein, B. (2010). Netrins guide migration of distinct glial cells in the *Drosophila* embryo. *Development* 137, 1251-1262.

Wang, S., Jayaram, S., Hemphala, J., Senti, K., Tsarouhas, V., Jin, H. & Samakovlis, C. (2006). Septate- junction-dependent luminal deposition of chitin deacetylases restricts tube elongation in the *Drosophila* trachea. *Current Biology* 16, 180-185.

Wang, S., Tsahouras, V., Xylourgidis, N., Sabri, N., Tiklova, K., Nautiyal, N., Gallio, M. & Samakovlis, C. (2009). The tyrosine kinase Stitcher activates Grainy head and epidermal wound healing in Drosophila. *Nat Cell Biol* 11, 890-895.

Welte, M. A., Gross, S. P., Postner, M., Block, S. M. & Wieschaus, E. F. (1998). Developmental Regulation of Vesicle Transport in *Drosophila* Embryos: Forces and Kinetics. *Cell* 92, 547-557.

Wodarz, A., Hinz, U., Engelbert, M. & Knust, E. (1995). Expression of Crumbs confers apical character on plasma membrane domains of ectodermal epithelia of *Drosophila. Cell* 82, 67-76.

Wu, V. M., Schulte, J., Hirschi, A., Tepass, U. & Beitel, G. J. (2004). Sinuous is a *Drosophila* claudin required for septate junction organization and epithelial tube size control. *The Journal of Cell Biology* 164, 313-323.

Yi, P., Han, Z., Li, X. & Olson, E. (2006). The Mevalonate Pathway Controls Heart Formation in Drosophila by Isoprenylation of Gγl. *Science* 313, 1301-1303.

Yi, P., Johnson, A., Han, Z., Wu, J. & Olson, E. (2008). Heterotrimeric G Proteins Regulate a Noncanonical Function of Septate Junction Proteins to Maintain Cardiac Integrity in *Drosophila. Developmental Cell* 15, 704-713.

Zhan, Y., Maung, S. W., Shao, B. & Myat, M. M. (2010). The bHLH transcription factor, hairy, refines the terminal cell fate in the *Drosophila* embryonic trachea. *PLoS One* 5, e14134.

Zhang, J., Felder, A., Liu, Y., Guo, L., Lange, S., Dalton, N., Gu, Y., Peterson, K., Mizisin, A., Shelton, G. et al. (2010). Nesprin 1 is critical for nuclear positioning and anchorage. *Hum Mol Genet* 19, 329-41.

Zhang, Q., Skepper, J., Yang, F., Davies, J., Hegyi, L., Roberts, R., Weissberg, P., Ellis, J. & Shanahan, C. (2001). Nesprins: a novel family of spectrin-repeat-containing proteins that localize to the nuclear membrane in multiple tissues. *J Cell Sci* 114, 4485-98.

6

Phospho-Signaling at Oocyte Maturation and Fertilization: Set Up for Embryogenesis and Beyond Part II. Kinase Regulators and Substrates

A.K.M. Mahbub Hasan[1], Takashi Matsumoto[2],
Shigeru Kihira[2], Junpei Yoshida[2] and Ken-ichi Sato[2,*]
[1]*Laboratory of Gene Biology, Department of Biochemistry and Molecular Biology,*
University of Dhaka, Dhaka,
[2]*Laboratory of Cell Signaling and Development, Department of Molecular Biosciences,*
Faculty of Life Sciences, Kyoto Sangyo University,
Kamigamo-Motoyama, Kita-ku, Kyoto
[1]*Bangladesh*
[2]*Japan*

1. Introduction

This chapter is the sequel to the chapter entitled "Phospho-signaling at Oocyte Maturation and Fertilization: Set Up for Embryogenesis and Beyond Part I. Protein Kinases" by Mahbub Hasan et al.

2. Kinase regulators and substrates in oocyte maturation, fertilization and activation of development

2.1 Actin

Filamentous actin or **F-actin** is a major component of stress fibers and involved in cellular architecture. Its dynamic rearrangement supports not only cellular morphology but also intracellular signal transduction that regulate cell-cell or cell-extracellular matrix interactions, cell motility, and proliferation. Several lines of evidence demonstrate that, in several organisms, oocyte cortical cytoskeleton involving F-actin network undergoes a dynamic rearrangement during meiosis/oocyte maturation and that this is often involving phosphorylation of actin and/or actin-interacting proteins (e.g. ADF/coffilin, see below) catalyzed by PKC (in Tubifex, *Xenopus*) (Capco et al. 1992; Shimizu 1997). In *Drosophila*, PKC phosphorylation of a tumor suppressor protein-homolog named Lgl (lethal (2) giant larvae) is responsible for actin-dependent oocyte polarity formation (Tian and Deng 2008). In mammalian oocytes (rat), F-actin has been implicated in tyrosine kinase-dependent

---

rearrangement of cortical structures (Meng et al. 2006). In unfertilized rat eggs, F-action is in association with PKC and RACKS and thought to suppress the cortical granule to exocytose, and after fertilization, PKC-dependent phosphorylation releases the actin suppression and cortical granule exocytosis occurs (Eliyahu et al. 2005).

2.2 ADF/coffilin

Actin-depolymerizing factor (ADF)/coffilin are an evolutionarily conserved F-actin-binding protein, whose function is essential for cortical actin cytoskeleton. It is well known that the actin-binding ability of ADF/coffilin can be regulated by its phosphorylation and dephosphorylation (Bamburg et al. 1999). This type of regulation of ADF/coffilin has been reported in maturing oocytes of starfish, where active transport of MPF from nucleus to cytoplasm is required for oocyte maturation (Santella et al. 2003), and dividing embryos of Xenopus, where cytokinesis involves the function of ADF/coffilin (Abe et al. 1996; Chiu et al. 2010; Tanaka et al. 2005). In the former case, MPF has been identified as a kinase for ADF/coffilin. In the latter case, protein phosphatase Slingshot is involved in Rho-dependent inactivation of ADF/coffilin, thereby promotes the rearrangement of actin cytoskeleton essential for cytokinesis.

2.3 ASIP/PAR-3

ASIP/PAR-3 (atypical PKC isotype-specific interacting protein/partitioning defective 3) is a PDZ-domain-containing adaptor protein that has been initially identified as a downstream element of PAR-6 in early embryos of the nematode C. elegans (Watts et al. 1996). Further studies have demonstrate the importance of PAR3 as an atypical PKC (aPKC)-interacting protein functioning in establishing asymmetric cell division and polarized cell structures in C. elegans and Drosophila embryos, and mammalian epithelial cells (Joberty et al. 2000). In Xenopus immature oocytes, ASIP/PAR-3 is shown to localize to animal hemisphere in association with aPKC, and upon hormone-induced oocyte maturation, aPKC undergoes kinase activity-dependent re-localization. These results suggest a potential role of ASIP/PAR-3 as a regulator and/or substrate of aPKC (Nakaya et al. 2000). Although phosphorylation of Ser-827 in ASIP/PAR-3 by aPKC has been shown in mammalian somatic cell systems (Hirose et al. 2002), its occurrence in oocyte/egg system is not yet demonstrated.

2.4 Astrin

Astrin is a spindle-associated non-motor protein that regulates mitotic cell cycle progression. In the meiosis of mouse oocytes, where centrioles are missing but multiple microtubule-organizing centers (MTOCs) are present, proper lining and segregation of homologous chromosomes and sister chromatids require the precise regulation of MTOCs by centrosomal protein kinases such as Aurora kinase and PLK1. It has been shown that inhibition of Astrin function by RNAi-mediated knockdown or overexpression of a coiled-coil domain of Astrin results in a defect in spindle disorganization, chromosome misalignment and meiosis progression arrest (Yuan et al. 2009). As Astrin localizes to the spindle apparatus, it is suggested that Astrin is a substrate of Aurora/PLK1. In support with this idea, site-directed mutation of Thr-24, Ser-66 or Ser-447, potential PLK1

Phospho-Signaling at Oocyte Maturation and Fertilization: Set Up for Embryogenesis
and Beyond Part II. Kinase Regulators and Substrates

91

phosphorylation sites in Astrin, causes oocyte meiotic arrest at metaphase I with highly disordered spindles and disorganized chromosomes (Yuan et al. 2009).

2.5 Bad

Bad is a member of BH3 (Bcl-2 homology 3) family proteins, the other members of which include Bax, Bak, Bik, Bid and Hrk. While Bcl-2, a firstly identified BH3 and other BH domain (BH1 and BH2)-containing protein, and its relative proteins (e.g. Bcl-xL) act as anti-apoptosis components, Bad and other BH3-only proteins participate in pro-apoptotic cellular functions (e.g. activation of caspases) (Danial 2008; Lutz 2000). Most of these anti-apoptotic or pro-apoptotic proteins localize to mitochondrial outer membranes and function as a sensor of intracellular damage as well as a trigger of mitochondrial death/survival pathway. Several species ranging from nematode, *Drosophila* and sea invertebrates to vertebrates including mammals undergo germline or ovarian/postovulatory oocyte apoptosis in an age-dependent or -independent manner (Buszczak and Cooley 2000; Chiba 2004; Morita and Tilly 1999). In particular, Bad has recently been identified as a factor for phospho-dependent mechanism of egg apoptosis in *Xenopus* (Du Pasquier et al. 2011). Bad in ovarian oocytes at the first meiotic propahse is negatively regulated by inhibitory phosphorylation on Ser-112 and Ser-136 by unknown mechanism (maybe PKA phosphorylation). Upon oocyte maturation, Bad becomes further phosphorylated on Ser-128 in a CDK- and JNK-dependent manner. The Ser-128 phosphorylation, if it exceeds the extent of those of Ser-112/Ser-136 phosphorylations during a long period of oocyte maturation in the absence of fertilizing sperm, will allow Bad to trigger a mitochondrial apoptotic pathway involving cytochrome c release and caspase activation. Whether normal process of oocyte maturation and fertilization involves anti-apoptotic mechanism is not known.

2.6 Brain-derived neurotrophic factor (BDNF)

BDNF is a member of neurotropic family of growth factors that include nerve growth factor (NGF). Its cellular functions are exerted by cell surface receptors such as TrkB, a tyrosine kinase/receptor, and p75 low-affinity NGF receptor (Chao and Hempstead 1995). In mammals including human, ovarian BDNF has been implicated in oogenesis, oocyte maturation, and pre-implantation embryogenesis (Kawamura et al. 2005; Zhang et al. 2010). In vitro maturation of mouse oocytes in the presence of cumulus cells is accompanied by BDNF-dependent activation of Akt/PKB and MAPK and its maintenance has been demonstrated (Zhang et al. 2010). Pharmacological experiments suggest that the Akt/PKB activation involves TrkB function (TrkB-PI3K-PIP$_3$ pathway), while the MAPK does not.

2.7 Bub1/BubR1

Bub1 and **BubR1** (Mad3 in yeast, worms and plants) are multidomain-containing protein-serine/threonine kinases that have been characterized as components of the mitotic checkpoint of spindle assembly (Bolanos-Garcia and Blundell 2011). In mouse oocytes, BubR1 is shown to act as a spindle assembly checkpoint protein in the first meiotic arrest (Homer et al. 2009; Jones and Holt 2010; Schwab et al. 2001; Wei et al. 2010). In maturing *Xenopus* oocytes, Bub1 is activated by MAPK-dependent p90[Rsk] phosphorylation, and is suggested to be involved in spindle assembly checkpoint and, in collaboration with cdk2/cyclin E complex, cytostatic arrest of the meiosis II (Schwab et al. 2001; Tunquist et al. 2002). Precise mechanism of the cytostatic arrest, i.e. inhibition of anaphase-promoting complex, is not known, because a substrate of Bub1 has not yet been identified. In mammals, first meiotic anaphase also seems to be regulated by Bub1-dependent mechanism (McGuinness et al. 2009).

2.8 Calcineurin

Calcineurin is a protein serine/threonine-specific phosphatase that can be up-regulated by the binding of Ca^{2+}/calmodulin (Pallen and Wang 1985), another target of which is CaMKII. In *Xenopus* eggs and cell-free egg extracts, Ca^{2+}-dependent exit of meiosis II involves transient activation of calcineurin. When the activation of calcineurin is blocked, inactivation of MPF by means of cyclin degradation does not occur and sperm nuclei remains condensed. In addition, cortical contraction of the pigmented granules in the animal hemisphere is also blocked. On the other hand, if the activity of calcineurin is artificially kept up-regulated for a prolonged period, growth of sperm aster is inhibited and fusion of the female and male pronuclei is also inhibited. It has been shown that calcineurin dephosphorylates Cdc20, a key regulator of the anaphase-promoting factor that is a substrate of MAPK (Mochida and Hunt 2007; Nishiyama et al. 2007). These results highlight a requirement of calcineurin for Ca^{2+}-dependent inactivation of cytostatic factor and for the onset of the mitotic cell cycle in the early embryos.

2.9 Caspase 2

Caspase 2 is a member of caspase family, which regulates and/or triggers the apoptotic cell death in response to a wide variety of extracellular and intracellular signals. It has been shown that in caspase 2-deficient mice, excess number of ovarian oocytes is a major cause, suggesting that caspase 2 is involved in ovarian oocyte apoptosis. Oocytes deficient in caspase 2 expression also exhibit a marked resistance to cell death induced by chemicals (Bergeron et al. 1998; Morita and Tilly 1999). Further insight into the roles of caspase 2 in the control of oocyte survival has been demonstrated by the studies with use of cell-free extracts prepared form *Xenopus* eggs. In this system, glucose-6-phosphate has been identified as an important component to drive continual operation of the pentose phosphate pathway that prolongs cell survival. In addition, NADPH generation by this pathway is critical for promoting CaMKII-dependent inhibitory phosphorylation of caspase 2 (Nutt et al. 2005). As CaMKII is known as a crucial component that inactivates CSF activity in frog and mammals, it is intriguing whether the CaMKII-caspase 2 axis also functions at fertilization.

2.10 Cdc20/Fizzy

Cdc20 is an activator of anaphase-promoting complex (APC) that directs the onset and progression of the meiotic and mitotic cell cycle (Chung and Chen 2003; Rudner and Murray 2000; Shteinberg et al. 1999; Tang et al. 2004; Weinstein 1997). In *Drosophila*, Cdc20-related gene Fizzy serves a similar function (Dawson et al. 1993; Pesin and Orr-Weaver 2008). The activity of Cdc20 is negatively regulated by phosphorylation on its serine and threonine residues: in case of *Xenopus* Cdc20, Ser-50, Thr-64, Thr-68 and Thr-79. In *Xenopus* maturing oocytes, phosphorylation of Cdc20 is catalyzed by MAPK, a component of cytostatic factor, and/or Bub1/BubR1 kinases, key regulators of spindle checkpoint, and it is involved in the maintenance of cytostatic factor activity that involves the inactivation of APC. Analyses using cell-free extracts prepared from unfertilized *Xenopus* eggs demonstrate that the phosphorylated form of Cdc20 is a target of calcineurin, whose phosphatase activity is transiently activated in response to Ca^{2+} signals (Mochida and Hunt 2007).

2.11 Cdc25 phosphatase (Cdc25A/B/C)

Cdc25 is a protein-tyrosine phosphatase that has been originally identified and characterized as a yeast cell cycle regulator (Fleig and Gould 1991). A major target of this phosphatase is the Cdc2 protein-serine/threonine kinase, its cyclin-associated form of which functions as MPF. Before oocyte maturation in vertebrates, the activity of Cdc2 protein is down-regulated by the absence of cyclin and by phosphorylation by Myt1/Wee1 dual-specificity kinases on Thr 14 and Tyr-15 residues. During oocyte maturation, however, both accumulation of newly synthesized cyclin as well as removal of the phosphates from Cdc2 ensures the Cdc2 activation (Karaiskou et al. 1998; Kim et al. 1999b; Oh et al. 2010; Perdiguero and Nebreda 2004; Perdiguero et al. 2003; Pirino et al. 2009; Qian et al. 2001; Rime et al. 1994; Zhang et al. 2008; Zhao et al. 2008). There are several types of Cdc25: e.g. Cdc25A, Cdc25B, and Cdc25C. PKA phosphorylation and activation of Cdc25B has been reported in mammals (Pirino et al. 2009). In *Xenopus*, Cdc25C is up-regulated by Plx1-mediated phosphorylation on Ser-287 (Qian et al. 2001). Other reports have shown that Xp38γ/SAPK (Perdiguero et al. 2003) and Greatwall kinase (Zhao et al. 2008) can be responsible for the stimulatory phosphorylation of Cdc25C. Cdc25A has been implicated in embryonic cell cycle regulation (Kim et al. 1999b).

2.12 Cdh1/Cort/Fzy

Cdh1 is an activator of anaphase promoting complex/cyclosome (APC/C), an E3 ubiquitin ligase that regulates the onset of anaphase during meiotic and mitotic cell cycle (Visintin et al. 1997). Several cell cycle regulators are subjected to Cdh1- and proteasome-dependent degradation, by which APC/C-dependent cell cycle progression through anaphase is triggered. In *Xenopus* egg cell-free extracts, Cdh1-dependent degradation of Aurora A kinase plays an important role in mitotic exit (Littlepage and Ruderman 2002). The Aurora A-Cdh1 interaction requires the phosphorylation of Aurora A on Ser-53 residue, which is a substrate of M-phase-activated kinase(s). On the other hand, APC-independent cellular function involving Cdh1 has also been suggested in *Xenopus* oocyte maturation (Papin et al. 2004). In immature mouse oocytes, where the meiotic cell cycle is paused at the prophase I, Emil-dependent mechanism of cdh1 inhibition (thereby inhibition of APC/C) functions for the MI arrest (Marangos et al. 2007). In *Drosophila* and *C. elegans*, Cdc20/Cdh1-related protein, Cort and Fzy, respectively, controls the meiotic cell cycle progression in a Cdh1-like manner (Kitagawa et al. 2002; Marangos et al. 2007; Swan and Schupbach 2007).

2.13 Cohesin/SCC1/Rec-8

Cohesin is a chromosome-binding protein that is involved in meiotic and mitotic assembly and segregation of sister chromatids (Heck 1997). In many vertebrate species, cell cycle progression through anaphase involves a proteolytic cleavage of cohesin, as catalyzed by separase and subsequent release of cohesin from the sister chromatids, so that the chromosomal segregation occurs. In *Xenopus*, however, proteolysis-independent release of cohesin from sister chromatids is working and it involves polo-like kinase phosphorylation of cohesin (Sumara et al. 2002). A similar phospho-dependent release of chromosome cohesion has been demonstrated in *C. elegans*, where the AIR-2 kinase (Aurora B kinase in this species) phosphorylation of the nematode cohesion Rec-8 (Rogers et al. 2002).

2.14 Crk adaptor protein (Crk/CRKL)

Crk is an SH2/SH3-containing adaptor protein that has been originally identified as an oncogene product (viral Crk or v-Crk) of avian sarcoma virus CT10 (Feller et al. 1994; Mayer et al. 1988; Mayer and Hanafusa 1990). Its SH2 domain-dependent phosphotyrosine-binding property and SH3 domain-dependent binding to proline-rich sequences in other molecules are required for malignant cell transformation. Three cellular homologues of v-Crk have been found in mammals: c-Crk I, c-Crk II, and c-Crk-like (CRKL). These cellular Crk family proteins have been identified as a major substrate of Bcr-Abl tyrosine kinase that causes chronic myeloid leukemia (CML) (Feller et al. 1998). Another aspect of Crk function has been demonstrated in the studies of *Xenopus* egg cell-free extract: apoptosis in aged egg extracts is shown to involve interaction between the SH2 domain of Crk and the tyrosine-phosphorylated form of Wee1 dual-specificity kinase (Evans et al. 1997; Smith et al. 2000). Further study has demonstrated that the SH3 domain of Crk is important for interacting with the nuclear export factor Crm1, an antagonistic factor for apoptosis in cell-free extract, and that mutually exclusive interaction between Crk and Crm1 or Wee1 in the nucleus regulates the onset of apoptosis.

2.15 Cyclin B

Cyclin is a family of CDK activator proteins, whose first example has been discovered in fertilized sea urchin eggs (Evans et al. 1983) and starfish maturing oocytes (Evans et al. 1983; Standart et al. 1987). Cyclin family consists of several proteins: cyclin A, B, D, E and others, and cyclin B are a component of MPF, another subunit of which is Cdc2/CDK1 serine/threonine-specific protein kinase (Hunt 1989; Maller 1990). In many species, hormone-induced MPF activity in maturing oocytes is generally dependent on *de novo* synthesis and accumulation of cyclin B (and subsequent phospho-dependent regulation of Cdc2/CDK1 by the actions of Wee1/Myt1 kinases and Cdc25 phosphatase is also important) (Gaffre et al. 2011). Fertilization triggers an ubiquitin/proteasome-dependent degradation of cyclin B that causes a rapid decrease of MPF activity (Edgecombe et al. 1991; Huo et al. 2004b; Lapasset et al. 2005; Lapasset et al. 2008; Meijer et al. 1989a; Meijer et al. 1991; Meijer et al. 1989b; Sakamoto et al. 1998). Other cyclins (e.g. cyclin A, D) serve a similar CDK-activating property, but have distinct physiological functions (e.g. G_1/S transition, spindle checkpoint) by interacting with a specific CDK member(s) (e.g. CDK2, CDK5).

2.16 *sn*-1,2-diacylglycerol (DG)

DG is one of two hydrolyzed products by phospholipase C of phosphatidylinositol 4,5-bisphosphate, another product of which is inositol 1,4,5-trisphosphate (IP_3). DG serve as a second messenger in a variety of extracellular signals such as hormones and neurotransmitters, and is well characterized as a direct activator for PKC, a family of

serine/threonine kinase (Nishizuka 1984; Nishizuka 1986). DG also acts as a substrate of DG
kinase that produces phosphatidic acid or PA, which has pleiotropic cellular functions. In
Xenopus eggs, fertilization promotes a rapid increase in intracellular DG concentration, a
large part of which seems to be due to phospholipase D (PLD)-mediated cleavage of
phosphatidylcholine (PC) (but not PIP_2). In support of this, choline, another product of PC
hydrolysis by PLD, is also accumulating in a similar time course of fertilization. Whether
DG is involved in the activation of egg PKC remains to be clarified (Stith et al. 1997).
Production of DG has also been examined in mouse eggs (Stith et al. 1997; Yu et al. 2008). In
this species, sperm-derived PLCζ seems to be mainly responsible for DG production and
subsequent PKC activation.

2.17 Initiation factor 4E-binding protein (4E-BP)

4E-BP is a binding protein for eukaryotic initiation factor 4E (eIF4E), an mRNA cap-binding
protein that facilitates the initiation of protein synthesis in association with eIF4F. The
interaction between 4E-BP and eIF4E depends on the phosphorylation state of 4E-BP: hypophosphorylated form of 4E-BP has an ability to bind to and inhibit eIF4E, whereas the
phosphorylated form of 4E-BP releases eIF4E so that eIF4E-eIF4F complex is formed and
promotes active translation of mRNA (Lasko 2003). In sea urchin eggs, fertilization is
accompanied by a rapid burst of protein synthesis. It has been shown that fertilization also
promotes a rapid decrease in 4E-BP as well as an increase in phosphorylated form of 4E-BP
(Cormier et al. 2001). Two-dimensional electrophoresis demonstrated that 4E-BP is
phosphorylated on multiple sites after fertilization. In mitotic sea urchin embryos, further
decrease in 4E-BP expression has been demonstrated and it is mediated by a rapamycin-
sensitive mechanism of proteolysis of 4E-BP (Salaun et al. 2003), suggesting that mTOR
(mammalian target of rapamycin)-like kinase is involved in the phosphorylation of 4E-BP. A
rapamycin-sensitive mechanism of global protein synthesis involving 4E-BP regulation (but
not translation of some proteins such as cyclin B and Mos, whose translational control
involves the phosphorylation of CPEB phosphorylation) has also been demonstrated in
maturing oocytes of starfish (Lapasset et al. 2008).

2.18 EGG-3/4/5

C. elegans **EGG-3** is a member of protein-tyrosine phosphatase-like (PTPL) family, whose
mutant egg undergoes fertilization normally but has a defect in polarized dispersal of F-
actin, formation of chitin eggshell, and production of polar bodies (Maruyama et al. 2007).
Although enzymatic substrate for EGG-3 has not yet been demonstrated (PTPL proteins are
supposed to be pseudo-phosphatase), its functional interaction with CHS-1, which is
required for deposition of egg shell, plays a role for proper distribution of MBK-2 kinase
that regulates degradation of maternal proteins and egg-to-embryo transition (Nishi and Lin
2005; Qu et al. 2006; Qu et al. 2007; Stitzel et al. 2007; Stitzel et al. 2006). Other members of
PTPL family such as EGG-4 and EGG-5 have also been characterized as components of
meiotic cell cycle progression and egg-to-embryo transition. These two EGG proteins have
no phosphatase activity, however, interact with YTY motif of MBK-2 kinase, which is
autophosphorylated in the active kinase, and inhibit the kinase activity (Cheng et al. 2009;
Parry et al. 2009).

2.19 Emi1 and Emi2/xErp1

In vertebrate unfertilized eggs, cytostatic factor (CSF) is responsible for maintaining the meiotic cell cycle at MII (metaphase of second meiosis) (Masui 2000; Tunquist and Maller 2003). As a candidate of molecule involved in CSF activity, several kinase proteins have been suggested and evaluated (e.g. Mos, MAPK, Rsk). On the other hand, APC/C (anaphase promoting complex/cyclosome) has been identified an initiator of meiotic resumption (thus, as a disruptor of CSF-mediated arrest or a main target of CSF activity). **Emi1** has been identified first as a negative regulator of APC/C in *Xenopus* eggs and cell-free extracts (Reimann et al. 2001a; Reimann et al. 2001b; Reimann and Jackson 2002). Thereafter, an Emi1-related protein named **Emi2/xErp1** has been identified and characterized as an essential component of CSF inhibition of APC/C (Hansen et al. 2006; Liu and Maller 2005; Rauh et al. 2005; Tang et al. 2008; Tung et al. 2005; Wu et al. 2007a; Wu et al. 2007b). In the current scenario, CSF arrest by Emi2/xErp1 of APC/C involves recruitment of PP2A to the Rsk-phosphorylated Emi2/xErp1 (this phosphorylation has stabilizing effect on Emi2/xErp1) and its phosphatase action on other phosphates in Emi2/xErp1 catalyzed by Cdc2/cyclin B complex (this phosphorylation weakens Emi2/xErp1). After fertilization, CaMKII and Plx1 phosphorylation promotes ubiquitin-dependent proteolysis of Emi2/xErp1, thereby APC/C is released from the inhibitory interaction with Emi2/xErp1 (Wu and Kornbluth 2008).

2.20 FKHRL1/FOXO3a

FKHRL (forkhead in rhabdomyosarcoma) is a transcription factor, whose activation has been implicated in the onset of apoptosis and Akt phosphorylation (on Thr-24, Ser-256, and Ser-319) leads to suppression of its function (Brunet et al. 1999; Tang et al. 1999). Its genetic loss or ablation can be a trigger of carcinogenesis, thus FKHRL is a tumor suppressor (Gallego Melcon and Sanchez de Toledo Codina 2007). Akt-dependent phosphorylation of FKHRL1 has been demonstrated in follicular oocytes that receive stem cell factor (SCF) for mammalian oocyte development (Reddy et al. 2005). SCF is a ligand for c-Kit receptor/tyrosine kinase that, upon its ligand-induced activation, promotes sequential activation of PI3K, PDK, and Akt. Thus, follicular development of oocytes involves the suppression of pro-apoptotic signal transduction by FKHRL1. In support of this, FKHRL1 gene-deficient mice exhibited excessive activation from primordial to primary follicles as well as enlarged oocyte sizes (Reddy et al. 2005). A similar pathway involving FOXO3a, a rat homologue of FKHRL transcription factor, has been shown in rat oocytes (Liu et al. 2009).

2.21 XGef

XGef is a *Xenopus* homologue of mammalian guanine nucleotide exchanging factor, RhoGEF that activates Rho-family small GTP-binding protein such as Cdc42. XGef has been initially identified as a CPEB-binding protein and in fact, it has been shown that XGef is involved in polyadenylation and translation of Mos mRNA during oocyte maturation (Reverte et al. 2003). GEF activity of XGef is required for Mos synthesis. In addition, interaction between XGef is responsible for an increase in CPEB phosphorylation during oocyte maturation, which is important for CPEB activation (Martinez et al. 2005). Further studies have shown that MAPK interacts with XGef and acts as a kinase of CPEB on Thr-22,

Thr-164, Ser-184, and Ser-248 (Keady et al. 2007). These phosphorylation sites seem to be required for another and most important phosphorylation event on CPEB: Ser-174 phosphorylation (maybe catalyzed by XRINGO/CDK1 kinase complex) (Kuo et al. 2011).

2.22 Grb2/7/10/14

Grb is a growth factor receptor-bound protein family that has one or more phosphotyrosine-binding and proline-rich interacting domains (i.e. SH2 and SH3 domains) and plays crucial roles in tyrosine kinase receptor-dependent signal transduction (Rozakis-Adcock et al. 1993). There are several Grb family members (e.g. Grb2), most well known of which is Grb2, whose *Drosophila* homologue is *drk* (Olivier et al. 1993). Grb2/*drk* directly interacts to receptor/tyrosine kinase with phosphotyrosine residue(s) (e.g. EGFR in mammals, *sevenless* in *Drosophila*). Because Grb2 interacts constitutively with Sos (*son of sevenless* in *Drosophila*), a guanine nucleotide-exchanging factor (GEF) for Ras, its recruitment to the plasma membranes leads to Ras activation and subsequent MAPK cascade propagation. In *Xenopus* oocytes expressing fibroblast growth factor receptor/kinase (FGFR), some Grb family members (Grb7, Grb10, and Grb14) have been implicated in tyrosine kinase-dependent signal transduction (Cailliau et al. 2003). Microinjection of Grb2 into immature *Xenopus* oocytes has been shown to cause oocyte maturation in a Ras-dependent manner (Browaeys-Poly et al. 2007; Cailliau et al. 2001). In this unusual, but interesting oocyte maturation system, SH2 domains and SH3 domain of Grb2 interact with tyrosine-phosphorylated lipovitellin 1 and PLCγ, respectively. Whether hormone-induced oocyte maturation involves Grb protein is not yet clear.

2.23 Heparin-binding and EGF-like growth factor (HB-EGF)

HB-EGF is a member of EGFR/Erb/HER ligand family, other members of which include EGF, transforming growth factor α, and heregulin. HB-EGF is initially expressed as a membrane-associated precursor and its mature form is secreted outside the cells is done by extracellular shedding as mediated by matrix metalloproteinases (MMPs). HB-EGF participates in several biological processes, including heart development and maintenance, skin wound healing, eyelid formation, progression of atherosclerosis and tumor formation (Miyamoto et al. 2006). In mammals, implantation of early embryos have been shown to involve the action of HB-EGF secreted from the surrounding epithelium as well as those autocrined (Lim and Dey 2009). In this system, HB-EGF exerts its biological functions through activation of intracellular Ca^{2+}-dependent pathways and MAPK cascade. Human trophoblast survival, where anti-apoptosis in low oxygen environment is a key event, has been shown to involve HB-EGF function (Armant et al. 2006). In other species such chicken and fish, expression of HB-EGF in oocytes is supposed to be required for ovarian follicle cell proliferation (Tse and Ge 2009; Wang et al. 2007).

2.24 Heterogenous nuclear ribonucleoprotein K (hnRNP K)

hnRNP K is a K homology (KH) domain-containing RNA-binding protein of the HnRNP family, other KH-containing RNA-binding proteins of which include hnRNP E1/E2 and Sam68 (Bomsztyk et al. 2004; Dreyfuss et al. 2002; Mattick 2004). hnRNP K binds to RNA through its three KH domains and serves multiple functions related to transcription and posttranscriptional regulation of mRNAs (e.g. splicing, translation). In *Xenopus* unfertilized eggs, hnRNP K is phosphorylated on serine and/or threonine residue(s). This phosphorylation seems to be done by MAPK, because a MAPKK inhibitor U0126, but not other inhibitors for MPF (Cdc2/cyclin B) and PKA, diminishes the signals. Consistently, fertilization results in a rapid decrease of the MAPK phosphorylation of hnRNP K. At the same time, hnRNP K becomes tyrosine-phosphorylated, most likely because of sperm-induced Src activation (Iwasaki et al. 2008). These MAPK and Src phosphorylation of hnRNP K has also been demonstrated in mammalian cell systems, in which RNA-binding property (i.e. inhibition of translation) of hnRNP K is up-regulated by MAPK and down-regulated by Src (Habelhah et al. 2001; Ostareck-Lederer et al. 2002). In *Xenopus* eggs and embryos (before mid-blastula transition, where zygotic transcription is activated), maternal mRNAs will be subjected to active protein synthesis to support embryonic development. Data obtained so far suggest that hnRNP K is involved in the suppression and release of specific subset of maternal mRNAs for its active translation (Iwasaki et al. 2008).

2.25 Heterotrimeric and monomeric GTP-binding proteins

G-proteins constitute a large family of proteins that includes small G-proteins and trimeric G-proteins, each of which act as a transducer for extracellular and/or intracellular signals (Gilman 1987; Kaziro et al. 1991). In the case of small G-proteins, a monomeric G-protein (e.g. Ras) is regulated by cell surface receptor-mediated modulation of GAP (GTPase-activating protein) and GEF (guanine nucleotide exchanging factor) activities, and the GTP-bound, active form interacts with effector molecules (e.g. Raf kinase) and regulates cellular functions. Trimeric G-proteins (e.g. Gi, Gs) consist of three subunits: α, β, and γ. Before activation, these three subunits containing GDP-bound form of a subunit are present in a tight complex. Upon activation of cognate cell surface receptors, they become dissociated and each of the subunit (GTP-bound form of α subunit and β/γ complex) exerts its cellular function. In some species, introduction of non-hydrolysable GTPγS or expression of G-protein-coupled cell surface receptor and its ligand activation, which promotes a constitutive activation of (mainly heterotrimeric) G-proteins, is shown to cause egg activation-like phenomena such as repetitive increase in intracellular Ca^{2+} concentration (in mammals) (Swann et al. 1989), cortical reactions (in *Xenopus*) (Kline et al. 1991), and DNA synthesis (in starfish) (Shilling et al. 1994). While involvement of some specific G-proteins (e.g. Gq) in the process of sperm-induced egg activation have been negatively evaluated (Runft et al. 1999; Williams et al. 1998), the fact that the *Xenopus* egg membrane-associated Src activity can be directly stimulated by GTPγS suggests that one or more unknown G-protein(s) serve as a signal transducer of gamete interaction (Sato et al. 2003; Shilling et al. 1994; Swann et al. 1989). Involvement of trimeric G-proteins in oocyte maturation is much more convincing in some species (Mehlmann 2005). Starfish and mouse oocyte meiotic arrest and/or maturation is shown to involve G-protein that directs PI3K-dependent or independent mechanism of Akt/MAPK/MPF/PKA activities (Han et al. 2006; Kalinowski et al. 2004; Kishimoto 2011; Mehlmann et al. 2004; Okumura et al. 2002). *Xenopus* oocyte maturation also seems to involve progesterone-induced membrane receptor activation that leads to modulation of G-protein (maybe Gs, not Gi)/adenylate cyclase pathway (Gallo et al. 1995; Kalinowski et al. 2003).

2.26 Histone H3

Histone is a family of basic polypeptides with ~130 amino acids and has been well characterized as DNA-binding proteins. Nucleosome, a complex of DNA-histones, is organized by an octamer of histone H2A, H2B, H3, and H4. Posttranslational modifications such as acetylation, methylation, and phosphorylation regulate the DNA-binding property of histones including H3. In some mammalian species, phosphorylation of H3 by aurora kinase and an adjacent dimethylated lysine residue are coordinately involved in chromosomal condensation during oocyte maturation (Bui et al. 2007; Ding et al. 2011; Eberlin et al. 2008; Gu et al. 2008; Jelinkova and Kubelka 2006; Maton et al. 2003; Swain et al. 2007; Wang et al. 2006).

2.27 Inositol trisphosphate receptor (IP3R)

Fertilization induces oscillation of inositol 1,4,5-trisphosphate receptor (**IP3R**)-dependent intracellular Ca^{2+} that is responsible for initiating oocyte maturation, egg activation and early embryogenesis. Three isoforms of IP3R have been detected. IP3R is dynamically regulated during meiotic maturation and is required for fertilization induced Ca^{2+} release in *Xenopus* (Kume et al. 1997; Runft et al. 1999). Developmentally regulated type 1 IP3R is up-regulated in oocytes at fertilization and down-regulated after fertilization and this down-regulation is mediated by degradation in proteasome pathway in mouse (Fissore et al. 1999; Jellerette et al. 2000; Parrington et al. 1998; Wakai et al. 2011) and bovine (Malcuit et al. 2005). IP3R1 is phosphorylated during both maturation and the first cell cycle mediated by M-phase kinases e.g. MAPK/ERK2 or polo-like kinase 1 and this is vital for IP3R function in optimum Ca^{2+} release at fertilization in *Xenopus,* mouse and pig (Ito et al. 2008; Ito et al. 2010; Lee et al. 2006; Sun et al. 2009; Vanderheyden et al. 2009). Type 1 IP3R is differentially distributed during human oocyte maturation through GV to MII stage and after fertilization in both peripheral and central in the zygotes and early 2-4-cell embryos and in perinuclear in the 6-8-cell embryos (Goud et al. 1999).

2.28 Insulin

Insulin is a peptide hormone and is crucial for follicular cell growth and development. The addition of insulin to the serum- and hormone-free maturation medium though does not improve the maturation but improves the fertilization rate of bovine oocytes in vitro (Matsui et al. 1995). Artificially induced impaired insulin secretion had a lower percentage of zygotes and a higher percentage of unfertilized and degenerated oocytes in mouse (Vesela et al. 1995). Mouse oocyte has the insulin receptor-beta and highly elevated insulin influences oocyte meiosis, chromatin remodeling, and embryonic developmental competence (Acevedo et al. 2007). Insulin did not activate MPF might be primarily due to the inability of the peptide to activate Ras and to stimulate Mos synthesis in *Xenopus* stage IV but successfully induced maturation of stage VI oocyte (Chesnel et al. 1997). Binding of insulin was revealed in oocytes, granulosa and theca internal cells of healthy pre-antral and antral follicles implying its function in these cells of swine (Quesnel 1999). Insulin increased the developmental potential of porcine oocytes and embryo (Lee et al. 2005). In insulin induced carp oocyte maturation, PI3K is an initial component of the signal transduction pathway, which proceeds, MAPK, and MPF activation (Paul et al. 2009).

2.29 Insulin-like growth factor -1 (IGF-1)

Insulin-like growth factor-1 (**IGF-1**) is primarily synthesized in liver and secreted in circulation that mediate endocrine signal important for the early embryonic development. In *in vitro* reconstructed horse oocytes, IGF-1 induced a bigger accumulation of MAPK (especially ERK2) in the cytoplasm that undergoes nuclear remodeling like a normal embryo following somatic cell nuclear transfer (Li et al. 2004). IGF-1 acts differentially to induce oocyte maturation competence but not meiotic resumption by IGF-1 in white bass (Weber and Sullivan 2005) and white perch (Weber et al. 2007). IGF-1 as like insulin also mediates its action through the activity of IRS-1 in *Xenopus* oocyte maturation (Chuang et al. 1993b). IGF-1 induced mammalian oocyte maturation and subsequently the embryo development e.g. in bovine (Bonilla et al. 2011; Stefanello et al. 2006; Wasielak and Bogacki 2007), mouse (Inzunza et al. 2010) and even human (Coppola et al. 2009).

2.30 Insulin receptor substrate-1 (IRS-1)

Insulin and insulin-like growth factor-1 (IGF-1) receptors (IR and IGFR-1) possess tyrosine-kinase enzymatic activity that is essential for signal transduction to mediate the putative effects of these hormones on oocyte maturation, fetal growth and development. This causes rapid tyrosine phosphorylation of a high-molecular-weight substrate termed insulin receptor substrate-1 (**IRS-1**), a docking protein that can bind with Src homology 2 domain containing molecules e.g. PI 3-kinase, Grb2. Insulin-induced maturation of *Xenopus* oocytes involve the activation of IRS-1 and PI 3-kinase where activation of PI 3-kinase might act upstream of mitogen-activated protein kinase activation and p70 S6K activation (Chuang et al. 1994; Chuang et al. 1993a; Chuang et al. 1993b; Liu et al. 1995; Yamamoto-Honda et al. 1996). IRS-1 is expressed maternally and constantly during *Xenopus* embryogenesis and is important for eye development (Bugner et al. 2011).

2.31 Integrin β1

Integrins are a family of cell surface receptors that mediate cell-cell and cell-matrix interactions in different cellular systems. Variety of integrins is differentially expressed during development, consistent with diverse roles for integrins in embryogenesis. **Integrin β1** (this subunit can interact with α6) is present on the mouse egg surface that increases the rate of sperm attachment but does not alter the total number of sperm that can attach or fuse to the egg (Baessler et al. 2009; Tarone et al. 1993). Integrin α6β1 in association with tetraspanin CD151 and CD9 complex do function in human and mouse gamete fusion (Ziyyat et al. 2006). In *Xenopus*, integrin β1 is present on the oocyte membrane throughout oogenesis and during maturation it is localized in several membrane vesicles in the cytoplasm might be to provide the material source for the rapid membrane formation during cleavage (Muller et al. 1993). Even integrin α6β1 might serve as potential clinical marker for evaluating sperm quality in men (Reddy et al. 2003).

2.32 Interleukin-7 (IL-7)

Interleukin-7 (**IL-7**, pre-B-cell growth factor) is playing its role not only as immunomodulator but also in the beginning of development. IL-7 in together with IL-8 inhibited the gamete interaction of hamster egg and sperm (Lambert et al. 1992). The role of

IL-7 was tested in differentiation during embryonic development e.g. in mouse: development of thymus (Wiles et al. 1992) and lymph node (Coles et al. 2006). IL-7 could be also a good marker of the embryo quality for implantation (Achour-Frydman et al. 2010). In rat granulosa cell culture of early antral and preovulatory follicles, IL-7 stimulated the phosphorylation of AKT, glycogen synthase kinase (GSK3B), and STAT5 proteins in a time- and dose-dependent manner (Cheng et al. 2011). It is concluded that oocyte-derived IL-7 act on neighboring granulosa cells as a survival factor and promote the nuclear maturation of pre-ovulatory oocytes through activation of the PIK3/AKT pathway (Cheng et al. 2011).

2.33 Lipovitellin (LV)

LV1 and **LV2** are components of crystallized yolk platelet in vertebrate oocytes, eggs, and embryos. Precursor protein of LVs, vitellogenin, is synthesized in a highly phosphorylated form in liver of adult and transferred to ovarian tissue, where growing oocytes actively incorporate vitellogenin through the action of specific oocyte membrane receptors (Bergink and Wallace 1974). The incorporated vitellogenin is subjected to partial proteolysis so that LV2 and other fragments such as lipovitellin 1, phosvitin, and pp25 are formed (Finn 2007). A similar set of yolk-associated proteins is also found in invertebrates including insect (e.g. vitelline). It is well known that phosvitin and pp25 are highly serine/threonine- phosphorylated proteins that serve as an energy source of oogenesis and early embryogenesis. On the other hand, tyrosine phosphorylation of LV1 (Browaeys-Poly et al. 2007) and LV2 (Kushima et al. 2011) has recently been demonstrated in *Xenopus*. In particular, tyrosine phosphorylation of LV2 is unusually stable during oogenesis, oocyte maturation, and early embryogenesis until the removal of yolk-associated materials from swimming tadpole (Kushima et al. 2011). Possible function of tyrosine-phosphorylated form of *Xenopus* LV1 and LV2 so far suggested is oocyte maturation (Browaeys-Poly et al. 2007; Kushima et al. 2011), although it's upstream (liver or oocyte) kinase and downstream cellular function is uncertain.

2.34 Maskin/Cytoplasmic polyadenylation element (CPE)-binding protein (CPEB)/TACC3/p82

Maskin is a cytoplasmic polyadenylation element-binding protein-associated factor. Dormant state of maternal mRNAs in immature oocytes is maintained by an abortive interaction of this protein with the eukaryotic initiation factors 4E and 4G. Phosphorylation of maskin promotes the dissociation of this interaction, thereby allows the dormant mRNAs to be translated actively. Aurora phosphorylation of maskin is reported to be involved in protein synthesis in maturing clam and *Xenopus* oocytes and in centrosome-dependent microtubule assembly at mitosis (Kinoshita et al. 2005; Pascreau et al. 2005).

2.35 Myosin regulatory light chain (MRLC)

Myosin regulatory light chain (**MRLC**) or, in short, myosin light chain (MLC) is a component of myosin that regulates the function of actin and actin filaments (see above) through the binding to the actin molecule. Unfertilized eggs of sea urchin undergo cortical contraction in response to calyculin A, an inhibitor for protein phosphates. The results suggest that an egg protein(s), in its phosphorylated form(s), is capable of inducing cortical contraction in this system. As a candidate phosphoprotein for this phenomenon, MRLC has been identified (Asano and Mabuchi 2001). Further biochemical experiments have demonstrated that CK2 (casein kinase 2) is a responsible kinase for the phosphorylation of MRLC (Komaba et al. 2001). Phosphorylation of MRLC in sea urchin eggs occurs on Ser-19 and Thr-18 residues, both of which are stimulatory phosphorylation sites (Asano and Mabuchi 2001). On the other hand, MRLC has also been identified as a phosphoprotein in cell-free extracts prepared form sea urchin eggs. In this system, phosphorylation of MRLC occurs at mitotic phase of cell cycle on Ser-1/2 and Thr-9, all of which are canonical PKC sites, and it is suggested that MPF is the responsible kinase (Totsukawa et al. 1996). In *Drosophila*, phosphorylation of Ser-21 of MRLC-homologue (*sqh*, spaghetti squash gene product) has been implicated as an important event for oogenesis (Jordan and Karess 1997).

2.36 Na$^+$/H$^+$ antiporter/exchanger

On fertilization there are marked changes in the cytoplasmic ionic concentration e.g. Ca^{2+}, H$^+$, are necessary and sufficient to constitute the egg activation and beyond. A second messenger type substance that stimulates protein kinase C linked the activation of the Na$^+$/H$^+$ exchange to the calcium transient and ultimately the protein synthesis is increased and the cytoplasmic alkalinization occur in sea urchin eggs (Swann and Whitaker 1985). In sea urchin eggs, though the **Na$^+$/H$^+$ exchanger** is regulated by PKC or Ca^{2+}/CaMK activities but fertilization mediated activation of this exchanger is Ca^{2+}, CaM-dependent (Shen 1989). G proteins activated Na$^+$/H$^+$ antiporter mediated by PKA and/or PKC in *Xenopus* oocytes (Busch 1997; Busch et al. 1995). A typical Na$^+$/H$^+$ exchanger mediated increased intracellular pH though activate the surf clam oocytes but is neither sufficient nor required for GVBD (Dube and Eckberg 1997). The function of Na$^+$/H$^+$ exchanger has also been described even for later stage of development e.g. blastocyst of mouse (Barr et al. 1998), bovine embryos (Lane and Bavister 1999) and human pre-implantation embryos (Phillips et al. 2000).

2.37 OMA-1

In *C. elegans*, two CCCH-type zinc finger proteins **OMA-1** and OMA-2 are expressed specifically in maturing oocytes and are functionally redundant during maturation. Both Oma-1 and Oma-2 mutant oocytes arrest at a defined point in prophase I and the removal of Myt1-like kinase Wee-1.3 results the release of prophase I arrest (Detwiler et al. 2001). As WEE-1.3 functions as a negative regulator, OMA-1 and OMA-2 either function upstream of WEE-1.3 or in parallel with WEE-1.3 as positive regulators of prophase progression (Detwiler et al. 2001). OMA-1 protein is largely reduced because of rapid degradation after the first mitotic division and this is necessary for the early embryonic development by regulating the temporal degradation of maternal proteins in early *C. elegans* embryos (Lin 2003; Shimada et al. 2006; Shirayama et al. 2006). OMA-1 is directly phosphorylated (Thr-239) by DYRK kinase MBK-2 that facilitates subsequent phosphorylation (Thr-339) by another kinase GSK-3 and these precisely timed phosphorylation events are important for its function in 1 cell embryo and degradation after first mitosis (Nishi and Lin 2005).

2.38 p53

The **p53** protein family includes three transcription factors-p53, p63 and p73 that play roles in both cancer and normal development (Levine et al. 2011). Mostly stable p53 protein is

Phospho-Signaling at Oocyte Maturation and Fertilization: Set Up for Embryogenesis
and Beyond Part II. Kinase Regulators and Substrates

103

synthesized during late oogenesis and stage VI oocyte and even after fertilization at least until the tadpole stage during *Xenopus* development (Tchang et al. 1993). After fertilization, part of the largely stored p53 is imported into the nucleus and associates both with decondensed DNA and the nuclear lamina envelope but not with any replication complexes during *Xenopus* early development (Tchang and Mechali 1999). In the absence of TPX2 (targeting protein for Xklp2), p53 can inhibit Aurora A, a serine/threonine kinase, activity (Eyers and Maller 2004). TPX2 is required for Aurora A activation and for p53 synthesis and phosphorylation during *Xenopus* oocyte maturation (Pascreau et al. 2009). The tumor suppressor protein p53 regulates the efficiency of human reproduction. The p53 allele encoding proline at 72 (Pro72) was found to be significantly higher (P=0.003) over the allele encoding arginine (Arg72) among women experiencing recurrent implantation failure (Kang et al. 2009; Kay et al. 2006; Levine et al. 2011)

2.39 p95

Several studies showed that in mammals, egg-specific extracellular matrix zona pellucida component ZP3 regulates an essential event in sperm function. Mouse zona pellucida glycoprotein ZP3 regulates acrosomal exocytosis by aggregating its corresponding receptors located in the mouse sperm plasma membrane e.g. a protein **p95** that might serve as a substrate for a tyrosine kinase in response to zona pellucida binding or itself act as tyrosine kinase (Saling 1991). A phosphotyrosine containing receptor tyrosine kinase was identified in human sperm that is similar to mouse sperm protein, p95, having tyrosine kinase activity and human ZP3 stimulate the tyrosine kinase activity of this protein (Burks et al. 1995; Naz and Ahmad 1994). Acrosome reaction was induced with increased tyrosine phosphorylation of p95 epitope only in capacitated human spermatozoa (Brewis et al. 1998).

2.40 Paxillin

Paxillin is a prominent focal adhesion docking protein that regulates somatic and germ cell signaling. Paxillin was shown as one of the major tyrosine kinase substrates during rat chick embryogenesis (Turner 1991) and regulator of Rho and Rac signaling during *Drosophila* development (Chen et al. 2005). It was described that paxillin is required for synthesis and activation of Mos (the germ cell Raf homolog), that promotes MEK and subsequently Erk signaling and then possibly Erk mediate the phosphorylation of paxillin required for steroid (testosterone)-induced *Xenopus* oocyte maturation (Rasar et al. 2006). In prostate cancer cell, EGFR-induced Erk activation requires Src-mediated phosphorylation of paxillin but paxillin was not involved in PKC-induced Erk signal (Sen et al. 2010). Erk-mediated phosphorylation of paxillin was necessary for both EGFR- and PKC-mediated cellular proliferation indicate that paxillin serves as a specific upstream regulator of Erk in response to receptor-tyrosine kinase activity but as a general regulator of downstream Erk actions regardless of agonist (Sen et al. 2010).

2.41 Peptidylarginine deiminase (PAD)

Peptidylarginine deiminase (PAD) catalyzes the post-translational modification of protein converting the arginine to citrulline in the presence of calcium ions. PAD is present in the cortical granules of mouse oocytes, is released extracellularly during the cortical reaction, and remains associated as a peripheral membrane protein until the blastocyst stage (Liu et al. 2005). In mouse peptidylarginine deiminase-like protein termed ePAD (p75) was expressed in immature oocyte, mature egg, and until the blastocyst stage of embryonic development (Wright et al. 2003). Peptidylarginine deiminase 6 (PAD6) is uniquely expressed in male and female germ cells but the inactivation of PAD6 gene leads to female infertility whereas male fertility is not affected (Esposito et al. 2007) and its transcript is detectable at embryonic day 16.5 in mouse (Choi et al. 2010). Mouse oocyte cytoplasmic sheet-associated PADI6 undergoes developmental change in phosphorylation that might be linked to interaction between PAD16-YWHA during oocyte maturation (Snow et al. 2008). PADI6-deficient mice are also infertile might be due to disruption of development beyond the two-cell stage (Snow et al. 2008).

2.42 Phosphodiesterase 3A (PDE3A)

Intracellular concentration of the second messenger cAMP is the key signaling molecules in the control of oocyte meiotic resumption mediated by the activity of phosphodiesterases (PDEs). cAMP blocks meiotic maturation of oocytes of a broad spectrum of species and cyclic nucleotide phosphodiesterase 3A (PDE3A) is primarily responsible for oocyte cAMP hydrolysis. The PDE3A activity in the regulation of oocyte maturation of several species has been studied extensively e.g. in rodent (Wiersma et al. 1998), rat (Richard et al. 2001), mouse (Masciarelli et al. 2004; Nogueira et al. 2003b; Nogueira et al. 2005), monkey (Jensen et al. 2005), porcine (Sasseville et al. 2006; Sasseville et al. 2007), bovine (Mayes and Sirard 2002; Thomas et al. 2002), and human (Nogueira et al. 2003a). Various PDE3 inhibitors were used like org9935, cilostamide, or milrinone. PDE3 activity is required for insulin/insulin-like growth factor-1 stimulation of *Xenopus* oocyte meiotic resumption. It should be note that the activation of PDE3A by PKB/Akt-mediated phosphorylation potentiates the *Xenopus* and mouse oocytes maturation (Han et al. 2006).

2.43 pp25 and phosvitin

Functions of multiple vitellogenin (VgA, VgB, and VgC)-derived yolk products, e.g. lipovitellin/**phosvitin** were described during oocyte maturation and early embryos in various species, e.g. barfin flounder, *Verasper moseri*, a marine teleost (Matsubara et al. 1999; Sawaguchi et al. 2006), red seabream (*Pagrus major*), another marine teleost and gray mullet (*Mugil cephalus*) (Amano et al. 2008). A substrate **pp25** for protein serine/threonine kinases was derived from the precursor of pp43 that is consisting of a portion of the *Xenopus* VgB1 protein (Xi et al. 2003). pp25 may have a role as an inhibitory modulator of some protein phosphorylation mediated by CKII and PKC in *Xenopus* oocytes and embryos (Sugimoto and Hashimoto 2006). A differentially distributed pp25 was shown to localize at the surface just below the plasma membrane in oocyte and in embryogenesis a transition from beneath the outer surface of each germ layer to endoderm during tail budding from where it gradually decreased and disappeared at the tadpole stage in *Xenopus* (Nakamura et al. 2007).

2.44 Protein methyl transferase 5 (PRMT5)

Distinct protein/DNA methylation patterns were observed in developmental stages during genomic reorganization. The protein methylase activity was measured at mesenchymal

blastula and at young gastrula of sea urchin embryonic development and lysine of histones
H3 and H4 are the main target (Branno et al. 1983). A Janus-2 (JAK-2) binding protein, JBP1,
acts as an arginine methyl transferase and is now designated as **PRMT5**. In *Xenopus* oocytes,
PRMT5 inhibited the oncogenic/transformed p21^{Ras} mediated maturation but not the
insulin mediated maturation that involve the wild-type p21^{Ras} (Chie et al. 2003). Decreased
level of methylated H3K79 was observed soon after fertilization and the hypomethylated
state was maintained at interphase (before the blastocyst stage) and variation in methylation
was observed at M phase (Ooga et al. 2008) in mouse. DNA methyltransferase-1 might work
during the late stage of oocyte differentiation, maturation and early embryonic development
in mammals e.g. cow (Lodde et al. 2009).

2.45 Proline-rich inositol phosphate 5-phosphatase (PIPP)

Different types of inositol polyphosphate 5-phosphatases (IPP) selectively remove the
phosphate from the 5-position of the inositol ring from both soluble and lipid substrates, i.e.,
inositol 1,4,5-trisphosphate, inositol 1,3,4,5-tetrakisphosphate, phosphatidylinositol 4,5-
bisphosphate or phosphatidylinositol 3,4,5-trisphosphate and they have various protein
modules probably responsible for specific cell organelle localization or recruitment e.g. SH2
domain, SH3-binding motif, proline-rich sequences, etc. (Erneux et al. 1998; Kong et al. 2000;
Mochizuki and Takenawa 1999). They demonstrate the restricted substrate specificity and
act downstream of various receptors by removing a phosphate. Proline-rich IPP (**PIPP**) had
been studied in PI3K pathway for early development of fertilized mouse eggs. PIPP might
affect development of fertilized mouse eggs by inhibition of level of phosphorylated Akt at
Ser-473 and subsequent inhibition of downstream signal cascades resulting reduced
cleavage rate of fertilized mouse eggs (Deng et al. 2011). In embryonic day 15.5 mice, SHIP2
a homologue of SHIP1 was strongly expressed in the liver, specific regions of the central
nervous system, the thymus, the lung, and the cartilage perichondrium (Schurmans et al.
1999).

2.46 Protein phosphatase 1/2A (PP1/PP2A)

Numerous protein kinases and phosphatases have important functions during mitosis and
meiosis. Protein phosphatase (PP) 1 (**PP1**) and 2A (**PP2A**) that preferentially
dephosphorylate the β- and α-subunit of phosphorylase kinase had been identified in
starfish oocyte (Pondaven and Cohen 1987). With the similar mechanism involved in
mammals and *Drosophila*, PP4, a centrosomal protein, involved in the recruitment of
pericentriolar material components to the centrosome from prophase to telophase, but not
during interphase, and is essential for the activation of microtubule nucleation that promote
spindle formation in *C. elegans* (Sumiyoshi et al. 2002). When the normal physiological
function of PP1 and PP2A was blocked, premature separation of sister chromatids during
meiosis I and aneuploidy in mouse oocytes was observed (Mailhes et al. 2003). In *Xenopus*
oocyte, PP2A negatively regulates Cdc2 activation whereas Aurora-A activation is indirectly
controlled by Cdc2 activity independent of either PP1 or PP2A activity (Maton et al. 2005).
Constant cyclin B levels are maintained during a CSF arrest through the regulation of Emi2
activity that inhibits the anaphase-promoting complex (APC), an E3 ubiquitin ligase that
targets cyclin B for degradation in vertebrates like *Xenopus* (Wu et al. 2007b). Rsk or Cdc2-
mediated phosphorylation of Emi2 was antagonized by PP2A, which could bind to Emi2

and promote Emi2-APC interactions results CSF arrest (Wu et al. 2007a; Wu et al. 2007b). Cdk1/cyclin B (MPF) induced active Gwl promotes PP2A (B55 is the regulatory subunit) inhibition to enter and maintenance the M phase that would otherwise remove MPF-driven phosphorylations (Castilho et al. 2009; Vigneron et al. 2009).

2.47 Protein tyrosine phosphatase (PTP)

In the early steps of embryogenesis both the protein tyrosine phosphorylation and the protein tyrosine phosphatase (**PTP**) regulated activities are involved. In *Xenopus* MPF and progesterone but not insulin-induced oocyte maturation was retarded by PTPase 1B action (Tonks et al. 1990) whereas non receptor PTP13 activate the oocyte maturation (Nedachi and Conti 2004). PTP exert its role by different mechanism for example, PTP regulate the oocyte maturation in pig (Kim et al. 1999a), receptor-type PTP regulate Fyn in zebrafish egg fertilization (Wu and Kinsey 2002), Src homology-2 domain containing PTP (SHP2) regulate normal human trophoblast proliferation (Forbes et al. 2009), and pseudo-PTP (lack at least one key residue in the catalytic site) regulate oocyte-embryo transition in nematode (Heighington and Kipreos 2009) and antagonist of PTP reduced GVBD and MAPK/MPF activities in sea water treated marine nemertean worms oocytes (Stricker and Smythe 2006). Receptor type PTP and PTP are essential for convergence and extension cell movements to shape the body axis during vertebrate gastrulation e.g. for zebrafish in a signaling pathway parallel to non-canonical Wnt and upstream of Fyn, Yes and RhoA (van Eekelen et al. 2010).

2.48 Pumilio1/2

In *Xenopus*, the cytoplasmic polyadenylation element (CPE) in the 3'-untranslated region (UTR) of cyclin B1 mRNA is responsible for both the translational repression (masking) and activation (unmasking) of the mRNA where CPE is bound by a CPE-binding (CPEB) protein (Hake and Richter 1994; Hodgman et al. 2001; Mendez and Richter 2001). *Xenopus* **pumilio** (Pum) in coordination with CPEB-maskin complex acts as a specific regulator for timing translational activation of cyclin B1 mRNA first as repressor in mature oocyte by binding and as activator by its release from phosphorylated CPEB during oocyte maturation (Nakahata et al. 2003). Usually nemo-like kinase (NLK) that acts downstream of Mos, phosphorylate Pum1, Pum2 and CPEB and this phosphorylation is proceeded with translational activation of cyclin B1 mRNA stored in oocytes for maturation (Ota et al. 2011a; Ota et al. 2011b).

2.49 p21[Ras]

In *Xenopus* oocytes, transformed/active **p21[Ras]** increased the level of total cell protein phosphorylation that culminated with germinal vesicle breakdown (GVBD) in the absence of protein synthesis and the same pattern of phosphorylation was observed by hormone either progesterone or insulin treatment (Nebreda et al. 1993). Activated p21[Ras] and GTPase-activating protein (GAP) complex may promote MAPK activity by tyrosine phosphorylation followed by the activation of S6-kinase II (Nebreda et al. 1993; Pomerance et al. 1992). Later it was shown that Ras-GAP activity is required for Cdc2 activation and Mos induction independent of MAPK activation (Pomerance et al. 1996). It should be note that active Ras increased MAPK and S6K activities and sensitized the

Phospho-Signaling at Oocyte Maturation and Fertilization: Set Up for Embryogenesis
and Beyond Part II. Kinase Regulators and Substrates

107

oocytes to insulin-stimulated maturation via IRS-1 (Chuang et al. 1994). T-Cell Origin protein Kinase (TOPK) and the nuclear kinase, DYRK1A are attractive candidates in insulin mediated wild-type p21[Ras]-induced oocyte maturation independent of MAPK (Qu et al. 2006; Qu et al. 2007). Phospholipase D (PLD) activity induced MAPK and S6K II activity might constitute a relevant step in Ras-induced GVBD in *Xenopus* oocytes was also reported (Carnero and Lacal 1995). p21[Ras] did not appear to be ubiquitous in the rat conceptus prior to gastrulation but was found in embryos from 6.5 to 12 days of age (Brewer and Brown 1992).

2.50 Phosphatidylinositol 3-kinase (PI3K)

PI3K is a lipid kinase that phosphorylates 3'-position in the inositol ling structure of inositol phospholipids (e.g. phosphatidylinositol 4,5-bisphosphate). Inactive PI3K consists of a heterodimer of one catalytic subunit (e.g. p110) and one regulatory subunit (e.g. p85), a latter of which is known to be tyrosine-phosphorylated in response to a variety of extracellular signals (Vanhaesebroeck et al. 1997). The tyrosine-phosphorylated regulatory subunit releases the catalytic subunit so that PI3K becomes enzymatically active. Involvement of PI3K in oocyte maturation and fertilization has been examined with the use of specific inhibitors such as LY294002 and Wortmannin as well as expression of native or mutant PI3K proteins (Chuang et al. 1993a; Hoshino and Sato 2008; Hoshino et al. 2004; Mammadova et al. 2009). In starfish oocyte, 1-methyladenine-induced oocyte maturation involves a sequential activation of the hormone receptor on the cell surface, G-proteins attached to the receptor, and PI3K. The activated PI3K promotes Akt kinase activation through the production of PIP$_3$ and stimulation of PIP$_3$-dependent protein kinase PDK1 (Kishimoto 2011). In oocytes of *Xenopus* or other frog species, PI3K is suggested to be a component of progesterone-induced oocyte maturation (Bagowski et al. 2001; Ota et al. 2008). However, wortmannin promotes oocyte maturation in the absence of hormonal signal (Carnero and Lacal 1998), suggesting the possibility that this drug targets unknown factor(s) other than PI3K or that, as opposed to the case in starfish, PI3K is negative regulator of oocyte maturation. On the other hand, LY294002 has been shown to block sperm-induced egg activation (Mammadova et al. 2009). LY294002 also blocks sperm-induced Src activation and Ca^{2+} release, suggesting that PIP$_3$ production by PI3K plays a role in fertilization. Interestingly, however, tyrosine phosphorylation of p85 subunit of PI3K is not detected, suggesting that alternative pathway for PI3K activation (e.g. recruitment to membrane microdomains) is working in this system.

2.51 Phospholipase Cγ (PLCγ)

PLCγ is a member of PLC family proteins (other members are PLCβ, PLCδ, PLCε, PLCζ etc.) that hydrolyzes phosphatidylinositol 4,5-bisphosphate into DG and IP$_3$, both of which are second messenger to promote PKC activation and intracellular Ca^{2+} mobilization, respectively (Rhee 2001). PLCγ is the first example of non-tyrosine kinase protein, whose structure contains SH2 and SH3 domains (Stahl et al. 1988). PLCγ is also unique in its regulatory mechanism, where tyrosine phosphorylation of the protein can up-regulate the enzyme activity. Under this background, function of PLCγ in oocyte maturation and fertilization has been analyzed extensively in relation to tyrosine kinase signaling. In fact, tyrosine kinase-dependent activation of PLCγ at fertilization has been demonstrated in some vertebrate (e.g. fish, frog) and invertebrate species (e.g. ascidian, sea urchin, starfish) (Carroll et al. 1999; Carroll et al. 1997; Giusti et al. 1999; Giusti et al. 2000; Mehlmann et al. 1998; Runft et al. 2004; Runft and Jaffe 2000; Runft et al. 2002; Runft et al. 1999; Sato et al. 2002a; Sato et al. 2001; Sato et al. 2003; Sato et al. 2000b; Shearer et al. 1999; Tokmakov et al. 2002). It should be noted that Src-dependent activation of PLCγ involves a new function of PLCγ as GEF for small G-protein Ras (Bivona et al. 2003), suggesting that other means of cellular function contributes to egg activation in these species. On the other hand, Ca^{2+} release associated with mammalian fertilization does not seem to involve tyrosine kinase activity and PLCγ activation, probably because sperm-derived PLCζ activity is necessary and sufficient for sperm-induced Ca^{2+} release in these species (Kurokawa et al. 2004; Parrington et al. 2002; Saunders et al. 2002).

2.52 RNA polymerase II large subunit

RNA polymerase II (also called RNAP II or Pol II), a complex of twelve subunits (p550) is an enzyme that catalyzes the transcription of DNA to synthesize precursors of mRNA and most snRNA and microRNA (Kornberg 1999; Sims et al. 2004). A large subunit of RNAPII (p220) was shown to be phosphorylated at the onset of wheat germination that moderately increase the RNA polymerase activity (Mazus et al. 1980). In *C. elegans*, embryonically transcribed gene products are required for gastrulation initiation where a large subunit of RNAPII is involved (Powell-Coffman et al. 1996). In *Xenopus*, the largest subunit of RNA polymerase II (RPB1) accumulates in large quantities from previtellogenic early diplotene oocytes up to fully grown oocytes where the C-terminal domain (CTD) was essentially hypophosphorylated in growing oocytes from stage IV to VI (Bellier et al. 1997). Upon maturation, RPB1 is hyperphosphorylated dramatically and abruptly but dephosphorylated within 1 h after fertilization (Bellier et al. 1997). Metaphase II-arrested oocytes showed a much stronger CTD kinase activity than that of prophase stage VI and this kinase activity were attributed to the activated MAPK i.e. RPB1 could be a substrate of MAPKs (e.g. p42) during *Xenopus* oocyte maturation (Bellier et al. 1997).

2.53 Receptor for activated C kinase (RACK)

PKC, serine/threonine kinase, is a pivotal enzyme in a variety of signal transduction pathways that includes the maturation through actin cytoskeleton rearrangement and cortical granules exocytosis (CGE) to early stages of embryogenesis. The translocation of PKC is facilitated by receptor for activated C kinase (**RACK**). Activation of PKC exposes the RACK-binding site, enabling the association of the enzyme with its anchoring RACK (Ron and Mochly-Rosen 1995). Inhibition of binding the PKC to RACK blocks the function of PKC (Ron et al. 1995). During the activation of MII eggs, PKCα, βII and γ individually and RACK1 together with both PKCα and PKCβII translocate to the egg cortex (Haberman et al. 2011). The association of PKC and actin with RACK1 is known to be involved in CGE. Upon egg activation, increased level of RACK1 shuttles activated PKCs to the egg cortex, thus facilitating CGE (Haberman et al. 2011). The phytohormone abscisic acid promoted the expression level of RACK that is regulated by Gα-protein and plays an important role in a basic cellular process as well as in rice embryogenesis and germination (Komatsu et al. 2005).

2.54 Rho

The **Rho** family of small GTPases is known to organize and maintain the actin filament-dependent cytoskeleton, and rho is involved in the control mechanism of cytokinesis. Actin-depolymerizing factor (ADF)/coffilin, a key regulator for actin dynamics during cytokinesis, is suppressed and reactivated by phosphorylation and dephosphorylation respectively. Rho-induced dephosphorylation of ADF/coffilin is dependent on the XSSH (*Xenopus* homologue of Slingshot phosphatase) activation that is caused by increase in the amount of F-actin induced by Rho signaling (Tanaka et al. 2005). XSSH may reorganize actin filaments through dephosphorylation and reactivation of ADF/coffilin at early stage of contractile ring formation during *Xenopus* cleavage (Tanaka et al. 2005). In sea urchin egg, Rho is synthesized early in oogenesis in soluble form, associates with cortical granules in the end of maturation and after insemination secreted by cortical granules exocytosis and retained in the fertilization membrane indicate the involvement of Rho in Ca^{2+}-regulated exocytosis or actin reorganization that accompany the egg activation (Covian-Nares et al. 2004; Cuellar-Mata et al. 2000; Manzo et al. 2003). In ascidians Rho proteins are involved in egg deformation, ooplasmic segregation and cytokinesis downstream of the Ca^{2+} transients (Yoshida et al. 2003).

2.55 Ribosomal S6

In *Xenopus* oocytes 40S ribosomal protein **S6** becomes phosphorylated by S6K on serine residues in response to hormones or growth factors and following microinjection of the tyrosine-specific protein kinases associated with Rous sarcoma virus or Abelson murine leukemia virus. S6 is minimally phosphorylated in unstimulated oocytes and in progesterone induced *Xenopus* oocyte maturation: phosphorylation of S6 precedes germinal vesicle breakdown (GVBD) and is maximal at the time when 50% of the oocytes have undergone GVBD (Erikson and Maller 1985; Hanocq-Quertier and Baltus 1981; Nielsen et al. 1982). In *Xenopus* oocytes, Ras (p21, have GTPase activity) proteins activate the pathway linked to S6 phosphorylation and that PKC has a synergistic effect on the Ras-mediated pathway (Kamata and Kung 1990). Microinjection of purified pp60[v-Src] into *Xenopus* caused the phosphorylation of S6 and accelerated the time course of progesterone-induced oocyte maturation (Spivack et al. 1984).

2.56 RINGO

RINGO/Speedy (Rapid Inducer of G2/M transition in Oocytes) proteins can bind to and directly stimulate CDKs (CDK1 and CDK2) that regulate cell cycle transition although they do not have amino acid sequence homology with cyclins. In *Xenopus* oocytes RINGO (XRINGO) accumulates transiently during meiosis I entry and this process is directly stimulated by several kinases, including PKA and GSK3β, and contributes to the maintenance of G2 arrest (Gutierrez et al. 2006). Later XRINGO is down-regulated/degraded after meiosis I that is mediated by the ubiquitin ligase Siah-2, which probably requires phosphorylation of XRINGO on Ser-243 and important for the omission of S phase at the meiosis-I-meiosis-II transition in *Xenopus* oocytes and finally trigger G2/M progression (Gutierrez et al. 2006; Karaiskou et al. 2001). p42 MAPK (ERK2) activity and RINGO accumulation are also required for activating phosphorylation of CPEB by Cdk1.

RINGO/Speedy, is necessary for CPEB-directed polyadenylation-induced translation of Mos and cyclin B1 mRNAs in maturing *Xenopus* oocytes (Padmanabhan and Richter 2006). Recently, it was shown that XGef (a Rho family guanine nucleotide exchange factor) is involved in XRINGO/CDK1-mediated activation of CPEB and that an XGef/XRINGO/ERK2/CPEB complex forms in ovo to facilitate the maturation process (Kuo et al. 2011). In mammals for example in porcine RINGO A2 (SPDYA2) speed up the oocyte maturation (Kume et al. 2007) and in mouse RINGO efficiently triggers meiosis resumption of oocytes and induces cell cycle arrest in embryos (Terret et al. 2001).

2.57 Sam68 adaptor protein (Sam68)

Sam68 is a KH domain-containing, STAR (signal transduction and activation of RNA) family RNA-binding protein that has been originally identified as a mitosis-specific Src-phosphorylated protein of 68 kDa (Taylor et al. 1995; Taylor and Shalloway 1994). Sam68 has also a proline-rich sequence that would interact with SH3 domain-containing proteins, linking its possible function to Src-dependent signal transduction pathways. The RNA-binding ability of Sam68 contributes to, like hnRNP K, another KH-containing RNA-binding protein, posttranscriptional regulation of mRNAs (e.g. splicing, translation). While its physiological function in spermatogenesis has been well known to date (Sette et al. 2010), roles of Sam68 in the oocyte and/or egg system have just recently been shown in mammalian species: Sam68-deficient female mice are severely subfertile (Bianchi et al. 2010). Further studies demonstrated that Sam68 directly binds the mRNAs for the follicle-stimulating hormone (FSH) and the luteinizing hormone (LH) receptors (FSHR and LHR) and is involved in proper expression of these transcripts in pre-ovulatory follicles in adult ovary. Whether these Sam68 functions involve phosphorylation of Sam68 is not known.

2.58 Separase

The cysteine protease named **separase** is widely expressed in unicellular and multicellular organisms and is involved in a timely cleavage of the sister chromatid protein cohesins/SCC1 so that the separation of sister chromatids is made possible in the anaphase. The activity of separase can be negatively regulated by two mechanisms: one is the binding of securin, and the other is Cdc2-dependent phosphorylation on Ser-1126 and subsequent phospho-dependent binding of cyclin B (Nagao and Yanagida 2002; Nasmyth et al. 2000; Stemmann et al. 2001). In meiotic cell cycles in *Xenopus* oocytes, phospho-dependent inhibition of separase seems to occur: progesterone-induced oocyte maturation promotes firstly an accumulation of *Xenopus* homolog of securin, and then it undergoes degradation at the meiotic anaphase I and II in an APC/C-dependent manner (Fan et al. 2006; Holland and Taylor 2006). Mutation studies of the phosphorylation site in separase demonstrated that phospho-dependent regulation of this enzyme also works in germ cell developmental stages and early embryonic (8-cell and 16-cell) stages (Huang et al. 2009).

2.59 SHB

The adaptor protein **SHB** (Src homology 2 domain-containing adapter protein B) mediates certain responses in platelet-derived growth factor (PDGF) receptor-, fibroblast growth factor (FGF) receptor-, neural growth factor (NGF) receptor-, T cell (TC) receptor-,

interleukin-2 (IL-2) receptor- and focal adhesion kinase- (FAK) signaling where in some cells
the Src-like Fyn-related kinase (FRK/RAK) act upstream of SHB (Cross et al. 2002; Karlsson
et al. 1998; Karlsson et al. 1995; Welsh et al. 1998).The absence of SHB enhanced ERK
(extracellular-signal regulated kinase) and RSK (ribosomal S6K) signaling in mouse oocytes
increasing the ribosomal protein S6 phosphorylation and activation (Calounova et al. 2010).
SHB regulates normal oocyte and follicle development and that perturbation of SHB
signaling causes defective meiosis I and early embryo development in mouse (Calounova et
al. 2010). The SHB protein is required for normal maturation of mesoderm and efficient
multilineage differentiation during in vitro differentiation of embryonic stem cells (Kriz et
al. 2006; Kriz et al. 2003).

2.60 Shc adaptor protein (Shc)

Src homology and collagen (**Shc**) is an SH2-containing adaptor protein that has been
identified as a mammalian proto-oncogene, whose overexpression in fibroblast cells leads to
the malignant transformation (McGlade et al. 1992; Pelicci et al. 1992; Rozakis-Adcock et al.
1992). Shc consists of three isoforms (i.e. p46, p52, and p66) produced by alternative
transcription and translation from one transcript and all isoforms also have an additional
phosphotyrosine-binding domain in its amino-terminal region, named PTB domain. In some
receptor/tyrosine kinase-mediated signal transduction pathway, Shc is recruited to the
phosphotyrosine clusters of the activated receptor proteins, phosphorylated on its tyrosine
residues (e.g. in mammals, Tyr-239/240 for Myc activation, Tyr-317 for MAPK/Fos
activation), and recruit other SH2 and/or SH3-containing proteins (e.g. Grb2) to elicit
downstream signaling cascade. In *Xenopus*, insulin-dependent oocyte maturation and egg
fertilization seem to involve tyrosine kinase-dependent function of Shc (Aoto et al. 1999;
Chesnel et al. 2003). Because two of three isoforms of Shc (p52 and p66) has been shown to
be a direct activator of Src tyrosine kinase (Sato et al. 2002b), it is interesting to examine
whether Shc-dependent Src activity contributes to these physiological events.

2.61 SNT/FRS2

Membrane anchored adaptor protein Suc1-associated neurotrophic target-1 or -2/fibroblast
growth factor receptor substrate-2 or (**SNT-1 or -2/FRS2**), is implicated in the transmission
of extracellular signals from several growth factor receptors e.g. fibroblast growth factor
receptors (FGFRs) and neurotrophin receptors (Trks) through their N-terminal
phosphotyrosine binding (PTB) domains to the mitogen-activated protein (MAP) kinase
signaling cascade during embryogenesis. SNT-1 physically associates with the Src-like
kinase Laloo, and SNT-1 activity is required for mesoderm induction by Laloo in *Xenopus*
(Akagi et al. 2002; Hama et al. 2001). Activated FGFR and FRS2 induced Mek/MAPK
activity for germinal vesicle breakdown (GVBD) and substantial H1 kinase activity might be
through PI3 kinase activation for *Xenopus* oocyte maturation but not by progesterone (Mood
et al. 2002). During progesterone-induced oocyte maturation Mek/MAPK activity is critical
for the induction and/or maintenance of H1 kinase activity (Mood et al. 2002).

2.62 Sperm receptor/p350

During fertilization, sperm must first bind in a species-specific manner to the eggs thick
extracellular coat, the zona pellucida or vitelline envelope and then undergo a form of

cellular exocytosis, the acrosome reaction. Little is known about sperm-binding proteins in egg envelope of vertebrate/invertebrate species. In sea urchin the sperm receptor is phosphorylated by an egg cortical tyrosine kinase in response to sperm or purified ligand (bindin) binding within 20 sec (Abassi and Foltz 1994). In sea urchin egg, a protein (**p350**) was isolated as sperm receptor with the egg plasma membrane-vitelline layer complexes (Giusti et al. 1997) and another report have shown that EBR1 gene product serves a species-specific sperm-interacting protein on the egg vitelline envelope (Kamei and Glabe 2003). In Ascidians (*Halocynthia roretzi*), the sperm-egg binding is mediated by the molecular interaction between HrUrabin, a glycosylphosphatidylinositol-anchored CRISP (cysteine-rich secretory protein)-like protein on the sperm surface and HrVC70 on the polymorphic vitelline coat, but that HrUrabin per se is unlikely to be a direct allorecognition protein (Urayama et al. 2008). In *Xenopus* egg, gp69/64 glycoproteins are two glycoforms in the vitelline envelope and have the same number of N-linked oligosaccharide chains but differ in the extent of O-glycosylation, might serve as sperm receptor (Tian et al. 1999). In *bufo*, gp75 is expressed by previtellogenic oocytes and follicle cells and can be considered as a sperm receptor that undergoes N-terminal proteolysis during fertilization (Scarpeci et al. 2008). mZP3, a zona pellucida glycoprotein that serve as sperm receptor is unique to mammalian eggs, from mice to humans, although related glycoproteins are found in vitelline envelopes of a variety of non-mammalian eggs, from fish to birds (Wassarman and Litscher 2001).

2.63 STAT1/3

Signal transducer and activator of transcription (**STAT**) proteins are transcription factors that play the important roles in fertility and early embryonic development. STAT1 and STAT3 are known to interact with each other and the heterodimer complex enters the nucleus and controls the expression of specific genes. Several studies have reported the association of JAK/STAT signaling pathway with fertility traits in cattle. Genotype combinations of STAT1 and STAT3 are found to promote fertilization and embryonic survival in Holstein cattle (Khatib et al. 2009). Leptin that is secreted from granulosa and follicular cells through the binding of leptin receptor can trigger the phosphorylation of STAT3 during mouse oocyte maturation (Matsuoka et al. 1999). JAK-STAT signaling crucially contributes to early embryonic patterning (Baumer et al. 2011). It was reported that *Drosophila* STAT (STAT92E) in conjunction with Zelda (Zld; Zinc-finger early *Drosophila* activator), plays an important role in the transcription of the zygotic genome at the onset of embryonic development (Tsurumi et al. 2011).

2.64 Stomatin-like protein-2 (SLP-2/STML-2)

Stomatin is an integral membrane protein, which is widely expressed in many cell types. **Stomatin-like protein-2** (SLP-2; p42), a novel and unusual stomatin homologue, has been implicated in interaction with erythrocyte cytoskeleton and presumably with other integral membrane proteins. SLP-2 is overexpressed in human esophageal squamous cell carcinoma, lung cancer, laryngeal cancer, and endometrial adenocarcinoma (Zhang et al. 2006). SLP-2 is a mitochondrial protein, interact with the mitochondrial fusion mediator mitofusin 2 (Mfn2) and might be participate in mitochondrial fusion (Hajek et al. 2007). On the other hand, human erythrocytes and T-cells express plasma membrane-associated SLP-2, where it seems to act as a transmembrane signaling involving protein phosphorylation (Kirchhof et al. 2008; Wang and Morrow 2000). In *Xenopus* eggs, a 40-kDa SLP-2-like protein has been identified as a membrane microdomain-associated protein that becomes tyrosine-phosphorylated by Src in vitro and in vivo (our unpublished results), suggesting that it is a component of sperm-induced tyrosine kinase signaling at fertilization.

2.65 Transcription factor IIIA

In *Xenopus* oocytes, transcription factor IIIA (**TFIIIA**), was isolated from the cytoplasmic 7 S ribonucleoprotein complex and is phosphorylated on Ser by CKII (Westmark et al. 2002). Expression of the TFIIIA gene is differentially regulated in oogenesis, early embryos and in somatic cells in *Xenopus*. The incorporation of histone H1 into chromatin during *Xenopus* embryogenesis directs the specific repression of the TFIIIA-activated transcription of 5S rRNA genes (Bouvet et al. 1994). Phospho-form of TFIIIA may allow the factor to act as repressor for oocyte-type 5S rRNA genes (Ghose et al. 2004). TFIIIA favorably binds to the somatic nucleosome whereas H1 preferentially binds to the oocyte nucleosome, excluding TFIIIA binding in *Xenopus* oocyte (Panetta et al. 1998).

2.66 TPX2

TPX2, targeting protein for *Xenopus* kinesin-like protein (Xklp2), has multiple functions during mitosis, including microtubule nucleation around the chromosomes and the targeting of Xklp2 and Aurora A, a serine/threonine kinase, to the spindle. At the physiological conditions, TPX2 is essential for microtubule nucleation around chromatin (Brunet et al. 2004). TPX2 is required for spindle assembly and spindle pole integrity in mouse oocyte maturation (Brunet et al. 2008). In *Xenopus* oocyte, activation of the centrosomal Aurora A by TPX2 is required during spindle assembly (Sardon et al. 2008). Localized Aurora A kinase activity is required to target the factors involved in microtubule (MT) nucleation and stabilization to the centrosome, therefore promoting the formation of a MT aster (Sardon et al. 2008). In *Xenopus*, TPX2 is required for nearly all Aurora A activation and for full p53 synthesis and phosphorylation during oocyte maturation (Pascreau et al. 2009).

2.67 Tr-kit

The c-kit, a tyrosine kinase receptor, is consists of an extracellular ligand binding domain and an intracellular kinase domain. With the onset of meiosis c-kit expression ceases, but a truncated c-kit product, **Tr-kit**, is specifically expressed in post-meiotic stages of spermatogenesis, and is accumulated in mature spermatozoa (Rossi et al. 2000). Fyn is localized in the cortex region underneath the plasma membrane in mouse oocytes. The interaction of Tr-kit with Fyn, make the Fyn active and that phosphorylate PLCγ1 with the result of Ca^{2+} oscillation (Sette et al. 2002). The truncated c-kit protein is present in primary tumors and shows a correlation between Tr-kit expression and activation of the Src pathway in the advanced stages of human prostate cancer (Paronetto et al. 2004). Recently it was shown that Tr-kit is present in the equatorial region of human spermatozoa, which are the first sperm components that enter into the oocyte cytoplasm after fusion with the egg (Muciaccia et al. 2010).

2.68 Tubulin β

Several studies were carried out to reveal the function of **tubulin** in some species oocytes to embryo because the spindle of vertebrate eggs must remain stable and well organized during the second meiotic arrest. The transition of tubulin from the quiescent oocyte state to that competent to form spindle microtubules may involve the changes in the availability of microtubule and qualitative changes in tubulin mRNAs occurred between the early blastula and hatched blastula stages in sea urchin embryos (Alexandraki and Ruderman 1985). Tubulin β1 mRNA is evenly distributed during early embryogenesis but in later stages of embryogenesis is predominantly expressed in neural derivatives whereas tubulin β3 mRNA is restricted to the mesoderm in *Drosophila* (Gasch et al. 1988). Vg1 RBP is associated with microtubules and co-precipitated by heterologous, polymerized tubulin in *Xenopus* oocytes (Elisha et al. 1995). It was shown recently that Fyn and tubulin are closely associated where Fyn can phosphorylate tubulin and thus SFKs mediate significant functions during the organization of the MII spindle that involves possibly microtubules in rat eggs (Talmor-Cohen et al. 2004). Similarly, well-organized microtubule formation increased the GVBD and MII development in mouse oocytes (Mohammadi Roushandeh and Habibi Roudkenar 2009).

2.69 Ubiquitin-proteasome pathway

The **ubiquitin-proteasome** pathway (Schonfelder et al. 2006) is involved in the degradation of proteins e.g. cyclin B, a regulatory subunit of MPF that are related to oocyte meiotic maturation, fertilization and embryogenesis. Proteasome (26S) catalyzes the ATP- and ubiquitin-dependent degradation of Mos in an early stage of meiotic maturation of *Xenopus* oocytes and egg activation (Aizawa et al. 1996; Ishida et al. 1993). *Xenopus* RINGO/Speedy, a direct activator of Cdk1 and Cdk2, is limitedly processed by UPP to maintenance of G2 arrest and fully degraded by the ubiquitin ligase Siah-2 during MI-MII transition (Gutierrez et al. 2006). UPP is important for oocyte meiotic maturation, fertilization, and early embryonic mitosis and may play its roles by regulating cyclin B1 degradation and MAPK/p90[Rsk] phosphorylation in pig (Huo et al. 2004a; Sun et al. 2004) and in mouse (Huo et al. 2004b; Karabinova et al. 2011; Tan et al. 2005a). UPP is required for meiotic maturation of rat oocyte (Tan et al. 2005b). In gold fish, cyclin B degradation is initiated by the ATP-dependent and ubiquitin-independent proteolytic activity of 26S proteasome and then the cyclin to be ubiquitinated for further destruction by ubiquitin-dependent activity of the 26S proteasome that leads to MPF inactivation (Tokumoto et al. 1997).

2.70 Uroplakin Ib/III (UPIb/UPIII)

Uroplakins (UP; UPIa, UPIb, UPII, UPIIIa and UPIIIb) were first identified in highly differentiated somatic cells plasma membrane called asymmetric unit membrane (AUM), which is believed to play a protective role. Recently, they were identified in genital tract (Kalma et al. 2009; Shapiro et al. 2000) and germ cells and their function has been described in *Xenopus* fertilization (Mahbub Hasan et al. 2011; Sakakibara et al. 2005; Sato et al. 2006), pathogen infection (Thumbikat et al. 2009a; Thumbikat et al. 2009b) and cancer (Matsumoto et al. 2008). In *Xenopus*, UPIIIa a single transmembrane protein is tyrosine phosphorylated transiently in the cytosolic domain by a tyrosine kinase Src and this tyrosine phosphorylation is required for sperm mediated egg activation. UPIIIa was shaded in the extracellular domain by cathepsin B like activity that is present in sperm and this activity are essential for egg activation and fertilization (Mahbub Hasan et al. 2005; Mizote et al. 1999). UPIIIa can serve as sperm receptor as the antibody against the extracellular domain of UPIIIa inhibited the fertilization (Sakakibara et al. 2005). UPIIIa is an interactive partner of UPIb, a tetraspanin and their interaction is required to negatively regulate the Src activity (Mahbub Hasan et al. 2007).

2.71 Vg1RBP

Xenopus **Vg1RBP** (RNA binding protein), also known as Vera or IMP3, is a member of the highly conserved IMP family of four KH (hnRNP K-homologous)-domain RNA binding proteins, with roles in RNA localization, translational control, RNA stability, and cell motility. *Xenopus* Vg1 mRNA is localized to the vegetal cortex during oogenesis for the regulation of germ layer formation and germ cell development where proteins e.g. Vg1RBP/Vera that specifically recognize the vegetal localization element (VLE) within the 3' untranslated region. It is reported that multiple KH domains are important in mediating RNA-protein and protein-protein interactions in the formation of a stable complex of Vg1RBP and Vg1 mRNA (Git and Standart 2002). PTB/hnRNP I (ribonucleo protein) is required for remodeling of the interaction between Vg1 mRNA and Vg1RBP/Vera in *Xenopus* oocytes (Lewis et al. 2008). Vg1RBP undergoes regulated phosphorylation by Erk2 MAPK during meiotic maturation in *Xenopus* (Git et al. 2009).

2.72 XEEK

The PAR-4 and PAR-1 kinases are necessary for the formation of the anterior-posterior (A-P) axis in *C. elegans*. The *Drosophila* PAR-4 homologue, LKB1, is required for the early A-P polarity of the oocyte, and for the repolarization of the oocyte cytoskeleton that defines the embryonic A-P axis in *Drosophila* (Martin and St Johnston 2003) and in mouse (Szczepanska and Maleszewski 2005). PKA phosphorylates *Drosophila* LKB1 on a conserved site that is important for its activity(Martin and St Johnston 2003). **LKB1/XEEK1** (*Xenopus* egg and embryo kinase 1) is found to exist in a complex with GSK3 and PKC, a known kinase for GSK3 and to regulate GSK3 phosphorylation resulting in increased Wnt-catenin signal in *Xenopus* embryonic development and mammalian cells (Clements and Kimelman 2003; Ossipova et al. 2003).

2.73 Xp95

In *Xenopus* oocytes, a protein **Xp95** is tyrosine-phosphorylated from the first through the second meiotic divisions during progesterone-induced oocyte maturation. The Xp95 protein sequence exhibited homology to mouse Rhophilin, budding yeast Bro1, and *Aspergillus* PalA, all of which are important in signal transduction (Che et al. 1999). Src kinase mediated phosphorylation of Xp95 was increased during oocyte maturation (Che et al. 1999). Xp95 is phosphorylated at multiple sites within the N-terminal half of the proline-rich domain (PRD) during *Xenopus* oocyte maturation and the phosphorylation may both positively and negatively modulate their interaction with partner proteins at different stage of cell cycle (Dejournett et al. 2007). Human homologue of Xp95, termed Hp95, induces G1 phase arrest in confluent HeLa cells when overexpressed (Wu et al. 2001).

2.74 Tyrosine 3-monooxygenase/tryptophan 5-monooxygenase activation protein (YWHA)/14-3-3

The tyrosine 3-monooxygenase/tryptophan 5-monooxygenase activation protein family (YWHA; also known as 14-3-3) are involved in the regulation of many intracellular processes. PKB, PKC and JNK target 14-3-3 to phosphorylate at different sites (Aitken 2006). YWHA might play the role regulating peptidylarginine deiminase type VI (PADI6), that undergo a dramatic developmental change in phosphorylation during mouse oocyte maturation until two cell stage (Snow et al. 2008). 14-3-3 protein binds to Cdc25C and inhibits dephosphorylation of Ser-287 by PP2A, allowing the arrest in the meiotic metaphase II in *Xenopus* oocytes (Hutchins et al. 2002). If 14-3-3 binding to Cdc25 is prevented while nuclear export is inhibited, the coordinate nuclear accumulation of Cdc25 that dephosphorylates Cdc2-cyclin B1 to make it active, which promotes oocyte maturation (Yang et al. 1999).

3. Conclusion

Since the discovery in the late 1800's of the gamete membrane interaction and fusion as an initial and indispensable process for the beginning of life, i.e. fertilization, a number of research have dealt with the molecular and cellular basis of fertilization. In this chapter, we have reviewed the structure and function of key molecules likely involved in the phospho-signaling at oocyte maturation, sperm-egg interaction and subsequent events for activation of development, collectively called "egg activation". This work is an updated version of the review paper that we published in 2000 (Sato et al. 2000a), and thus a special focus point in this chapter is the kinases (both tyrosine kinases and serine/threonine kinases, total number of 53) and their regulators and/or substrates expressed in oocytes/eggs and/or early embryos of animal species (including some algae, total number of 74). We have compiled the currently available knowledge in the molecular level to explore the general as well as the species-specific features of oocyte maturation and fertilization, which is widely employed as an only-one strategy to give rise to a newborn in the bisexual reproduction system. It seems that number of kinases and their regulators/substrates will still be growing from day to day, and we may miss some important molecules in this chapter: we would continue to update that information not cited here in a future. Although the phospho-signaling system is just one kind of the post-translational modifications of cellular proteins, other kinds of steps e.g. transcriptional regulations or post-transcriptional modifications would also contribute to oocyte maturation and fertilization. We hope that this chapter could be helpful and enthusiastic for the readers in any kind of research field that deals with molecular (in particular, cellular proteins') network involved in physiological and/or pathological features of biological system.

4. Acknowledgements

We apologize to those whose work was not cited or insufficiently cited. This work is supported by a Grant-in-Aid on Innovative Areas (22112522), and a grant for Private University Strategic Research Foundation Support Program (S0801060) from the Ministry of Education, Culture, Sports, Science and Technology, Japan to K.S.

5. References

Abassi YA, Foltz KR. 1994. Tyrosine phosphorylation of the egg receptor for sperm at fertilization. Dev Biol 164(2):430-443.

Abe H, Obinata T, Minamide LS, Bamburg JR. 1996. Xenopus laevis actin-depolymerizing factor/cofilin: a phosphorylation-regulated protein essential for development. J Cell Biol 132(5):871-885.

Acevedo N, Ding J, Smith GD. 2007. Insulin signaling in mouse oocytes. Biol Reprod 77(5):872-879.

Achour-Frydman N, Ledee N, Fallet C. 2010. [Secrets of proportions in follicular liquid]. J Gynecol Obstet Biol Reprod (Paris) 39(1 Suppl):2-4.

Aitken A. 2006. 14-3-3 proteins: a historic overview. Semin Cancer Biol 16(3):162-172.

Aizawa H, Kawahara H, Tanaka K, Yokosawa H. 1996. Activation of the proteasome during Xenopus egg activation implies a link between proteasome activation and intracellular calcium release. Biochem Biophys Res Commun 218(1):224-228.

Akagi K, Kyun Park E, Mood K, Daar IO. 2002. Docking protein SNT1 is a critical mediator of fibroblast growth factor signaling during Xenopus embryonic development. Dev Dyn 223(2):216-228.

Alexandraki D, Ruderman JV. 1985. Expression of alpha- and beta-tubulin genes during development of sea urchin embryos. Dev Biol 109(2):436-451.

Amano H, Fujita T, Hiramatsu N, Kagawa H, Matsubara T, Sullivan CV, Hara A. 2008. Multiple vitellogenin-derived yolk proteins in gray mullet (Mugil cephalus): disparate proteolytic patterns associated with ovarian follicle maturation. Mol Reprod Dev 75(8):1307-1317.

Aoto M, Sato K, Takeba S, Horiuchi Y, Iwasaki T, Tokmakov AA, Fukami Y. 1999. A 58-kDa Shc protein is present in Xenopus eggs and is phosphorylated on tyrosine residues upon egg activation. Biochem Biophys Res Commun 258(2):265-270.

Armant DR, Kilburn BA, Petkova A, Edwin SS, Duniec-Dmuchowski ZM, Edwards HJ, Romero R, Leach RE. 2006. Human trophoblast survival at low oxygen concentrations requires metalloproteinase-mediated shedding of heparin-binding EGF-like growth factor. Development 133(4):751-759.

Asano Y, Mabuchi I. 2001. Calyculin-A, an inhibitor for protein phosphatases, induces cortical contraction in unfertilized sea urchin eggs. Cell Motil Cytoskeleton 48(4):245-261.

Baessler KA, Lee Y, Sampson NS. 2009. Beta1 integrin is an adhesion protein for sperm binding to eggs. ACS Chem Biol 4(5):357-366.

Bagowski CP, Myers JW, Ferrell JE, Jr. 2001. The classical progesterone receptor associates with p42 MAPK and is involved in phosphatidylinositol 3-kinase signaling in Xenopus oocytes. J Biol Chem 276(40):37708-37714.

Bamburg JR, McGough A, Ono S. 1999. Putting a new twist on actin: ADF/cofilins modulate actin dynamics. Trends Cell Biol 9(9):364-370.

Barr KJ, Garrill A, Jones DH, Orlowski J, Kidder GM. 1998. Contributions of Na+/H+ exchanger isoforms to preimplantation development of the mouse. Mol Reprod Dev 50(2):146-153.

Baumer D, Trauner J, Hollfelder D, Cerny A, Schoppmeier M. 2011. JAK-STAT signalling is required throughout telotrophic oogenesis and short-germ embryogenesis of the beetle Tribolium. Dev Biol 350(1):169-182.

Bellier S, Dubois MF, Nishida E, Almouzni G, Bensaude O. 1997. Phosphorylation of the RNA polymerase II largest subunit during Xenopus laevis oocyte maturation. Mol Cell Biol 17(3):1434-1440.

Bergeron L, Perez GI, Macdonald G, Shi L, Sun Y, Jurisicova A, Varmuza S, Latham KE, Flaws JA, Salter JC, Hara H, Moskowitz MA, Li E, Greenberg A, Tilly JL, Yuan J. 1998. Defects in regulation of apoptosis in caspase-2-deficient mice. Genes Dev 12(9):1304-1314.

Bergink EW, Wallace RA. 1974. Precursor-product relationship between amphibian vitellogenin and the yolk proteins, lipovitellin and phosvitin. J Biol Chem 249(9):2897-2903.

Bianchi E, Barbagallo F, Valeri C, Geremia R, Salustri A, De Felici M, Sette C. 2010. Ablation of the Sam68 gene impairs female fertility and gonadotropin-dependent follicle development. Hum Mol Genet 19(24):4886-4894.

Bivona TG, Perez De Castro I, Ahearn IM, Grana TM, Chiu VK, Lockyer PJ, Cullen PJ, Pellicer A, Cox AD, Philips MR. 2003. Phospholipase Cgamma activates Ras on the Golgi apparatus by means of RasGRP1. Nature 424(6949):694-698.

Bolanos-Garcia VM, Blundell TL. 2011. BUB1 and BUBR1: multifaceted kinases of the cell cycle. Trends Biochem Sci 36(3):141-150.

Bomsztyk K, Denisenko O, Ostrowski J. 2004. hnRNP K: one protein multiple processes. Bioessays 26(6):629-638.

Bonilla AQ, Oliveira LJ, Ozawa M, Newsom EM, Lucy MC, Hansen PJ. 2011. Developmental changes in thermoprotective actions of insulin-like growth factor-1 on the preimplantation bovine embryo. Mol Cell Endocrinol 332(1-2):170-179.

Bouvet P, Dimitrov S, Wolffe AP. 1994. Specific regulation of Xenopus chromosomal 5S rRNA gene transcription in vivo by histone H1. Genes Dev 8(10):1147-1159.

Branno M, De Franciscis V, Tosi L. 1983. In vitro methylation of histones in sea urchin nuclei during early embryogenesis. Biochim Biophys Acta 741(1):136-142.

Brewer LM, Brown NA. 1992. Distribution of p21ras in postimplantation rat embryos. Anat Rec 234(3):443-451.

Brewis IA, Clayton R, Browes CE, Martin M, Barratt CL, Hornby DP, Moore HD. 1998. Tyrosine phosphorylation of a 95 kDa protein and induction of the acrosome reaction in human spermatozoa by recombinant human zona pellucida glycoprotein 3. Mol Hum Reprod 4(12):1136-1144.

Browaeys-Poly E, Broutin I, Antoine AF, Marin M, Lescuyer A, Vilain JP, Ducruix A, Cailliau K. 2007. A non-canonical Grb2-PLC-gamma1-Sos cascade triggered by lipovitellin 1, an apolipoprotein B homologue. Cell Signal 19(12):2540-2548.

Brunet A, Bonni A, Zigmond MJ, Lin MZ, Juo P, Hu LS, Anderson MJ, Arden KC, Blenis J, Greenberg ME. 1999. Akt promotes cell survival by phosphorylating and inhibiting a Forkhead transcription factor. Cell 96(6):857-868.

Brunet S, Dumont J, Lee KW, Kinoshita K, Hikal P, Gruss OJ, Maro B, Verlhac MH. 2008. Meiotic regulation of TPX2 protein levels governs cell cycle progression in mouse oocytes. PLoS One 3(10):e3338.

Brunet S, Sardon T, Zimmerman T, Wittmann T, Pepperkok R, Karsenti E, Vernos I. 2004. Characterization of the TPX2 domains involved in microtubule nucleation and spindle assembly in Xenopus egg extracts. Mol Biol Cell 15(12):5318-5328.

Bugner V, Aurhammer T, Kuhl M. 2011. Xenopus laevis insulin receptor substrate IRS-1 is important for eye development. Dev Dyn 240(7):1705-1715.

Bui HT, Van Thuan N, Kishigami S, Wakayama S, Hikichi T, Ohta H, Mizutani E, Yamaoka E, Wakayama T, Miyano T. 2007. Regulation of chromatin and chromosome morphology by histone H3 modifications in pig oocytes. Reproduction 133(2):371-382.

Burks DJ, Carballada R, Moore HD, Saling PM. 1995. Interaction of a tyrosine kinase from human sperm with the zona pellucida at fertilization. Science 269(5220):83-86.

Busch S. 1997. Cloning and sequencing of the cDNA encoding for a Na+/H+ exchanger from Xenopus laevis oocytes (X1-NHE). Biochim Biophys Acta 1325(1):13-16.

Busch S, Wieland T, Esche H, Jakobs KH, Siffert W. 1995. G protein regulation of the Na+/H+ antiporter in Xenopus laevis oocytes. Involvement of protein kinases A and C. J Biol Chem 270(30):17898-17901.

Buszczak M, Cooley L. 2000. Eggs to die for: cell death during Drosophila oogenesis. Cell Death Differ 7(11):1071-1074.

Cailliau K, Browaeys-Poly E, Broutin-L'Hermite I, Nioche P, Garbay C, Ducruix A, Vilain JP. 2001. Grb2 promotes reinitiation of meiosis in Xenopus oocytes. Cell Signal 13(1):51-55.

Cailliau K, Le Marcis V, Bereziat V, Perdereau D, Cariou B, Vilain JP, Burnol AF, Browaeys-Poly E. 2003. Inhibition of FGF receptor signalling in Xenopus oocytes: differential effect of Grb7, Grb10 and Grb14. FEBS Lett 548(1-3):43-48.

Calounova G, Livera G, Zhang XQ, Liu K, Gosden RG, Welsh M. 2010. The Src homology 2 domain-containing adapter protein B (SHB) regulates mouse oocyte maturation. PLoS One 5(6):e11155.

Capco DG, Tutnick JM, Bement WM. 1992. The role of protein kinase C in reorganization of the cortical cytoskeleton during the transition from oocyte to fertilization-competent egg. J Exp Zool 264(4):395-405.

Carnero A, Lacal JC. 1995. Activation of intracellular kinases in Xenopus oocytes by p21ras and phospholipases: a comparative study. Mol Cell Biol 15(2):1094-1101.

Carnero A, Lacal JC. 1998. Wortmannin, an inhibitor of phosphatidyl-inositol 3-kinase, induces oocyte maturation through a MPF-MAPK-dependent pathway. FEBS Lett 422(2):155-159.

Carroll DJ, Albay DT, Terasaki M, Jaffe LA, Foltz KR. 1999. Identification of PLCgamma-dependent and -independent events during fertilization of sea urchin eggs. Dev Biol 206(2):232-247.

Carroll DJ, Ramarao CS, Mehlmann LM, Roche S, Terasaki M, Jaffe LA. 1997. Calcium release at fertilization in starfish eggs is mediated by phospholipase Cgamma. J Cell Biol 138(6):1303-1311.

Castilho PV, Williams BC, Mochida S, Zhao Y, Goldberg ML. 2009. The M phase kinase Greatwall (Gwl) promotes inactivation of PP2A/B55delta, a phosphatase directed against CDK phosphosites. Mol Biol Cell 20(22):4777-4789.

Chao MV, Hempstead BL. 1995. p75 and Trk: a two-receptor system. Trends Neurosci 18(7):321-326.

Che S, El-Hodiri HM, Wu CF, Nelman-Gonzalez M, Weil MM, Etkin LD, Clark RB, Kuang J. 1999. Identification and cloning of xp95, a putative signal transduction protein in Xenopus oocytes. J Biol Chem 274(9):5522-5531.

Chen GC, Turano B, Ruest PJ, Hagel M, Settleman J, Thomas SM. 2005. Regulation of Rho and Rac signaling to the actin cytoskeleton by paxillin during Drosophila development. Mol Cell Biol 25(3):979-987.

Cheng KC, Klancer R, Singson A, Seydoux G. 2009. Regulation of MBK-2/DYRK by CDK-1 and the pseudophosphatases EGG-4 and EGG-5 during the oocyte-to-embryo transition. Cell 139(3):560-572.

Cheng Y, Yata A, Klein C, Cho JH, Deguchi M, Hsueh AJ. 2011. Oocyte-expressed interleukin 7 suppresses granulosa cell apoptosis and promotes oocyte maturation in rats. Biol Reprod 84(4):707-714.

Chesnel F, Bonnec G, Tardivel A, Boujard D. 1997. Comparative effects of insulin on the activation of the Raf/Mos-dependent MAP kinase cascade in vitellogenic versus postvitellogenic Xenopus oocytes. Dev Biol 188(1):122-133.

Chesnel F, Heligon C, Richard-Parpaillon L, Boujard D. 2003. Molecular cloning and characterization of an adaptor protein Shc isoform from Xenopus laevis oocytes. Biol Cell 95(5):311-320.

Chiba K. 2004. MI arrest and apoptosis in starfish oocytes. Zoolog Sci 21(12):1193.

Chie L, Cook JR, Chung D, Hoffmann R, Yang Z, Kim Y, Pestka S, Pincus MR. 2003. A protein methyl transferase, PRMT5, selectively blocks oncogenic ras-p21 mitogenic signal transduction. Ann Clin Lab Sci 33(2):200-207.

Chiu TT, Patel N, Shaw AE, Bamburg JR, Klip A. 2010. Arp2/3- and cofilin-coordinated actin dynamics is required for insulin-mediated GLUT4 translocation to the surface of muscle cells. Mol Biol Cell 21(20):3529-3539.

Choi M, Lee OH, Jeon S, Park M, Lee DR, Ko JJ, Yoon TK, Rajkovic A, Choi Y. 2010. The oocyte-specific transcription factor, Nobox, regulates the expression of Pad6, a peptidylarginine deiminase in the oocyte. FEBS Lett 584(16):3629-3634.

Chuang LM, Hausdorff SF, Myers MG, Jr., White MF, Birnbaum MJ, Kahn CR. 1994. Interactive roles of Ras, insulin receptor substrate-1, and proteins with Src homology-2 domains in insulin signaling in Xenopus oocytes. J Biol Chem 269(44):27645-27649.

Chuang LM, Myers MG, Jr., Backer JM, Shoelson SE, White MF, Birnbaum MJ, Kahn CR. 1993a. Insulin-stimulated oocyte maturation requires insulin receptor substrate 1 and interaction with the SH2 domains of phosphatidylinositol 3-kinase. Mol Cell Biol 13(11):6653-6660.

Chuang LM, Myers MG, Jr., Seidner GA, Birnbaum MJ, White MF, Kahn CR. 1993b. Insulin receptor substrate 1 mediates insulin and insulin-like growth factor I-stimulated maturation of Xenopus oocytes. Proc Natl Acad Sci U S A 90(11):5172-5175.

Chung E, Chen RH. 2003. Phosphorylation of Cdc20 is required for its inhibition by the spindle checkpoint. Nat Cell Biol 5(8):748-753.

Clements WK, Kimelman D. 2003. Wnt signalling gets XEEKy. Nat Cell Biol 5(10):861-863.

Coles MC, Veiga-Fernandes H, Foster KE, Norton T, Pagakis SN, Seddon B, Kioussis D. 2006. Role of T and NK cells and IL7/IL7r interactions during neonatal maturation of lymph nodes. Proc Natl Acad Sci U S A 103(36):13457-13462.

Coppola D, Ouban A, Gilbert-Barness E. 2009. Expression of the insulin-like growth factor receptor 1 during human embryogenesis. Fetal Pediatr Pathol 28(2):47-54.

Cormier P, Pyronnet S, Morales J, Mulner-Lorillon O, Sonenberg N, Belle R. 2001. eIF4E association with 4E-BP decreases rapidly following fertilization in sea urchin. Dev Biol 232(2):275-283.

Covian-Nares F, Martinez-Cadena G, Lopez-Godinez J, Voronina E, Wessel GM, Garcia-Soto J. 2004. A Rho-signaling pathway mediates cortical granule translocation in the sea urchin oocyte. Mech Dev 121(3):225-235.

Cross MJ, Lu L, Magnusson P, Nyqvist D, Holmqvist K, Welsh M, Claesson-Welsh L. 2002. The Shb adaptor protein binds to tyrosine 766 in the FGFR-1 and regulates the Ras/MEK/MAPK pathway via FRS2 phosphorylation in endothelial cells. Mol Biol Cell 13(8):2881-2893.

Cuellar-Mata P, Martinez-Cadena G, Lopez-Godinez J, Obregon A, Garcia-Soto J. 2000. The GTP-binding protein RhoA localizes to the cortical granules of Strongylocentrotus purpuratas sea urchin egg and is secreted during fertilization. Eur J Cell Biol 79(2):81-91.

Danial NN. 2008. BAD: undertaker by night, candyman by day. Oncogene 27 Suppl 1:S53-70.

Dawson IA, Roth S, Akam M, Artavanis-Tsakonas S. 1993. Mutations of the fizzy locus cause metaphase arrest in Drosophila melanogaster embryos. Development 117(1):359-376.

Dejournett RE, Kobayashi R, Pan S, Wu C, Etkin LD, Clark RB, Bogler O, Kuang J. 2007. Phosphorylation of the proline-rich domain of Xp95 modulates Xp95 interaction with partner proteins. Biochem J 401(2):521-531.

Deng X, Feng C, Wang EH, Zhu YQ, Cui C, Zong ZH, Li GS, Liu C, Meng J, Yu BZ. 2011. Influence of proline-rich inositol polyphosphate 5-phosphatase, on early development of fertilized mouse eggs, via inhibition of phosphorylation of Akt. Cell Prolif 44(2):156-165.

Detwiler MR, Reuben M, Li X, Rogers E, Lin R. 2001. Two zinc finger proteins, OMA-1 and OMA-2, are redundantly required for oocyte maturation in C. elegans. Dev Cell 1(2):187-199.

Ding J, Swain JE, Smith GD. 2011. Aurora kinase-A regulates microtubule organizing center (MTOC) localization, chromosome dynamics, and histone-H3 phosphorylation in mouse oocytes. Mol Reprod Dev 78(2):80-90.

Dreyfuss G, Kim VN, Kataoka N. 2002. Messenger-RNA-binding proteins and the messages they carry. Nat Rev Mol Cell Biol 3(3):195-205.

Du Pasquier D, Dupre A, Jessus C. 2011. Unfertilized Xenopus eggs die by bad-dependent apoptosis under the control of Cdk1 and JNK. PLoS One 6(8):e23672.

Dube F, Eckberg WR. 1997. Intracellular pH increase driven by an Na+/H+ exchanger upon activation of surf clam oocytes. Dev Biol 190(1):41-54.

Eberlin A, Grauffel C, Oulad-Abdelghani M, Robert F, Torres-Padilla ME, Lambrot R, Spehner D, Ponce-Perez L, Wurtz JM, Stote RH, Kimmins S, Schultz P, Dejaegere A,

Tora L. 2008. Histone H3 tails containing dimethylated lysine and adjacent phosphorylated serine modifications adopt a specific conformation during mitosis and meiosis. Mol Cell Biol 28(5):1739-1754.

Edgecombe M, Patel R, Whitaker M. 1991. A cyclin-abundance cycle-independent p34cdc2 tyrosine phosphorylation cycle in early sea urchin embryos. EMBO J 10(12):3769-3775.

Elisha Z, Havin L, Ringel I, Yisraeli JK. 1995. Vg1 RNA binding protein mediates the association of Vg1 RNA with microtubules in Xenopus oocytes. EMBO J 14(20):5109-5114.

Eliyahu E, Tsaadon A, Shtraizent N, Shalgi R. 2005. The involvement of protein kinase C and actin filaments in cortical granule exocytosis in the rat. Reproduction 129(2):161-170.

Erikson E, Maller JL. 1985. A protein kinase from Xenopus eggs specific for ribosomal protein S6. Proc Natl Acad Sci U S A 82(3):742-746.

Erneux C, Govaerts C, Communi D, Pesesse X. 1998. The diversity and possible functions of the inositol polyphosphate 5-phosphatases. Biochim Biophys Acta 1436(1-2):185-199.

Esposito G, Vitale AM, Leijten FP, Strik AM, Koonen-Reemst AM, Yurttas P, Robben TJ, Coonrod S, Gossen JA. 2007. Peptidylarginine deiminase (PAD) 6 is essential for oocyte cytoskeletal sheet formation and female fertility. Mol Cell Endocrinol 273(1-2):25-31.

Evans EK, Lu W, Strum SL, Mayer BJ, Kornbluth S. 1997. Crk is required for apoptosis in Xenopus egg extracts. EMBO J 16(2):230-241.

Evans T, Rosenthal ET, Youngblom J, Distel D, Hunt T. 1983. Cyclin: a protein specified by maternal mRNA in sea urchin eggs that is destroyed at each cleavage division. Cell 33(2):389-396.

Eyers PA, Maller JL. 2004. Regulation of Xenopus Aurora A activation by TPX2. J Biol Chem 279(10):9008-9015.

Fan HY, Sun QY, Zou H. 2006. Regulation of Separase in meiosis: Separase is activated at the metaphase I-II transition in Xenopus oocytes during meiosis. Cell Cycle 5(2):198-204.

Feller SM, Posern G, Voss J, Kardinal C, Sakkab D, Zheng J, Knudsen BS. 1998. Physiological signals and oncogenesis mediated through Crk family adapter proteins. J Cell Physiol 177(4):535-552.

Feller SM, Ren R, Hanafusa H, Baltimore D. 1994. SH2 and SH3 domains as molecular adhesives: the interactions of Crk and Abl. Trends Biochem Sci 19(11):453-458.

Finn RN. 2007. Vertebrate yolk complexes and the functional implications of phosvitins and other subdomains in vitellogenins. Biol Reprod 76(6):926-935.

Fissore RA, Longo FJ, Anderson E, Parys JB, Ducibella T. 1999. Differential distribution of inositol trisphosphate receptor isoforms in mouse oocytes. Biol Reprod 60(1):49-57.

Fleig UN, Gould KL. 1991. Regulation of cdc2 activity in Schizosaccharomyces pombe: the role of phosphorylation. Semin Cell Biol 2(4):195-204.

Forbes K, West G, Garside R, Aplin JD, Westwood M. 2009. The protein-tyrosine phosphatase, SRC homology-2 domain containing protein tyrosine phosphatase-2,

Phospho-Signaling at Oocyte Maturation and Fertilization: Set Up for Embryogenesis
and Beyond Part II. Kinase Regulators and Substrates

123

is a crucial mediator of exogenous insulin-like growth factor signaling to human trophoblast. Endocrinology 150(10):4744-4754.

Gaffre M, Martoriati A, Belhachemi N, Chambon JP, Houliston E, Jessus C, Karaiskou A. 2011. A critical balance between Cyclin B synthesis and Myt1 activity controls meiosis entry in Xenopus oocytes. Development 138(17):3735-3744.

Gallego Melcon S, Sanchez de Toledo Codina J. 2007. Molecular biology of rhabdomyosarcoma. Clin Transl Oncol 9(7):415-419.

Gallo CJ, Hand AR, Jones TL, Jaffe LA. 1995. Stimulation of Xenopus oocyte maturation by inhibition of the G-protein alpha S subunit, a component of the plasma membrane and yolk platelet membranes. J Cell Biol 130(2):275-284.

Gasch A, Hinz U, Leiss D, Renkawitz-Pohl R. 1988. The expression of beta 1 and beta 3 tubulin genes of Drosophila melanogaster is spatially regulated during embryogenesis. Mol Gen Genet 211(1):8-16.

Ghose R, Malik M, Huber PW. 2004. Restricted specificity of Xenopus TFIIIA for transcription of somatic 5S rRNA genes. Mol Cell Biol 24(6):2467-2477.

Gilman AG. 1987. G proteins: transducers of receptor-generated signals. Annu Rev Biochem 56:615-649.

Git A, Allison R, Perdiguero E, Nebreda AR, Houliston E, Standart N. 2009. Vg1RBP phosphorylation by Erk2 MAP kinase correlates with the cortical release of Vg1 mRNA during meiotic maturation of Xenopus oocytes. RNA 15(6):1121-1133.

Git A, Standart N. 2002. The KH domains of Xenopus Vg1RBP mediate RNA binding and self-association. RNA 8(10):1319-1333.

Giusti AF, Carroll DJ, Abassi YA, Foltz KR. 1999. Evidence that a starfish egg Src family tyrosine kinase associates with PLC-gamma1 SH2 domains at fertilization. Dev Biol 208(1):189-199.

Giusti AF, Hoang KM, Foltz KR. 1997. Surface localization of the sea urchin egg receptor for sperm. Dev Biol 184(1):10-24.

Giusti AF, Xu W, Hinkle B, Terasaki M, Jaffe LA. 2000. Evidence that fertilization activates starfish eggs by sequential activation of a Src-like kinase and phospholipase cgamma. J Biol Chem 275(22):16788-16794.

Goud PT, Goud AP, Van Oostveldt P, Dhont M. 1999. Presence and dynamic redistribution of type I inositol 1,4,5-trisphosphate receptors in human oocytes and embryos during in-vitro maturation, fertilization and early cleavage divisions. Mol Hum Reprod 5(5):441-451.

Gu L, Wang Q, Wang CM, Hong Y, Sun SG, Yang SY, Wang JG, Hou Y, Sun QY, Liu WQ. 2008. Distribution and expression of phosphorylated histone H3 during porcine oocyte maturation. Mol Reprod Dev 75(1):143-149.

Gutierrez GJ, Vogtlin A, Castro A, Ferby I, Salvagiotto G, Ronai Z, Lorca T, Nebreda AR. 2006. Meiotic regulation of the CDK activator RINGO/Speedy by ubiquitin-proteasome-mediated processing and degradation. Nat Cell Biol 8(10):1084-1094.

Habelhah H, Shah K, Huang L, Ostareck-Lederer A, Burlingame AL, Shokat KM, Hentze MW, Ronai Z. 2001. ERK phosphorylation drives cytoplasmic accumulation of hnRNP-K and inhibition of mRNA translation. Nat Cell Biol 3(3):325-330.

Haberman Y, Alon LT, Eliyahu E, Shalgi R. 2011. Receptor for activated C kinase (RACK) and protein kinase C (PKC) in egg activation. Theriogenology 75(1):80-89.

Hajek P, Chomyn A, Attardi G. 2007. Identification of a novel mitochondrial complex containing mitofusin 2 and stomatin-like protein 2. J Biol Chem 282(8):5670-5681.

Hake LE, Richter JD. 1994. CPEB is a specificity factor that mediates cytoplasmic polyadenylation during Xenopus oocyte maturation. Cell 79(4):617-627.

Hama J, Xu H, Goldfarb M, Weinstein DC. 2001. SNT-1/FRS2alpha physically interacts with Laloo and mediates mesoderm induction by fibroblast growth factor. Mech Dev 109(2):195-204.

Han SJ, Vaccari S, Nedachi T, Andersen CB, Kovacina KS, Roth RA, Conti M. 2006. Protein kinase B/Akt phosphorylation of PDE3A and its role in mammalian oocyte maturation. EMBO J 25(24):5716-5725.

Hanocq-Quertier J, Baltus E. 1981. Phosphorylation of ribosomal proteins during maturation of Xenopus laevis oocytes. Eur J Biochem 120(2):351-355.

Hansen DV, Tung JJ, Jackson PK. 2006. CaMKII and polo-like kinase 1 sequentially phosphorylate the cytostatic factor Emi2/XErp1 to trigger its destruction and meiotic exit. Proc Natl Acad Sci U S A 103(3):608-613.

Heck MM. 1997. Condensins, cohesins, and chromosome architecture: how to make and break a mitotic chromosome. Cell 91(1):5-8.

Heighington CS, Kipreos ET. 2009. Embryogenesis: Degenerate phosphatases control the oocyte-to-embryo transition. Curr Biol 19(20):R939-941.

Hirose T, Izumi Y, Nagashima Y, Tamai-Nagai Y, Kurihara H, Sakai T, Suzuki Y, Yamanaka T, Suzuki A, Mizuno K, Ohno S. 2002. Involvement of ASIP/PAR-3 in the promotion of epithelial tight junction formation. J Cell Sci 115(Pt 12):2485-2495.

Hodgman R, Tay J, Mendez R, Richter JD. 2001. CPEB phosphorylation and cytoplasmic polyadenylation are catalyzed by the kinase IAK1/Eg2 in maturing mouse oocytes. Development 128(14):2815-2822.

Holland AJ, Taylor SS. 2006. Cyclin-B1-mediated inhibition of excess separase is required for timely chromosome disjunction. J Cell Sci 119(Pt 16):3325-3336.

Homer H, Gui L, Carroll J. 2009. A spindle assembly checkpoint protein functions in prophase I arrest and prometaphase progression. Science 326(5955):991-994.

Hoshino Y, Sato E. 2008. Protein kinase B (PKB/Akt) is required for the completion of meiosis in mouse oocytes. Dev Biol 314(1):215-223.

Hoshino Y, Yokoo M, Yoshida N, Sasada H, Matsumoto H, Sato E. 2004. Phosphatidylinositol 3-kinase and Akt participate in the FSH-induced meiotic maturation of mouse oocytes. Mol Reprod Dev 69(1):77-86.

Huang X, Andreu-Vieyra CV, Wang M, Cooney AJ, Matzuk MM, Zhang P. 2009. Preimplantation mouse embryos depend on inhibitory phosphorylation of separase to prevent chromosome missegregation. Mol Cell Biol 29(6):1498-1505.

Hunt T. 1989. Maturation promoting factor, cyclin and the control of M-phase. Curr Opin Cell Biol 1(2):268-274.

Huo LJ, Fan HY, Liang CG, Yu LZ, Zhong ZS, Chen DY, Sun QY. 2004a. Regulation of ubiquitin-proteasome pathway on pig oocyte meiotic maturation and fertilization. Biol Reprod 71(3):853-862.

Huo LJ, Fan HY, Zhong ZS, Chen DY, Schatten H, Sun QY. 2004b. Ubiquitin-proteasome pathway modulates mouse oocyte meiotic maturation and fertilization via regulation of MAPK cascade and cyclin B1 degradation. Mech Dev 121(10):1275-1287.

Hutchins JR, Dikovskaya D, Clarke PR. 2002. Dephosphorylation of the inhibitory phosphorylation site S287 in Xenopus Cdc25C by protein phosphatase-2A is inhibited by 14-3-3 binding. FEBS Lett 528(1-3):267-271.

Inzunza J, Danielsson O, Lalitkumar PG, Larsson O, Axelson M, Tohonen V, Danielsson KG, Stavreus-Evers A. 2010. Selective insulin-like growth factor-I antagonist inhibits mouse embryo development in a dose-dependent manner. Fertil Steril 93(8):2621-2626.

Ishida N, Tanaka K, Tamura T, Nishizawa M, Okazaki K, Sagata N, Ichihara A. 1993. Mos is degraded by the 26S proteasome in a ubiquitin-dependent fashion. FEBS Lett 324(3):345-348.

Ito J, Yoon SY, Lee B, Vanderheyden V, Vermassen E, Wojcikiewicz R, Alfandari D, De Smedt H, Parys JB, Fissore RA. 2008. Inositol 1,4,5-trisphosphate receptor 1, a widespread Ca2+ channel, is a novel substrate of polo-like kinase 1 in eggs. Dev Biol 320(2):402-413.

Ito J, Yoshida T, Kasai Y, Wakai T, Parys JB, Fissore RA, Kashiwazaki N. 2010. Phosphorylation of inositol 1,4,5-triphosphate receptor 1 during in vitro maturation of porcine oocytes. Anim Sci J 81(1):34-41.

Iwasaki T, Koretomo Y, Fukuda T, Paronetto MP, Sette C, Fukami Y, Sato K. 2008. Expression, phosphorylation, and mRNA-binding of heterogeneous nuclear ribonucleoprotein K in Xenopus oocytes, eggs, and early embryos. Dev Growth Differ 50(1):23-40.

Jelinkova L, Kubelka M. 2006. Neither Aurora B activity nor histone H3 phosphorylation is essential for chromosome condensation during meiotic maturation of porcine oocytes. Biol Reprod 74(5):905-912.

Jellerette T, He CL, Wu H, Parys JB, Fissore RA. 2000. Down-regulation of the inositol 1,4,5-trisphosphate receptor in mouse eggs following fertilization or parthenogenetic activation. Dev Biol 223(2):238-250.

Jensen JT, Zelinski-Wooten MB, Schwinof KM, Vance JE, Stouffer RL. 2005. The phosphodiesterase 3 inhibitor ORG 9935 inhibits oocyte maturation during gonadotropin-stimulated ovarian cycles in rhesus macaques. Contraception 71(1):68-73.

Joberty G, Petersen C, Gao L, Macara IG. 2000. The cell-polarity protein Par6 links Par3 and atypical protein kinase C to Cdc42. Nat Cell Biol 2(8):531-539.

Jones KT, Holt JE. 2010. BubR1 highlights essential function of Cdh1 in mammalian oocytes. Cell Cycle 9(6):1029-1030.

Jordan P, Karess R. 1997. Myosin light chain-activating phosphorylation sites are required for oogenesis in Drosophila. J Cell Biol 139(7):1805-1819.

Kalinowski RR, Berlot CH, Jones TL, Ross LF, Jaffe LA, Mehlmann LM. 2004. Maintenance of meiotic prophase arrest in vertebrate oocytes by a Gs protein-mediated pathway. Dev Biol 267(1):1-13.

Kalinowski RR, Jaffe LA, Foltz KR, Giusti AF. 2003. A receptor linked to a Gi-family G-protein functions in initiating oocyte maturation in starfish but not frogs. Dev Biol 253(1):139-149.

Kalma Y, Granot I, Gnainsky Y, Or Y, Czernobilsky B, Dekel N, Barash A. 2009. Endometrial biopsy-induced gene modulation: first evidence for the expression of bladder-transmembranal uroplakin Ib in human endometrium. Fertil Steril 91(4):1042-1049, 1049 e1041-1049.

Kamata T, Kung HF. 1990. Modulation of maturation and ribosomal protein S6 phosphorylation in Xenopus oocytes by microinjection of oncogenic ras protein and protein kinase C. Mol Cell Biol 10(3):880-886.

Kamei N, Glabe CG. 2003. The species-specific egg receptor for sea urchin sperm adhesion is EBR1, a novel ADAMTS protein. Genes Dev. 17(20):2502-2507.

Kang HJ, Feng Z, Sun Y, Atwal G, Murphy ME, Rebbeck TR, Rosenwaks Z, Levine AJ, Hu W. 2009. Single-nucleotide polymorphisms in the p53 pathway regulate fertility in humans. Proc Natl Acad Sci U S A 106(24):9761-9766.

Karabinova P, Kubelka M, Susor A. 2011. Proteasomal degradation of ubiquitinated proteins in oocyte meiosis and fertilization in mammals. Cell Tissue Res.

Karaiskou A, Cayla X, Haccard O, Jessus C, Ozon R. 1998. MPF amplification in Xenopus oocyte extracts depends on a two-step activation of cdc25 phosphatase. Exp Cell Res 244(2):491-500.

Karaiskou A, Perez LH, Ferby I, Ozon R, Jessus C, Nebreda AR. 2001. Differential regulation of Cdc2 and Cdk2 by RINGO and cyclins. J Biol Chem 276(38):36028-36034.

Karlsson T, Kullander K, Welsh M. 1998. The Src homology 2 domain protein Shb transmits basic fibroblast growth factor- and nerve growth factor-dependent differentiation signals in PC12 cells. Cell Growth Differ 9(9):757-766.

Karlsson T, Songyang Z, Landgren E, Lavergne C, Di Fiore PP, Anafi M, Pawson T, Cantley LC, Claesson-Welsh L, Welsh M. 1995. Molecular interactions of the Src homology 2 domain protein Shb with phosphotyrosine residues, tyrosine kinase receptors and Src homology 3 domain proteins. Oncogene 10(8):1475-1483.

Kawamura K, Kawamura N, Mulders SM, Sollewijn Gelpke MD, Hsueh AJ. 2005. Ovarian brain-derived neurotrophic factor (BDNF) promotes the development of oocytes into preimplantation embryos. Proc Natl Acad Sci U S A 102(26):9206-9211.

Kay C, Jeyendran RS, Coulam CB. 2006. p53 tumour suppressor gene polymorphism is associated with recurrent implantation failure. Reprod Biomed Online 13(4):492-496.

Kaziro Y, Itoh H, Kozasa T, Nakafuku M, Satoh T. 1991. Structure and function of signal-transducing GTP-binding proteins. Annu Rev Biochem 60:349-400.

Keady BT, Kuo P, Martinez SE, Yuan L, Hake LE. 2007. MAPK interacts with XGef and is required for CPEB activation during meiosis in Xenopus oocytes. J Cell Sci 120(Pt 6):1093-1103.

Khatib H, Huang W, Mikheil D, Schutzkus V, Monson RL. 2009. Effects of signal transducer and activator of transcription (STAT) genes STAT1 and STAT3 genotypic combinations on fertilization and embryonic survival rates in Holstein cattle. J Dairy Sci 92(12):6186-6191.

Kim JH, Do HJ, Wang WH, Machaty Z, Han YM, Day BN, Prather RS. 1999a. A protein tyrosine phosphatase inhibitor, sodium orthovanadate, causes parthenogenetic activation of pig oocytes via an increase in protein tyrosine kinase activity. Biol Reprod 61(4):900-905.

Phospho-Signaling at Oocyte Maturation and Fertilization: Set Up for Embryogenesis
and Beyond Part II. Kinase Regulators and Substrates
127

Kim SH, Li C, Maller JL. 1999b. A maternal form of the phosphatase Cdc25A regulates early embryonic cell cycles in Xenopus laevis. Dev Biol 212(2):381-391.

Kinoshita K, Noetzel TL, Pelletier L, Mechtler K, Drechsel DN, Schwager A, Lee M, Raff JW, Hyman AA. 2005. Aurora A phosphorylation of TACC3/maskin is required for centrosome-dependent microtubule assembly in mitosis. J Cell Biol 170(7):1047-1055.

Kirchhof MG, Chau LA, Lemke CD, Vardhana S, Darlington PJ, Marquez ME, Taylor R, Rizkalla K, Blanca I, Dustin ML, Madrenas J. 2008. Modulation of T cell activation by stomatin-like protein 2. J Immunol 181(3):1927-1936.

Kishimoto T. 2011. A primer on meiotic resumption in starfish oocytes: The proposed signaling pathway triggered by maturation-inducing hormone. Mol Reprod Dev 78(10-11):704-707.

Kitagawa R, Law E, Tang L, Rose AM. 2002. The Cdc20 homolog, FZY-1, and its interacting protein, IFY-1, are required for proper chromosome segregation in Caenorhabditis elegans. Curr Biol 12(24):2118-2123.

Kline D, Kopf GS, Muncy LF, Jaffe LA. 1991. Evidence for the involvement of a pertussis toxin-insensitive G-protein in egg activation of the frog, Xenopus laevis. Dev Biol 143(2):218-229.

Komaba S, Hamao H, Murata-Hori M, Hosoya H. 2001. Identification of myosin II kinase from sea urchin eggs as protein kinase CK2. Gene 275(1):141-148.

Komatsu S, Abbasi F, Kobori E, Fujisawa Y, Kato H, Iwasaki Y. 2005. Proteomic analysis of rice embryo: an approach for investigating Galpha protein-regulated proteins. Proteomics 5(15):3932-3941.

Kong AM, Speed CJ, O'Malley CJ, Layton MJ, Meehan T, Loveland KL, Cheema S, Ooms LM, Mitchell CA. 2000. Cloning and characterization of a 72-kDa inositol-polyphosphate 5-phosphatase localized to the Golgi network. J Biol Chem 275(31):24052-24064.

Kornberg RD. 1999. Eukaryotic transcriptional control. Trends Cell Biol 9(12):M46-49.

Kriz V, Agren N, Lindholm CK, Lenell S, Saldeen J, Mares J, Welsh M. 2006. The SHB adapter protein is required for normal maturation of mesoderm during in vitro differentiation of embryonic stem cells. J Biol Chem 281(45):34484-34491.

Kriz V, Anneren C, Lai C, Karlsson J, Mares J, Welsh M. 2003. The SHB adapter protein is required for efficient multilineage differentiation of mouse embryonic stem cells. Exp Cell Res 286(1):40-56.

Kume S, Endo T, Nishimura Y, Kano K, Naito K. 2007. Porcine SPDYA2 (RINGO A2) stimulates CDC2 activity and accelerates meiotic maturation of porcine oocytes. Biol Reprod 76(3):440-447.

Kume S, Yamamoto A, Inoue T, Muto A, Okano H, Mikoshiba K. 1997. Developmental expression of the inositol 1,4,5-trisphosphate receptor and structural changes in the endoplasmic reticulum during oogenesis and meiotic maturation of Xenopus laevis. Dev Biol 182(2):228-239.

Kuo P, Runge E, Lu X, Hake LE. 2011. XGef influences XRINGO/CDK1 signaling and CPEB activation during Xenopus oocyte maturation. Differentiation 81(2):133-140.

Kurokawa M, Sato K, Fissore RA. 2004. Mammalian fertilization: from sperm factor to phospholipase Czeta. Biol Cell 96(1):37-45.

Kushima S, Mammadova G, Mahbub Hasan AK, Fukami Y, Sato K. 2011. Characterization of Lipovitellin 2 as a tyrosine-phosphorylated protein in oocytes, eggs and early embryos of Xenopus laevis. Zoolog Sci 28(8):550-559.

Lambert H, Collazo I, Steinleitner A. 1992. IL-7 and IL-8 inhibit gamete interaction in the zona-free hamster egg sperm penetration assay. Mediators Inflamm 1(1):67-69.

Lane M, Bavister BD. 1999. Regulation of intracellular pH in bovine oocytes and cleavage stage embryos. Mol Reprod Dev 54(4):396-401.

Lapasset L, Pradet-Balade B, Lozano JC, Peaucellier G, Picard A. 2005. Nuclear envelope breakdown may deliver an inhibitor of protein phosphatase 1 which triggers cyclin B translation in starfish oocytes. Dev Biol 285(1):200-210.

Lapasset L, Pradet-Balade B, Verge V, Lozano JC, Oulhen N, Cormier P, Peaucellier G. 2008. Cyclin B synthesis and rapamycin-sensitive regulation of protein synthesis during starfish oocyte meiotic divisions. Mol Reprod Dev 75(11):1617-1626.

Lasko P. 2003. Gene regulation at the RNA layer: RNA binding proteins in intercellular signaling networks. Sci STKE 2003(179):RE6.

Lee B, Vermassen E, Yoon SY, Vanderheyden V, Ito J, Alfandari D, De Smedt H, Parys JB, Fissore RA. 2006. Phosphorylation of IP3R1 and the regulation of [Ca2+]i responses at fertilization: a role for the MAP kinase pathway. Development 133(21):4355-4365.

Lee MS, Kang SK, Lee BC, Hwang WS. 2005. The beneficial effects of insulin and metformin on in vitro developmental potential of porcine oocytes and embryos. Biol Reprod 73(6):1264-1268.

Levine AJ, Tomasini R, McKeon FD, Mak TW, Melino G. 2011. The p53 family: guardians of maternal reproduction. Nat Rev Mol Cell Biol 12(4):259-265.

Lewis RA, Gagnon JA, Mowry KL. 2008. PTB/hnRNP I is required for RNP remodeling during RNA localization in Xenopus oocytes. Mol Cell Biol 28(2):678-686.

Li X, Dai Y, Allen WR. 2004. Influence of insulin-like growth factor-I on cytoplasmic maturation of horse oocytes in vitro and organization of the first cell cycle following nuclear transfer and parthenogenesis. Biol Reprod 71(4):1391-1396.

Lim HJ, Dey SK. 2009. HB-EGF: a unique mediator of embryo-uterine interactions during implantation. Exp Cell Res 315(4):619-626.

Lin R. 2003. A gain-of-function mutation in oma-1, a C. elegans gene required for oocyte maturation, results in delayed degradation of maternal proteins and embryonic lethality. Dev Biol 258(1):226-239.

Littlepage LE, Ruderman JV. 2002. Identification of a new APC/C recognition domain, the A box, which is required for the Cdh1-dependent destruction of the kinase Aurora-A during mitotic exit. Genes Dev 16(17):2274-2285.

Liu H, Luo LL, Qian YS, Fu YC, Sui XX, Geng YJ, Huang DN, Gao ST, Zhang RL. 2009. FOXO3a is involved in the apoptosis of naked oocytes and oocytes of primordial follicles from neonatal rat ovaries. Biochem Biophys Res Commun 381(4):722-727.

Liu J, Maller JL. 2005. Calcium elevation at fertilization coordinates phosphorylation of XErp1/Emi2 by Plx1 and CaMK II to release metaphase arrest by cytostatic factor. Curr Biol 15(16):1458-1468.

Liu M, Oh A, Calarco P, Yamada M, Coonrod SA, Talbot P. 2005. Peptidylarginine deiminase (PAD) is a mouse cortical granule protein that plays a role in preimplantation embryonic development. Reprod Biol Endocrinol 3:42.

Liu XJ, Sorisky A, Zhu L, Pawson T. 1995. Molecular cloning of an amphibian insulin receptor substrate 1-like cDNA and involvement of phosphatidylinositol 3-kinase in insulin-induced Xenopus oocyte maturation. Mol Cell Biol 15(7):3563-3570.

Lodde V, Modina SC, Franciosi F, Zuccari E, Tessaro I, Luciano AM. 2009. Localization of DNA methyltransferase-1 during oocyte differentiation, in vitro maturation and early embryonic development in cow. Eur J Histochem 53(4):199-207.

Lutz RJ. 2000. Role of the BH3 (Bcl-2 homology 3) domain in the regulation of apoptosis and Bcl-2-related proteins. Biochem Soc Trans 28(2):51-56.

Mahbub Hasan AK, Fukami Y, Sato KI. 2011. Gamete membrane microdomains and their associated molecules in fertilization signaling. Mol Reprod Dev. 78(10-11):814-830.

Mahbub Hasan AK, Ou Z, Sakakibara K, Hirahara S, Iwasaki T, Sato K, Fukami Y. 2007. Characterization of Xenopus egg membrane microdomains containing uroplakin Ib/III complex: roles of their molecular interactions for subcellular localization and signal transduction. Genes Cells 12(2):251-267.

Mahbub Hasan AK, Sato K, Sakakibara K, Ou Z, Iwasaki T, Ueda Y, Fukami Y. 2005. Uroplakin III, a novel Src substrate in Xenopus egg rafts, is a target for sperm protease essential for fertilization. Dev Biol 286(2):483-492.

Mailhes JB, Hilliard C, Fuseler JW, London SN. 2003. Okadaic acid, an inhibitor of protein phosphatase 1 and 2A, induces premature separation of sister chromatids during meiosis I and aneuploidy in mouse oocytes in vitro. Chromosome Res 11(6):619-631.

Malcuit C, Knott JG, He C, Wainwright T, Parys JB, Robl JM, Fissore RA. 2005. Fertilization and inositol 1,4,5-trisphosphate (IP3)-induced calcium release in type-1 inositol 1,4,5-trisphosphate receptor down-regulated bovine eggs. Biol Reprod 73(1):2-13.

Maller JL. 1990. Xenopus oocytes and the biochemistry of cell division. Biochemistry 29(13):3157-3166.

Mammadova G, Iwasaki T, Tokmakov AA, Fukami Y, Sato K. 2009. Evidence that phosphatidylinositol 3-kinase is involved in sperm-induced tyrosine kinase signaling in Xenopus egg fertilization. BMC Dev Biol 9:68.

Manzo S, Martinez-Cadena G, Lopez-Godinez J, Pedraza-Reyes M, Garcia-Soto J. 2003. A Rho GTPase controls the rate of protein synthesis in the sea urchin egg. Biochem Biophys Res Commun 310(3):685-690.

Marangos P, Verschuren EW, Chen R, Jackson PK, Carroll J. 2007. Prophase I arrest and progression to metaphase I in mouse oocytes are controlled by Emi1-dependent regulation of APC(Cdh1). J Cell Biol 176(1):65-75.

Martin SG, St Johnston D. 2003. A role for Drosophila LKB1 in anterior-posterior axis formation and epithelial polarity. Nature 421(6921):379-384.

Martinez SE, Yuan L, Lacza C, Ransom H, Mahon GM, Whitehead IP, Hake LE. 2005. XGef mediates early CPEB phosphorylation during Xenopus oocyte meiotic maturation. Mol Biol Cell 16(3):1152-1164.

Maruyama R, Velarde NV, Klancer R, Gordon S, Kadandale P, Parry JM, Hang JS, Rubin J, Stewart-Michaelis A, Schweinsberg P, Grant BD, Piano F, Sugimoto A, Singson A. 2007. EGG-3 regulates cell-surface and cortex rearrangements during egg activation in Caenorhabditis elegans. Curr Biol 17(18):1555-1560.

Masciarelli S, Horner K, Liu C, Park SH, Hinckley M, Hockman S, Nedachi T, Jin C, Conti M, Manganiello V. 2004. Cyclic nucleotide phosphodiesterase 3A-deficient mice as a model of female infertility. J Clin Invest 114(2):196-205.

Masui Y. 2000. The elusive cytostatic factor in the animal egg. Nat Rev Mol Cell Biol 1(3):228-232.

Maton G, Lorca T, Girault JA, Ozon R, Jessus C. 2005. Differential regulation of Cdc2 and Aurora-A in Xenopus oocytes: a crucial role of phosphatase 2A. J Cell Sci 118(Pt 11):2485-2494.

Maton G, Thibier C, Castro A, Lorca T, Prigent C, Jessus C. 2003. Cdc2-cyclin B triggers H3 kinase activation of Aurora-A in Xenopus oocytes. J Biol Chem 278(24):21439-21449.

Matsubara T, Ohkubo N, Andoh T, Sullivan CV, Hara A. 1999. Two forms of vitellogenin, yielding two distinct lipovitellins, play different roles during oocyte maturation and early development of barfin flounder, Verasper moseri, a marine teleost that spawns pelagic eggs. Dev Biol 213(1):18-32.

Matsui M, Takahashi Y, Hishinuma M, Kanagawa H. 1995. Effects of supplementation of the maturation media with insulin on in vitro maturation and in vitro fertilization of bovine oocytes. Jpn J Vet Res 43(3-4):145-153.

Matsumoto K, Satoh T, Irie A, Ishii J, Kuwao S, Iwamura M, Baba S. 2008. Loss expression of uroplakin III is associated with clinicopathologic features of aggressive bladder cancer. Urology 72(2):444-449.

Matsuoka T, Tahara M, Yokoi T, Masumoto N, Takeda T, Yamaguchi M, Tasaka K, Kurachi H, Murata Y. 1999. Tyrosine phosphorylation of STAT3 by leptin through leptin receptor in mouse metaphase 2 stage oocyte. Biochem Biophys Res Commun 256(3):480-484.

Mattick JS. 2004. RNA regulation: a new genetics? Nat Rev Genet 5(4):316-323.

Mayer BJ, Hamaguchi M, Hanafusa H. 1988. A novel viral oncogene with structural similarity to phospholipase C. Nature 332(6161):272-275.

Mayer BJ, Hanafusa H. 1990. Mutagenic analysis of the v-crk oncogene: requirement for SH2 and SH3 domains and correlation between increased cellular phosphotyrosine and transformation. J Virol 64(8):3581-3589.

Mayes MA, Sirard MA. 2002. Effect of type 3 and type 4 phosphodiesterase inhibitors on the maintenance of bovine oocytes in meiotic arrest. Biol Reprod 66(1):180-184.

Mazus B, Szurmak B, Buchowicz J. 1980. Phosphorylation in vitro and in vivo of the wheat embryo RNA polymerase II. Acta Biochim Pol 27(1):9-19.

McGlade J, Cheng A, Pelicci G, Pelicci PG, Pawson T. 1992. Shc proteins are phosphorylated and regulated by the v-Src and v-Fps protein-tyrosine kinases. Proc Natl Acad Sci U S A 89(19):8869-8873.

McGuinness BE, Anger M, Kouznetsova A, Gil-Bernabe AM, Helmhart W, Kudo NR, Wuensche A, Taylor S, Hoog C, Novak B, Nasmyth K. 2009. Regulation of APC/C activity in oocytes by a Bub1-dependent spindle assembly checkpoint. Curr Biol 19(5):369-380.

Mehlmann LM. 2005. Stops and starts in mammalian oocytes: recent advances in understanding the regulation of meiotic arrest and oocyte maturation. Reproduction 130(6):791-799.

Mehlmann LM, Carpenter G, Rhee SG, Jaffe LA. 1998. SH2 domain-mediated activation of
    phospholipase Cgamma is not required to initiate Ca2+ release at fertilization of
    mouse eggs. Dev Biol 203(1):221-232.
Mehlmann LM, Saeki Y, Tanaka S, Brennan TJ, Evsikov AV, Pendola FL, Knowles BB, Eppig
    JJ, Jaffe LA. 2004. The Gs-linked receptor GPR3 maintains meiotic arrest in
    mammalian oocytes. Science 306(5703):1947-1950.
Meijer L, Arion D, Golsteyn R, Pines J, Brizuela L, Hunt T, Beach D. 1989a. Cyclin is a
    component of the sea urchin egg M-phase specific histone H1 kinase. EMBO J
    8(8):2275-2282.
Meijer L, Azzi L, Wang JY. 1991. Cyclin B targets p34cdc2 for tyrosine phosphorylation.
    EMBO J 10(6):1545-1554.
Meijer L, Dostmann W, Genieser HG, Butt E, Jastorff B. 1989b. Starfish oocyte maturation:
    evidence for a cyclic AMP-dependent inhibitory pathway. Dev Biol 133(1):58-66.
Mendez R, Richter JD. 2001. Translational control by CPEB: a means to the end. Nat Rev Mol
    Cell Biol 2(7):521-529.
Meng XQ, Zheng KG, Yang Y, Jiang MX, Zhang YL, Sun QY, Li YL. 2006. Proline-rich
    tyrosine kinase2 is involved in F-actin organization during in vitro maturation of
    rat oocyte. Reproduction 132(6):859-867.
Miyamoto S, Yagi H, Yotsumoto F, Kawarabayashi T, Mekada E. 2006. Heparin-binding
    epidermal growth factor-like growth factor as a novel targeting molecule for cancer
    therapy. Cancer Sci 97(5):341-347.
Mizote A, Okamoto S, Iwao Y. 1999. Activation of Xenopus eggs by proteases: possible
    involvement of a sperm protease in fertilization. Dev Biol 208(1):79-92.
Mochida S, Hunt T. 2007. Calcineurin is required to release Xenopus egg extracts from
    meiotic M phase. Nature 449(7160):336-340.
Mochizuki Y, Takenawa T. 1999. Novel inositol polyphosphate 5-phosphatase localizes at
    membrane ruffles. J Biol Chem 274(51):36790-36795.
Mohammadi Roushandeh A, Habibi Roudkenar M. 2009. The influence of meiotic spindle
    configuration by cysteamine during in vitro maturation of mouse oocytes. Iran
    Biomed J 13(2):73-78.
Mood K, Friesel R, Daar IO. 2002. SNT1/FRS2 mediates germinal vesicle breakdown
    induced by an activated FGF receptor1 in Xenopus oocytes. J Biol Chem
    277(36):33196-33204.
Morita Y, Tilly JL. 1999. Oocyte apoptosis: like sand through an hourglass. Dev Biol 213(1):1-
    17.
Muciaccia B, Sette C, Paronetto MP, Barchi M, Pensini S, D'Agostino A, Gandini L, Geremia
    R, Stefanini M, Rossi P. 2010. Expression of a truncated form of KIT tyrosine kinase
    in human spermatozoa correlates with sperm DNA integrity. Hum Reprod
    25(9):2188-2202.
Muller AH, Gawantka V, Ding X, Hausen P. 1993. Maturation induced internalization of
    beta 1-integrin by Xenopus oocytes and formation of the maternal integrin pool.
    Mech Dev 42(1-2):77-88.
Nagao K, Yanagida M. 2002. Regulating sister chromatid separation by separase
    phosphorylation. Dev Cell 2(1):2-4.

Nakahata S, Kotani T, Mita K, Kawasaki T, Katsu Y, Nagahama Y, Yamashita M. 2003. Involvement of Xenopus Pumilio in the translational regulation that is specific to cyclin B1 mRNA during oocyte maturation. Mech Dev 120(8):865-880.

Nakamura H, Yoshitome S, Sugimoto I, Sado Y, Kawahara A, Ueno S, Miyahara T, Yoshida Y, Aoki-Yagi N, Hashimoto E. 2007. Cellular distribution of Mr 25,000 protein, a protein partially overlapping phosvitin and lipovitellin 2 in vitellogenin B1, and yolk proteins in Xenopus laevis oocytes and embryos. Comp Biochem Physiol A Mol Integr Physiol 148(3):621-628.

Nakaya M, Fukui A, Izumi Y, Akimoto K, Asashima M, Ohno S. 2000. Meiotic maturation induces animal-vegetal asymmetric distribution of aPKC and ASIP/PAR-3 in Xenopus oocytes. Development 127(23):5021-5031.

Nasmyth K, Peters JM, Uhlmann F. 2000. Splitting the chromosome: cutting the ties that bind sister chromatids. Science 288(5470):1379-1385.

Naz RK, Ahmad K. 1994. Molecular identities of human sperm proteins that bind human zona pellucida: nature of sperm-zona interaction, tyrosine kinase activity, and involvement of FA-1. Mol Reprod Dev 39(4):397-408.

Nebreda AR, Porras A, Santos E. 1993. p21ras-induced meiotic maturation of Xenopus oocytes in the absence of protein synthesis: MPF activation is preceded by activation of MAP and S6 kinases. Oncogene 8(2):467-477.

Nedachi T, Conti M. 2004. Potential role of protein tyrosine phosphatase nonreceptor type 13 in the control of oocyte meiotic maturation. Development 131(20):4987-4998.

Nielsen PJ, Thomas G, Maller JL. 1982. Increased phosphorylation of ribosomal protein S6 during meiotic maturation of Xenopus oocytes. Proc Natl Acad Sci U S A 79(9):2937-2941.

Nishi Y, Lin R. 2005. DYRK2 and GSK-3 phosphorylate and promote the timely degradation of OMA-1, a key regulator of the oocyte-to-embryo transition in C. elegans. Dev Biol 288(1):139-149.

Nishiyama T, Yoshizaki N, Kishimoto T, Ohsumi K. 2007. Transient activation of calcineurin is essential to initiate embryonic development in Xenopus laevis. Nature 449(7160):341-345.

Nishizuka Y. 1984. The role of protein kinase C in cell surface signal transduction and tumour promotion. Nature 308(5961):693-698.

Nishizuka Y. 1986. Studies and perspectives of protein kinase C. Science 233(4761):305-312.

Nogueira D, Albano C, Adriaenssens T, Cortvrindt R, Bourgain C, Devroey P, Smitz J. 2003a. Human oocytes reversibly arrested in prophase I by phosphodiesterase type 3 inhibitor in vitro. Biol Reprod 69(3):1042-1052.

Nogueira D, Cortvrindt R, De Matos DG, Vanhoutte L, Smitz J. 2003b. Effect of phosphodiesterase type 3 inhibitor on developmental competence of immature mouse oocytes in vitro. Biol Reprod 69(6):2045-2052.

Nogueira D, Cortvrindt R, Everaerdt B, Smitz J. 2005. Effects of long-term in vitro exposure to phosphodiesterase type-3 inhibitors on follicle and oocyte development. Reproduction 130(2):177-186.

Nutt LK, Margolis SS, Jensen M, Herman CE, Dunphy WG, Rathmell JC, Kornbluth S. 2005. Metabolic regulation of oocyte cell death through the CaMKII-mediated phosphorylation of caspase-2. Cell 123(1):89-103.

Oh JS, Han SJ, Conti M. 2010. Wee1B, Myt1, and Cdc25 function in distinct compartments of
the mouse oocyte to control meiotic resumption. J Cell Biol 188(2):199-207.

Okumura E, Fukuhara T, Yoshida H, Hanada Si S, Kozutsumi R, Mori M, Tachibana K,
Kishimoto T. 2002. Akt inhibits Myt1 in the signalling pathway that leads to meiotic
G2/M-phase transition. Nat Cell Biol 4(2):111-116.

Olivier JP, Raabe T, Henkemeyer M, Dickson B, Mbamalu G, Margolis B, Schlessinger J,
Hafen E, Pawson T. 1993. A Drosophila SH2-SH3 adaptor protein implicated in
coupling the sevenless tyrosine kinase to an activator of Ras guanine nucleotide
exchange, Sos. Cell 73(1):179-191.

Ooga M, Inoue A, Kageyama S, Akiyama T, Nagata M, Aoki F. 2008. Changes in H3K79
methylation during preimplantation development in mice. Biol Reprod 78(3):413-
424.

Ossipova O, Bardeesy N, DePinho RA, Green JB. 2003. LKB1 (XEEK1) regulates Wnt
signalling in vertebrate development. Nat Cell Biol 5(10):889-894.

Ostareck-Lederer A, Ostareck DH, Cans C, Neubauer G, Bomsztyk K, Superti-Furga G,
Hentze MW. 2002. c-Src-mediated phosphorylation of hnRNP K drives
translational activation of specifically silenced mRNAs. Mol Cell Biol 22(13):4535-
4543.

Ota R, Kotani T, Yamashita M. 2011a. Biochemical characterization of Pumilio1 and
Pumilio2 in Xenopus oocytes. J Biol Chem 286(4):2853-2863.

Ota R, Kotani T, Yamashita M. 2011b. Possible involvement of Nemo-like kinase 1 in
Xenopus oocyte maturation as a kinase responsible for Pumilio1, Pumilio2, and
CPEB phosphorylation. Biochemistry 50(25):5648-5659.

Ota R, Suwa K, Kotani T, Mita K, Yamashita M. 2008. Possible involvement of
phosphatidylinositol 3-kinase, but not protein kinase B or glycogen synthase kinase
3beta, in progesterone-induced oocyte maturation in the Japanese brown frog, Rana
japonica. Zoolog Sci 25(7):773-781.

Padmanabhan K, Richter JD. 2006. Regulated Pumilio-2 binding controls RINGO/Spy
mRNA translation and CPEB activation. Genes Dev 20(2):199-209.

Pallen CJ, Wang JH. 1985. A multifunctional calmodulin-stimulated phosphatase. Arch
Biochem Biophys 237(2):281-291.

Panetta G, Buttinelli M, Flaus A, Richmond TJ, Rhodes D. 1998. Differential nucleosome
positioning on Xenopus oocyte and somatic 5 S RNA genes determines both TFIIIA
and H1 binding: a mechanism for selective H1 repression. J Mol Biol 282(3):683-697.

Papin C, Rouget C, Lorca T, Castro A, Mandart E. 2004. XCdh1 is involved in progesterone-
induced oocyte maturation. Dev Biol 272(1):66-75.

Paronetto MP, Farini D, Sammarco I, Maturo G, Vespasiani G, Geremia R, Rossi P, Sette C.
2004. Expression of a truncated form of the c-Kit tyrosine kinase receptor and
activation of Src kinase in human prostatic cancer. Am J Pathol 164(4):1243-1251.

Parrington J, Brind S, De Smedt H, Gangeswaran R, Lai FA, Wojcikiewicz R, Carroll J. 1998.
Expression of inositol 1,4,5-trisphosphate receptors in mouse oocytes and early
embryos: the type I isoform is upregulated in oocytes and downregulated after
fertilization. Dev Biol 203(2):451-461.

Parrington J, Jones ML, Tunwell R, Devader C, Katan M, Swann K. 2002. Phospholipase C isoforms in mammalian spermatozoa: potential components of the sperm factor that causes Ca2+ release in eggs. Reproduction 123(1):31-39.

Parry JM, Velarde NV, Lefkovith AJ, Zegarek MH, Hang JS, Ohm J, Klancer R, Maruyama R, Druzhinina MK, Grant BD, Piano F, Singson A. 2009. EGG-4 and EGG-5 Link Events of the Oocyte-to-Embryo Transition with Meiotic Progression in C. elegans. Curr Biol 19(20):1752-1757.

Pascreau G, Delcros JG, Cremet JY, Prigent C, Arlot-Bonnemains Y. 2005. Phosphorylation of maskin by Aurora-A participates in the control of sequential protein synthesis during Xenopus laevis oocyte maturation. J Biol Chem 280(14):13415-13423.

Pascreau G, Eckerdt F, Lewellyn AL, Prigent C, Maller JL. 2009. Phosphorylation of p53 is regulated by TPX2-Aurora A in xenopus oocytes. J Biol Chem 284(9):5497-5505.

Paul S, Pramanick K, Kundu S, Bandyopadhyay A, Mukherjee D. 2009. Involvement of PI3 kinase and MAP kinase in IGF-I- and insulin-induced oocyte maturation in Cyprinus carpio. Mol Cell Endocrinol 309(1-2):93-100.

Pelicci G, Lanfrancone L, Grignani F, McGlade J, Cavallo F, Forni G, Nicoletti I, Pawson T, Pelicci PG. 1992. A novel transforming protein (SHC) with an SH2 domain is implicated in mitogenic signal transduction. Cell 70(1):93-104.

Perdiguero E, Nebreda AR. 2004. Regulation of Cdc25C activity during the meiotic G2/M transition. Cell Cycle 3(6):733-737.

Perdiguero E, Pillaire MJ, Bodart JF, Hennersdorf F, Frodin M, Duesbery NS, Alonso G, Nebreda AR. 2003. Xp38gamma/SAPK3 promotes meiotic G(2)/M transition in Xenopus oocytes and activates Cdc25C. EMBO J 22(21):5746-5756.

Pesin JA, Orr-Weaver TL. 2008. Regulation of APC/C activators in mitosis and meiosis. Annu Rev Cell Dev Biol 24:475-499.

Phillips KP, Leveille MC, Claman P, Baltz JM. 2000. Intracellular pH regulation in human preimplantation embryos. Hum Reprod 15(4):896-904.

Pirino G, Wescott MP, Donovan PJ. 2009. Protein kinase A regulates resumption of meiosis by phosphorylation of Cdc25B in mammalian oocytes. Cell Cycle 8(4):665-670.

Pomerance M, Schweighoffer F, Tocque B, Pierre M. 1992. Stimulation of mitogen-activated protein kinase by oncogenic Ras p21 in Xenopus oocytes. Requirement for Ras p21-GTPase-activating protein interaction. J Biol Chem 267(23):16155-16160.

Pomerance M, Thang MN, Tocque B, Pierre M. 1996. The Ras-GTPase-activating protein SH3 domain is required for Cdc2 activation and mos induction by oncogenic Ras in Xenopus oocytes independently of mitogen-activated protein kinase activation. Mol Cell Biol 16(6):3179-3186.

Pondaven P, Cohen P. 1987. Identification of protein phosphatases-1 and 2A and inhibitor-2 in oocytes of the starfish Asterias rubens and Marthasterias glacialis. Eur J Biochem 167(1):135-140.

Powell-Coffman JA, Knight J, Wood WB. 1996. Onset of C. elegans gastrulation is blocked by inhibition of embryonic transcription with an RNA polymerase antisense RNA. Dev Biol 178(2):472-483.

Qian YW, Erikson E, Taieb FE, Maller JL. 2001. The polo-like kinase Plx1 is required for activation of the phosphatase Cdc25C and cyclin B-Cdc2 in Xenopus oocytes. Mol Biol Cell 12(6):1791-1799.

Qu Y, Adler V, Chu T, Platica O, Michl J, Pestka S, Izotova L, Boutjdir M, Pincus MR. 2006.
Two dual specificity kinases are preferentially induced by wild-type rather than by
oncogenic RAS-P21 in Xenopus oocytes. Front Biosci 11:2420-2427.

Qu Y, Adler V, Izotova L, Pestka S, Bowne W, Michl J, Boutjdir M, Friedman FK, Pincus MR.
2007. The dual-specificity kinases, TOPK and DYRK1A, are critical for oocyte
maturation induced by wild-type--but not by oncogenic--ras-p21 protein. Front
Biosci 12:5089-5097.

Quesnel H. 1999. Localization of binding sites for IGF-I, insulin and GH in the sow ovary. J
Endocrinol 163(2):363-372.

Rasar M, DeFranco DB, Hammes SR. 2006. Paxillin regulates steroid-triggered meiotic
resumption in oocytes by enhancing an all-or-none positive feedback kinase loop. J
Biol Chem 281(51):39455-39464.

Rauh NR, Schmidt A, Bormann J, Nigg EA, Mayer TU. 2005. Calcium triggers exit from
meiosis II by targeting the APC/C inhibitor XErp1 for degradation. Nature
437(7061):1048-1052.

Reddy P, Shen L, Ren C, Boman K, Lundin E, Ottander U, Lindgren P, Liu YX, Sun QY, Liu
K. 2005. Activation of Akt (PKB) and suppression of FKHRL1 in mouse and rat
oocytes by stem cell factor during follicular activation and development. Dev Biol
281(2):160-170.

Reddy VR, Rajeev SK, Gupta V. 2003. Alpha 6 beta 1 Integrin is a potential clinical marker
for evaluating sperm quality in men. Fertil Steril 79 Suppl 3:1590-1596.

Reimann JD, Freed E, Hsu JY, Kramer ER, Peters JM, Jackson PK. 2001a. Emi1 is a mitotic
regulator that interacts with Cdc20 and inhibits the anaphase promoting complex.
Cell 105(5):645-655.

Reimann JD, Gardner BE, Margottin-Goguet F, Jackson PK. 2001b. Emi1 regulates the
anaphase-promoting complex by a different mechanism than Mad2 proteins. Genes
Dev 15(24):3278-3285.

Reimann JD, Jackson PK. 2002. Emi1 is required for cytostatic factor arrest in vertebrate
eggs. Nature 416(6883):850-854.

Reverte CG, Yuan L, Keady BT, Lacza C, Attfield KR, Mahon GM, Freeman B, Whitehead IP,
Hake LE. 2003. XGef is a CPEB-interacting protein involved in Xenopus oocyte
maturation. Dev Biol 255(2):383-398.

Rhee SG. 2001. Regulation of phosphoinositide-specific phospholipase C. Annu Rev
Biochem 70:281-312.

Richard FJ, Tsafriri A, Conti M. 2001. Role of phosphodiesterase type 3A in rat oocyte
maturation. Biol Reprod 65(5):1444-1451.

Rime H, Huchon D, De Smedt V, Thibier C, Galaktionov K, Jessus C, Ozon R. 1994.
Microinjection of Cdc25 protein phosphatase into Xenopus prophase oocyte
activates MPF and arrests meiosis at metaphase I. Biol Cell 82(1):11-22.

Rogers E, Bishop JD, Waddle JA, Schumacher JM, Lin R. 2002. The aurora kinase AIR-2
functions in the release of chromosome cohesion in Caenorhabditis elegans meiosis.
J Cell Biol 157(2):219-229.

Ron D, Luo J, Mochly-Rosen D. 1995. C2 region-derived peptides inhibit translocation and
function of beta protein kinase C in vivo. J Biol Chem 270(41):24180-24187.

Ron D, Mochly-Rosen D. 1995. An autoregulatory region in protein kinase C: the pseudoanchoring site. Proc Natl Acad Sci U S A 92(2):492-496.

Rossi P, Sette C, Dolci S, Geremia R. 2000. Role of c-kit in mammalian spermatogenesis. J Endocrinol Invest 23(9):609-615.

Rozakis-Adcock M, Fernley R, Wade J, Pawson T, Bowtell D. 1993. The SH2 and SH3 domains of mammalian Grb2 couple the EGF receptor to the Ras activator mSos1. Nature 363(6424):83-85.

Rozakis-Adcock M, McGlade J, Mbamalu G, Pelicci G, Daly R, Li W, Batzer A, Thomas S, Brugge J, Pelicci PG, et al. 1992. Association of the Shc and Grb2/Sem5 SH2-containing proteins is implicated in activation of the Ras pathway by tyrosine kinases. Nature 360(6405):689-692.

Rudner AD, Murray AW. 2000. Phosphorylation by Cdc28 activates the Cdc20-dependent activity of the anaphase-promoting complex. J Cell Biol 149(7):1377-1390.

Runft LL, Carroll DJ, Gillett J, Giusti AF, O'Neill FJ, Foltz KR. 2004. Identification of a starfish egg PLC-gamma that regulates Ca2+ release at fertilization. Dev Biol 269(1):220-236.

Runft LL, Jaffe LA. 2000. Sperm extract injection into ascidian eggs signals Ca(2+) release by the same pathway as fertilization. Development 127(15):3227-3236.

Runft LL, Jaffe LA, Mehlmann LM. 2002. Egg activation at fertilization: where it all begins. Dev Biol 245(2):237-254.

Runft LL, Watras J, Jaffe LA. 1999. Calcium release at fertilization of Xenopus eggs requires type I IP(3) receptors, but not SH2 domain-mediated activation of PLCgamma or G(q)-mediated activation of PLCbeta. Dev Biol 214(2):399-411.

Sakakibara K, Sato K, Yoshino K, Oshiro N, Hirahara S, Mahbub Hasan AK, Iwasaki T, Ueda Y, Iwao Y, Yonezawa K, Fukami Y. 2005. Molecular identification and characterization of Xenopus egg uroplakin III, an egg raft-associated transmembrane protein that is tyrosine-phosphorylated upon fertilization. J Biol Chem 280(15):15029-15037.

Sakamoto I, Takahara K, Yamashita M, Iwao Y. 1998. Changes in cyclin B during oocyte maturation and early embryonic cell cycle in the newt, Cynops pyrrhogaster: requirement of germinal vesicle for MPF activation. Dev Biol 195(1):60-69.

Salaun P, Pyronnet S, Morales J, Mulner-Lorillon O, Belle R, Sonenberg N, Cormier P. 2003. eIF4E/4E-BP dissociation and 4E-BP degradation in the first mitotic division of the sea urchin embryo. Dev Biol 255(2):428-439.

Saling PM. 1991. How the egg regulates sperm function during gamete interaction: facts and fantasies. Biol Reprod 44(2):246-251.

Santella L, Ercolano E, Lim D, Nusco GA, Moccia F. 2003. Activated M-phase-promoting factor (MPF) is exported from the nucleus of starfish oocytes to increase the sensitivity of the Ins(1,4,5)P3 receptors. Biochem Soc Trans 31(Pt 1):79-82.

Sardon T, Peset I, Petrova B, Vernos I. 2008. Dissecting the role of Aurora A during spindle assembly. EMBO J 27(19):2567-2579.

Sasseville M, Cote N, Guillemette C, Richard FJ. 2006. New insight into the role of phosphodiesterase 3A in porcine oocyte maturation. BMC Dev Biol 6:47.

Phospho-Signaling at Oocyte Maturation and Fertilization: Set Up for Embryogenesis
and Beyond Part II. Kinase Regulators and Substrates

137

Sasseville M, Cote N, Vigneault C, Guillemette C, Richard FJ. 2007. 3'5'-cyclic adenosine monophosphate-dependent up-regulation of phosphodiesterase type 3A in porcine cumulus cells. Endocrinology 148(4):1858-1867.

Sato K, Iwasaki T, Ogawa K, Konishi M, Tokmakov AA, Fukami Y. 2002a. Low density detergent-insoluble membrane of Xenopus eggs: subcellular microdomain for tyrosine kinase signaling in fertilization. Development 129(4):885-896.

Sato K, Nagao T, Kakumoto M, Kimoto M, Otsuki T, Iwasaki T, Tokmakov AA, Owada K, Fukami Y. 2002b. Adaptor protein Shc is an isoform-specific direct activator of the tyrosine kinase c-Src. J Biol Chem 277(33):29568-29576.

Sato K, Ogawa K, Tokmakov AA, Iwasaki T, Fukami Y. 2001. Hydrogen peroxide induces Src family tyrosine kinase-dependent activation of Xenopus eggs. Dev Growth Differ 43(1):55-72.

Sato K, Tokmakov AA, Fukami Y. 2000a. Fertilization signalling and protein-tyrosine kinases. Comp Biochem Physiol B Biochem Mol Biol 126(2):129-148.

Sato K, Tokmakov AA, He CL, Kurokawa M, Iwasaki T, Shirouzu M, Fissore RA, Yokoyama S, Fukami Y. 2003. Reconstitution of Src-dependent phospholipase Cgamma phosphorylation and transient calcium release by using membrane rafts and cell-free extracts from Xenopus eggs. J Biol Chem 278(40):38413-38420.

Sato K, Tokmakov AA, Iwasaki T, Fukami Y. 2000b. Tyrosine kinase-dependent activation of phospholipase Cgamma is required for calcium transient in Xenopus egg fertilization. Dev Biol 224(2):453-469.

Sato K, Yoshino K, Tokmakov AA, Iwasaki T, Yonezawa K, Fukami Y. 2006. Studying fertilization in cell-free extracts: focusing on membrane/lipid raft functions and proteomics. Methods Mol Biol 322:395-411.

Saunders CM, Larman MG, Parrington J, Cox LJ, Royse J, Blayney LM, Swann K, Lai FA. 2002. PLC zeta: a sperm-specific trigger of Ca(2+) oscillations in eggs and embryo development. Development 129(15):3533-3544.

Sawaguchi S, Ohkubo N, Matsubara T. 2006. Identification of two forms of vitellogenin-derived phosvitin and elucidation of their fate and roles during oocyte maturation in the barfin flounder, Verasper moseri. Zoolog Sci 23(11):1021-1029.

Scarpeci SL, Sanchez ML, Cabada MO. 2008. Cellular origin of the Bufo arenarum sperm receptor gp75, a ZP2 family member: its proteolysis after fertilization. Biol Cell 100(4):219-230.

Schonfelder EM, Knuppel T, Tasic V, Miljkovic P, Konrad M, Wuhl E, Antignac C, Bakkaloglu A, Schaefer F, Weber S. 2006. Mutations in Uroplakin IIIA are a rare cause of renal hypodysplasia in humans. Am J Kidney Dis 47(6):1004-1012.

Schurmans S, Carrio R, Behrends J, Pouillon V, Merino J, Clement S. 1999. The mouse SHIP2 (Inppl1) gene: complementary DNA, genomic structure, promoter analysis, and gene expression in the embryo and adult mouse. Genomics 62(2):260-271.

Schwab MS, Roberts BT, Gross SD, Tunquist BJ, Taieb FE, Lewellyn AL, Maller JL. 2001. Bub1 is activated by the protein kinase p90(Rsk) during Xenopus oocyte maturation. Curr Biol 11(3):141-150.

Sen A, O'Malley K, Wang Z, Raj GV, Defranco DB, Hammes SR. 2010. Paxillin regulates androgen- and epidermal growth factor-induced MAPK signaling and cell proliferation in prostate cancer cells. J Biol Chem 285(37):28787-28795.

Sette C, Messina V, Paronetto MP. 2010. Sam68: a new STAR in the male fertility firmament. J Androl 31(1):66-74.

Sette C, Paronetto MP, Barchi M, Bevilacqua A, Geremia R, Rossi P. 2002. Tr-kit-induced resumption of the cell cycle in mouse eggs requires activation of a Src-like kinase. EMBO J 21(20):5386-5395.

Shapiro E, Huang HY, Wu XR. 2000. Uroplakin and androgen receptor expression in the human fetal genital tract: insights into the development of the vagina. J Urol 164(3 Pt 2):1048-1051.

Shearer J, De Nadai C, Emily-Fenouil F, Gache C, Whitaker M, Ciapa B. 1999. Role of phospholipase Cgamma at fertilization and during mitosis in sea urchin eggs and embryos. Development 126(10):2273-2284.

Shen SS. 1989. Na+-H+ antiport during fertilization of the sea urchin egg is blocked by W-7 but is insensitive to K252a and H-7. Biochem Biophys Res Commun 161(3):1100-1108.

Shilling FM, Carroll DJ, Muslin AJ, Escobedo JA, Williams LT, Jaffe LA. 1994. Evidence for both tyrosine kinase and G-protein-coupled pathways leading to starfish egg activation. Dev Biol 162(2):590-599.

Shimada M, Yokosawa H, Kawahara H. 2006. OMA-1 is a P granules-associated protein that is required for germline specification in Caenorhabditis elegans embryos. Genes Cells 11(4):383-396.

Shimizu T. 1997. Reorganization of the cortical actin cytoskeleton during maturation division in the Tubifex egg: possible involvement of protein kinase C. Dev Biol 188(1):110-121.

Shirayama M, Soto MC, Ishidate T, Kim S, Nakamura K, Bei Y, van den Heuvel S, Mello CC. 2006. The Conserved Kinases CDK-1, GSK-3, KIN-19, and MBK-2 Promote OMA-1 Destruction to Regulate the Oocyte-to-Embryo Transition in C. elegans. Curr Biol 16(1):47-55.

Shteinberg M, Protopopov Y, Listovsky T, Brandeis M, Hershko A. 1999. Phosphorylation of the cyclosome is required for its stimulation by Fizzy/cdc20. Biochem Biophys Res Commun 260(1):193-198.

Sims RJ, 3rd, Mandal SS, Reinberg D. 2004. Recent highlights of RNA-polymerase-II-mediated transcription. Curr Opin Cell Biol 16(3):263-271.

Smith JJ, Evans EK, Murakami M, Moyer MB, Moseley MA, Woude GV, Kornbluth S. 2000. Wee1-regulated apoptosis mediated by the crk adaptor protein in Xenopus egg extracts. J Cell Biol 151(7):1391-1400.

Snow AJ, Puri P, Acker-Palmer A, Bouwmeester T, Vijayaraghavan S, Kline D. 2008. Phosphorylation-dependent interaction of tyrosine 3-monooxygenase/tryptophan 5-monooxygenase activation protein (YWHA) with PADI6 following oocyte maturation in mice. Biol Reprod 79(2):337-347.

Spivack JG, Erikson RL, Maller JL. 1984. Microinjection of pp60v-src into Xenopus oocytes increases phosphorylation of ribosomal protein S6 and accelerates the rate of progesterone-induced meiotic maturation. Mol Cell Biol 4(8):1631-1634.

Stahl ML, Ferenz CR, Kelleher KL, Kriz RW, Knopf JL. 1988. Sequence similarity of phospholipase C with the non-catalytic region of src. Nature 332(6161):269-272.

Standart N, Minshull J, Pines J, Hunt T. 1987. Cyclin synthesis, modification and destruction
    during meiotic maturation of the starfish oocyte. Dev Biol 124(1):248-258.
Stefanello JR, Barreta MH, Porciuncula PM, Arruda JN, Oliveira JF, Oliveira MA, Goncalves
    PB. 2006. Effect of angiotensin II with follicle cells and insulin-like growth factor-I
    or insulin on bovine oocyte maturation and embryo development. Theriogenology
    66(9):2068-2076.
Stemmann O, Zou H, Gerber SA, Gygi SP, Kirschner MW. 2001. Dual inhibition of sister
    chromatid separation at metaphase. Cell 107(6):715-726.
Stith BJ, Woronoff K, Espinoza R, Smart T. 1997. sn-1,2-diacylglycerol and choline increase
    after fertilization in Xenopus laevis. Mol Biol Cell 8(4):755-765.
Stitzel ML, Cheng KC, Seydoux G. 2007. Regulation of MBK-2/Dyrk kinase by dynamic
    cortical anchoring during the oocyte-to-zygote transition. Curr Biol 17(18):1545-
    1554.
Stitzel ML, Pellettieri J, Seydoux G. 2006. The C. elegans DYRK Kinase MBK-2 Marks Oocyte
    Proteins for Degradation in Response to Meiotic Maturation. Curr Biol 16(1):56-62.
Stricker SA, Smythe TL. 2006. Differing mechanisms of cAMP- versus seawater-induced
    oocyte maturation in marine nemertean worms II. The roles of tyrosine kinases and
    phosphatases. Mol Reprod Dev 73(12):1564-1577.
Sugimoto I, Hashimoto E. 2006. Modulation of protein phosphorylation by Mr 25,000
    protein partially overlapping phosvitin and lipovitellin 2 in Xenopus laevis
    vitellogenin B1 protein. Protein J 25(2):109-115.
Sumara I, Vorlaufer E, Stukenberg PT, Kelm O, Redemann N, Nigg EA, Peters JM. 2002. The
    dissociation of cohesin from chromosomes in prophase is regulated by Polo-like
    kinase. Mol Cell 9(3):515-525.
Sumiyoshi E, Sugimoto A, Yamamoto M. 2002. Protein phosphatase 4 is required for
    centrosome maturation in mitosis and sperm meiosis in C. elegans. J Cell Sci 115(Pt
    7):1403-1410.
Sun L, Haun S, Jones RC, Edmondson RD, Machaca K. 2009. Kinase-dependent regulation of
    inositol 1,4,5-trisphosphate-dependent Ca2+ release during oocyte maturation. J
    Biol Chem 284(30):20184-20196.
Sun QY, Fuchimoto D, Nagai T. 2004. Regulatory roles of ubiquitin-proteasome pathway in
    pig oocyte meiotic maturation and fertilization. Theriogenology 62(1-2):245-255.
Swain JE, Ding J, Brautigan DL, Villa-Moruzzi E, Smith GD. 2007. Proper chromatin
    condensation and maintenance of histone H3 phosphorylation during mouse
    oocyte meiosis requires protein phosphatase activity. Biol Reprod 76(4):628-638.
Swan A, Schupbach T. 2007. The Cdc20 (Fzy)/Cdh1-related protein, Cort, cooperates with
    Fzy in cyclin destruction and anaphase progression in meiosis I and II in
    Drosophila. Development 134(5):891-899.
Swann K, Igusa Y, Miyazaki S. 1989. Evidence for an inhibitory effect of protein kinase C on
    G-protein-mediated repetitive calcium transients in hamster eggs. EMBO J
    8(12):3711-3718.
Swann K, Whitaker M. 1985. Stimulation of the Na/H exchanger of sea urchin eggs by
    phorbol ester. Nature 314(6008):274-277.
Szczepanska K, Maleszewski M. 2005. LKB1/PAR4 protein is asymmetrically localized in
    mouse oocytes and associates with meiotic spindle. Gene Expr Patterns 6(1):86-93.

Talmor-Cohen A, Tomashov-Matar R, Tsai WB, Kinsey WH, Shalgi R. 2004. Fyn kinase-tubulin interaction during meiosis of rat eggs. Reproduction 128(4):387-393.

Tan X, Peng A, Wang Y, Tang Z. 2005a. The effects of proteasome inhibitor lactacystin on mouse oocyte meiosis and first cleavage. Sci China C Life Sci 48(3):287-294.

Tan X, Peng A, Wang YC, Wang Y, Sun QY. 2005b. Participation of the ubiquitin-proteasome pathway in rat oocyte activation. Zygote 13(1):87-95.

Tanaka K, Okubo Y, Abe H. 2005. Involvement of slingshot in the Rho-mediated dephosphorylation of ADF/cofilin during Xenopus cleavage. Zoolog Sci 22(9):971-984.

Tang ED, Nunez G, Barr FG, Guan KL. 1999. Negative regulation of the forkhead transcription factor FKHR by Akt. J Biol Chem 274(24):16741-16746.

Tang W, Wu JQ, Guo Y, Hansen DV, Perry JA, Freel CD, Nutt L, Jackson PK, Kornbluth S. 2008. Cdc2 and Mos regulate Emi2 stability to promote the meiosis I-meiosis II transition. Mol Biol Cell 19(8):3536-3543.

Tang Z, Shu H, Oncel D, Chen S, Yu H. 2004. Phosphorylation of Cdc20 by Bub1 provides a catalytic mechanism for APC/C inhibition by the spindle checkpoint. Mol Cell 16(3):387-397.

Tarone G, Russo MA, Hirsch E, Odorisio T, Altruda F, Silengo L, Siracusa G. 1993. Expression of beta 1 integrin complexes on the surface of unfertilized mouse oocyte. Development 117(4):1369-1375.

Taylor SJ, Anafi M, Pawson T, Shalloway D. 1995. Functional interaction between c-Src and its mitotic target, Sam 68. J Biol Chem 270(17):10120-10124.

Taylor SJ, Shalloway D. 1994. An RNA-binding protein associated with Src through its SH2 and SH3 domains in mitosis. Nature 368(6474):867-871.

Tchang F, Gusse M, Soussi T, Mechali M. 1993. Stabilization and expression of high levels of p53 during early development in Xenopus laevis. Dev Biol 159(1):163-172.

Tchang F, Mechali M. 1999. Nuclear import of p53 during Xenopus laevis early development in relation to DNA replication and DNA repair. Exp Cell Res 251(1):46-56.

Terret ME, Ferby I, Nebreda AR, Verlhac MH. 2001. RINGO efficiently triggers meiosis resumption in mouse oocytes and induces cell cycle arrest in embryos. Biol Cell 93(1-2):89-97.

Thomas RE, Armstrong DT, Gilchrist RB. 2002. Differential effects of specific phosphodiesterase isoenzyme inhibitors on bovine oocyte meiotic maturation. Dev Biol 244(2):215-225.

Thumbikat P, Berry RE, Schaeffer AJ, Klumpp DJ. 2009a. Differentiation-induced uroplakin III expression promotes urothelial cell death in response to uropathogenic E. coli. Microbes Infect 11(1):57-65.

Thumbikat P, Berry RE, Zhou G, Billips BK, Yaggie RE, Zaichuk T, Sun TT, Schaeffer AJ, Klumpp DJ. 2009b. Bacteria-induced uroplakin signaling mediates bladder response to infection. PLoS Pathog 5(5):e1000415.

Tian AG, Deng WM. 2008. Lgl and its phosphorylation by aPKC regulate oocyte polarity formation in Drosophila. Development 135(3):463-471.

Tian J, Gong H, Lennarz WJ. 1999. Xenopus laevis sperm receptor gp69/64 glycoprotein is a homolog of the mammalian sperm receptor ZP2. Proc Natl Acad Sci U S A 96(3):829-834.

Tokmakov AA, Sato KI, Iwasaki T, Fukami Y. 2002. Src kinase induces calcium release in
    Xenopus egg extracts via PLCgamma and IP3-dependent mechanism. Cell Calcium
    32(1):11-20.
Tokumoto T, Yamashita M, Tokumoto M, Katsu Y, Horiguchi R, Kajiura H, Nagahama Y.
    1997. Initiation of cyclin B degradation by the 26S proteasome upon egg activation.
    J Cell Biol 138(6):1313-1322.
Tonks NK, Cicirelli MF, Diltz CD, Krebs EG, Fischer EH. 1990. Effect of microinjection of a
    low-Mr human placenta protein tyrosine phosphatase on induction of meiotic cell
    division in Xenopus oocytes. Mol Cell Biol 10(2):458-463.
Totsukawa G, Himi-Nakamura E, Komatsu S, Iwata K, Tezuka A, Sakai H, Yazaki K,
    Hosoya H. 1996. Mitosis-specific phosphorylation of smooth muscle regulatory
    light chain of myosin II at Ser-1 and/or -2 and Thr-9 in sea urchin egg extract. Cell
    Struct Funct 21(6):475-482.
Tse AC, Ge W. 2009. Differential regulation of betacellulin and heparin-binding EGF-like
    growth factor in cultured zebrafish ovarian follicle cells by EGF family ligands.
    Comp Biochem Physiol A Mol Integr Physiol 153(1):13-17.
Tsurumi A, Xia F, Li J, Larson K, LaFrance R, Li WX. 2011. STAT is an essential activator of
    the zygotic genome in the early Drosophila embryo. PLoS Genet 7(5):e1002086.
Tung JJ, Hansen DV, Ban KH, Loktev AV, Summers MK, Adler JR, 3rd, Jackson PK. 2005. A
    role for the anaphase-promoting complex inhibitor Emi2/XErp1, a homolog of
    early mitotic inhibitor 1, in cytostatic factor arrest of Xenopus eggs. Proc Natl Acad
    Sci U S A 102(12):4318-4323.
Tunquist BJ, Maller JL. 2003. Under arrest: cytostatic factor (CSF)-mediated metaphase arrest
    in vertebrate eggs. Genes Dev 17(6):683-710.
Tunquist BJ, Schwab MS, Chen LG, Maller JL. 2002. The spindle checkpoint kinase bub1 and
    cyclin e/cdk2 both contribute to the establishment of meiotic metaphase arrest by
    cytostatic factor. Curr Biol 12(12):1027-1033.
Turner CE. 1991. Paxillin is a major phosphotyrosine-containing protein during embryonic
    development. J Cell Biol 115(1):201-207.
Urayama S, Harada Y, Nakagawa Y, Ban S, Akasaka M, Kawasaki N, Sawada H. 2008.
    Ascidian sperm glycosylphosphatidylinositol-anchored CRISP-like protein as a
    binding partner for an allorecognizable sperm receptor on the vitelline coat. J Biol
    Chem 283(31):21725-21733.
van Eekelen M, Runtuwene V, Overvoorde J, den Hertog J. 2010. RPTPalpha and
    PTPepsilon signaling via Fyn/Yes and RhoA is essential for zebrafish convergence
    and extension cell movements during gastrulation. Dev Biol 340(2):626-639.
Vanderheyden V, Wakai T, Bultynck G, De Smedt H, Parys JB, Fissore RA. 2009. Regulation
    of inositol 1,4,5-trisphosphate receptor type 1 function during oocyte maturation by
    MPM-2 phosphorylation. Cell Calcium 46(1):56-64.
Vanhaesebroeck B, Leevers SJ, Panayotou G, Waterfield MD. 1997. Phosphoinositide 3-
    kinases: a conserved family of signal transducers. Trends Biochem Sci 22(7):267-
    272.
Vesela J, Cikos S, Hlinka D, Rehak P, Baran V, Koppel J. 1995. Effects of impaired insulin
    secretion on the fertilization of mouse oocytes. Hum Reprod 10(12):3233-3236.

Vigneron S, Brioudes E, Burgess A, Labbe JC, Lorca T, Castro A. 2009. Greatwall maintains mitosis through regulation of PP2A. EMBO J 28(18):2786-2793.

Visintin R, Prinz S, Amon A. 1997. CDC20 and CDH1: a family of substrate-specific activators of APC-dependent proteolysis. Science 278(5337):460-463.

Wakai T, Vanderheyden V, Yoon SY, Cheon B, Zhang N, Parys JB, Fissore RA. 2011. Regulation of inositol 1,4,5-trisphosphate receptor function during mouse oocyte maturation. J Cell Physiol.

Wang Q, Wang CM, Ai JS, Xiong B, Yin S, Hou Y, Chen DY, Schatten H, Sun QY. 2006. Histone phosphorylation and pericentromeric histone modifications in oocyte meiosis. Cell Cycle 5(17):1974-1982.

Wang Y, Li J, Ying Wang C, Yan Kwok AH, Leung FC. 2007. Epidermal growth factor (EGF) receptor ligands in the chicken ovary: I. Evidence for heparin-binding EGF-like growth factor (HB-EGF) as a potential oocyte-derived signal to control granulosa cell proliferation and HB-EGF and kit ligand expression. Endocrinology 148(7):3426-3440.

Wang Y, Morrow JS. 2000. Identification and characterization of human SLP-2, a novel homologue of stomatin (band 7.2b) present in erythrocytes and other tissues. J Biol Chem 275(11):8062-8071.

Wasielak M, Bogacki M. 2007. Apoptosis inhibition by insulin-like growth factor (IGF)-I during in vitro maturation of bovine oocytes. J Reprod Dev 53(2):419-426.

Wassarman PM, Litscher ES. 2001. Multiple functions of mouse zona pellucida glycoprotein mZP3, the sperm receptor. Ital J Anat Embryol 106(2 Suppl 2):21-32.

Watts JL, Etemad-Moghadam B, Guo S, Boyd L, Draper BW, Mello CC, Priess JR, Kemphues KJ. 1996. par-6, a gene involved in the establishment of asymmetry in early C. elegans embryos, mediates the asymmetric localization of PAR-3. Development 122(10):3133-3140.

Weber GM, Moore AB, Sullivan CV. 2007. In vitro actions of insulin-like growth factor-I on ovarian follicle maturation in white perch (Morone americana). Gen Comp Endocrinol 151(2):180-187.

Weber GM, Sullivan CV. 2005. Insulin-like growth factor-I induces oocyte maturational competence but not meiotic resumption in white bass (Morone chrysops) follicles in vitro: evidence for rapid evolution of insulin-like growth factor action. Biol Reprod 72(5):1177-1186.

Wei L, Liang XW, Zhang QH, Li M, Yuan J, Li S, Sun SC, Ouyang YC, Schatten H, Sun QY. 2010. BubR1 is a spindle assembly checkpoint protein regulating meiotic cell cycle progression of mouse oocyte. Cell Cycle 9(6):1112-1121.

Weinstein J. 1997. Cell cycle-regulated expression, phosphorylation, and degradation of p55Cdc. A mammalian homolog of CDC20/Fizzy/slp1. J Biol Chem 272(45):28501-28511.

Welsh M, Songyang Z, Frantz JD, Trub T, Reedquist KA, Karlsson T, Miyazaki M, Cantley LC, Band H, Shoelson SE. 1998. Stimulation through the T cell receptor leads to interactions between SHB and several signaling proteins. Oncogene 16(7):891-901.

Westmark CJ, Ghose R, Huber PW. 2002. Phosphorylation of Xenopus transcription factor IIIA by an oocyte protein kinase CK2. Biochem J 362(Pt 2):375-382.

Wiersma A, Hirsch B, Tsafriri A, Hanssen RG, Van de Kant M, Kloosterboer HJ, Conti M, Hsueh AJ. 1998. Phosphodiesterase 3 inhibitors suppress oocyte maturation and consequent pregnancy without affecting ovulation and cyclicity in rodents. J Clin Invest 102(3):532-537.

Wiles MV, Ruiz P, Imhof BA. 1992. Interleukin-7 expression during mouse thymus development. Eur J Immunol 22(4):1037-1042.

Williams CJ, Mehlmann LM, Jaffe LA, Kopf GS, Schultz RM. 1998. Evidence that Gq family G proteins do not function in mouse egg activation at fertilization. Dev Biol 198(1):116-127.

Wright PW, Bolling LC, Calvert ME, Sarmento OF, Berkeley EV, Shea MC, Hao Z, Jayes FC, Bush LA, Shetty J, Shore AN, Reddi PP, Tung KS, Samy E, Allietta MM, Sherman NE, Herr JC, Coonrod SA. 2003. ePAD, an oocyte and early embryo-abundant peptidylarginine deiminase-like protein that localizes to egg cytoplasmic sheets. Dev Biol 256(1):73-88.

Wu JQ, Hansen DV, Guo Y, Wang MZ, Tang W, Freel CD, Tung JJ, Jackson PK, Kornbluth S. 2007a. Control of Emi2 activity and stability through Mos-mediated recruitment of PP2A. Proc Natl Acad Sci U S A 104(42):16564-16569.

Wu JQ, Kornbluth S. 2008. Across the meiotic divide - CSF activity in the post-Emi2/XErp1 era. J Cell Sci 121(Pt 21):3509-3514.

Wu Q, Guo Y, Yamada A, Perry JA, Wang MZ, Araki M, Freel CD, Tung JJ, Tang W, Margolis SS, Jackson PK, Yamano H, Asano M, Kornbluth S. 2007b. A role for Cdc2- and PP2A-mediated regulation of Emi2 in the maintenance of CSF arrest. Curr Biol 17(3):213-224.

Wu W, Kinsey WH. 2002. Role of PTPase(s) in regulating Fyn kinase at fertilization of the zebrafish egg. Dev Biol 247(2):286-294.

Wu Y, Pan S, Che S, He G, Nelman-Gonzalez M, Weil MM, Kuang J. 2001. Overexpression of Hp95 induces G1 phase arrest in confluent HeLa cells. Differentiation 67(4-5):139-153.

Xi J, Sugimoto I, Yoshitome S, Yasuda H, Ogura K, Mori N, Li Z, Ito S, Hashimoto E. 2003. Purification and characterization of Mr 43,000 protein similar to Mr 25,000 protein, a substrate for protein Ser/Thr kinases, identified as a part of Xenopus laevis vitellogenin B1. J Protein Chem 22(6):571-583.

Yamamoto-Honda R, Honda Z, Ueki K, Tobe K, Kaburagi Y, Takahashi Y, Tamemoto H, Suzuki T, Itoh K, Akanuma Y, Yazaki Y, Kadowaki T. 1996. Mutant of insulin receptor substrate-1 incapable of activating phosphatidylinositol 3-kinase did not mediate insulin-stimulated maturation of Xenopus laevis oocytes. J Biol Chem 271(45):28677-28681.

Yang J, Winkler K, Yoshida M, Kornbluth S. 1999. Maintenance of G2 arrest in the Xenopus oocyte: a role for 14-3-3-mediated inhibition of Cdc25 nuclear import. EMBO J 18(8):2174-2183.

Yoshida M, Horiuchi Y, Sensui N, Morisawa M. 2003. Signaling pathway from [Ca2+]i transients to ooplasmic segregation involves small GTPase rho in the ascidian egg. Dev Growth Differ 45(3):275-281.

Yu Y, Halet G, Lai FA, Swann K. 2008. Regulation of diacylglycerol production and protein kinase C stimulation during sperm- and PLCzeta-mediated mouse egg activation. Biol Cell 100(11):633-643.

Yuan J, Li M, Wei L, Yin S, Xiong B, Li S, Lin SL, Schatten H, Sun QY. 2009. Astrin regulates meiotic spindle organization, spindle pole tethering and cell cycle progression in mouse oocytes. Cell Cycle 8(20):3384-3395.

Zhang L, Ding F, Cao W, Liu Z, Liu W, Yu Z, Wu Y, Li W, Li Y. 2006. Stomatin-like protein 2 is overexpressed in cancer and involved in regulating cell growth and cell adhesion in human esophageal squamous cell carcinoma. Clin Cancer Res 12(5):1639-1646.

Zhang L, Liang Y, Liu Y, Xiong CL. 2010. The role of brain-derived neurotrophic factor in mouse oocyte maturation in vitro involves activation of protein kinase B. Theriogenology 73(8):1096-1103.

Zhang Y, Zhang Z, Xu XY, Li XS, Yu M, Yu AM, Zong ZH, Yu BZ. 2008. Protein kinase A modulates Cdc25B activity during meiotic resumption of mouse oocytes. Dev Dyn 237(12):3777-3786.

Zhao Y, Haccard O, Wang R, Yu J, Kuang J, Jessus C, Goldberg ML. 2008. Roles of Greatwall kinase in the regulation of cdc25 phosphatase. Mol Biol Cell 19(4):1317-1327.

Ziyyat A, Rubinstein E, Monier-Gavelle F, Barraud V, Kulski O, Prenant M, Boucheix C, Bomsel M, Wolf JP. 2006. CD9 controls the formation of clusters that contain tetraspanins and the integrin alpha 6 beta 1, which are involved in human and mouse gamete fusion. J Cell Sci 119(Pt 3):416-424.

Phospho-Signaling at Oocyte Maturation and Fertilization: Set Up for Embryogenesis and Beyond Part I. Protein Kinases

A.K.M. Mahbub Hasan[1], Takashi Matsumoto[2],
Shigeru Kihira[2], Junpei Yoshida[2] and Ken-ichi Sato[2,*]
*[1]Laboratory of Gene Biology, Department of Biochemistry and Molecular Biology,
University of Dhaka, Dhaka,
[2]Laboratory of Cell Signaling and Development, Department of Molecular Biosciences,
Faculty of Life Sciences, Kyoto Sangyo University,
Kamigamo-Motoyama, Kita-ku, Kyoto
[1]Bangladesh
[2]Japan*

1. Introduction

In the field of developmental biology, day by day data are accumulated to describe the molecular mechanisms involved in gamete cell production (oogenesis and spermatogenesis) and the sperm-egg interaction/fusion (fertilization) leading the formation of zygote to embryo (embryogenesis) that ultimately develop into a complete body. Here, we will review how *oocyte maturation*, sperm mediated *egg activation/fertilization* and early steps of *embryogenesis* are accomplished and regulated through protein phosphorylation(s) highlighting the participating molecules (e.g. protein kinases) (**this chapter**) and their regulators and substrates (**another chapter entitled "Part II. Kinase Regulators and Substrates"**). Meiosis is the process by which diploid germ-line cell reduces their number of chromosomes in half to generate haploid gamete and combine with opposite sex haploid gamete to create a genetically new, diploid individual. Oocyte maturation, which undergoes two meiotic cell cycles that arrest at several stages, has been studied extensively in many species of vertebrates and invertebrates. A lot of review articles on oocyte meiotic maturation of different species have been written (Kang and Han 2011; Liang et al. 2007; Machaca 2007; Madgwick and Jones 2007; Schmitt and Nebreda 2002a; Tripathi et al. 2010). In almost all vertebrates, oocyte meiotic cell cycle starts during fetal life (at 4-5 weeks) but arrest at first in diplotene stage of first meiotic prophase (before the metaphase I or MI) that may last for several months or years in follicular microenvironment depending on the species (Mehlmann and Jaffe 2005; Sirard 2001; Trounson et al. 2001; Wassmann et al. 2003). The progression of meiotic cell cycle is also arrested, in many but not all species, at stages of second meiotic metaphase II (MII) and/or metaphase-like arrest (MIII). During oocyte

* Corresponding Author

maturation different kinds of molecules e.g. second messengers, protein kinases, protein phosphatases and their regulator and/or substrate proteins are involved. Here, the molecular mechanisms involved in the arrest and resumption of these stages will be discussed briefly.

MPF (maturation or M-phase promoting factor), a serine/threonine kinase, is composed of a catalytic subunit cyclin-dependent kinase 1 (Cdc2/CDK1), and a regulatory subunit, cyclin B; are the key components in the maintenance of diplotene arrest. In activated MPF, dephosphorylated CDK1 is associated with cyclin B and both cyclin B synthesis and degradation is required for MPF activity (Clarke and Karsenti 1991; Ledan et al. 2001). Cyclin B is accumulated in diplotene-arrested oocytes due to the presence of early mitotic inhibitor1 (Emi1) that inhibits anaphase promoting complex/cyclosome (APC/C), an ubiquitin ligase complex responsible for the destruction of cyclin B (Marangos et al. 2007). In oocyte, the level of cGMP and cAMP are very high and they are secreted from cumulus and granulosa cells surrounding the oocyte and are essential for the maintenance of meiotic arrest at diplotene stage (Norris et al. 2009; Sirard and Bilodeau 1990b; Sun et al. 2009; Vaccari et al. 2008). The increased level of cGMP inactivates phosphodiesterase 3A (PDE3A) and prevents hydrolysis of cAMP thus further increase its level (Mayes and Sirard 2002; Tsafriri et al. 1996; Vaccari et al. 2008). In diplotene-arrested oocytes, high concentrations of cAMP activate protein kinase A (PKA), and activated PKA phosphorylates two CDK1 regulators such as cell division cycle 25 homologue B (Cdc25B) phosphatase (Pirino et al. 2009) and Wee1/Myt1 (myelin transcription factor 1) kinase (Han and Conti 2006; Stanford and Ruderman 2005). The inactivation of Cdc25B and activation of Wee1/Myt1 kinase ultimately inactivate MPF activity for the maintenance of meiotic arrest at diplotene stage (Han and Conti 2006; Potapova et al. 2009; Solc et al. 2010). Luteinizing hormone (LH) released from surrounding granulosa cells act indirectly on oocytes to resume diplotene arrest at the onset of puberty (Mehlmann 2005; Zhang et al. 2009). LH mediated MAPK activation in granulosa cells interrupts the cells-oocytes communications and the result is the decrease of cAMP and cGMP level in oocytes (Liang et al. 2007; Mehlmann 2005; Norris et al. 2009). Reduced level of intraoocyte cGMP causes the activation of PDE3A activity that further reduces the intra oocyte cAMP level (Tornell et al. 1991; Wang et al. 2008). Net reduction of cAMP in oocytes inhibits PKA actions and dephospho-form of Cdc25B phosphatase remains active (Han and Conti 2006). On the other hand, dephospho-form of Wee1/Myt1 kinase remains inactive (Han and Conti 2006; Liang et al. 2007; Mehlmann et al. 2002; Solc et al. 2010) and finally resumes the diplotene arrest that is morphologically characterized by germinal vesicle breakdown (GVBD).

Getting release from diplotene arrest, activated MAPK through proper organization of metaphase spindle makes the progression of MI when homologous chromosomes are segregated (Sirard and Bilodeau 1990a). Oocytes are arrested at MI until the entire sister chromatids properly attached to the bipolar spindle and aligned at the metaphase plate where spindle assembly checkpoint (SAC) proteins e.g. Mad2 (metaphase arrest deficient 2), Bub1, and Bub3 (budding uninhibited by benzimidazole 1 and 3) act for all the required activities (Hupalowska et al. 2008; Li et al. 2009; Niault et al. 2007; Wassmann et al. 2003). The SAC proteins for accurate homologous chromosome segregation and to delay anaphase onset target APC/C (Brunet and Maro 2005; Homer 2011; Wassmann et al. 2003). Formation of functional spindle, spindle migration correlates with the progressive increase and continuous MPF activity (Brunet and Maro 2005; Madgwick et al. 2004). Mos/MAPK activity is also important in microtubule reorganization and positioning of metaphase spindle to the oocyte cortex (Choi et al. 1996; Verlhac et al. 1996; Zhou et al. 1991). At the end of MI, MPF activity is declined and is characterized by first polar body extrusion. After completion of MI, oocytes undergo some cytoplasmic changes and progress to the arrest at MII with further high MPF activity until fertilization. Stabilization of MPF activity is maintained by CSF (cytostatic factor) activity, not a single molecule but a total activity (Madgwick and Jones 2007; Wu and Kornbluth 2008) and by Mos-mediated MAPK pathway (Perry and Verlhac 2008; Shoji et al. 2006). Emi1 and Emi2 are two members of Emi/Erp family of proteins that has also the CSF activity (Schmidt et al. 2006) and functions in MII arrest (Madgwick and Jones 2007; Schmidt et al. 2006; Shoji et al. 2006; Tang et al. 2008). Complex of dephosphorylated active Emi2 with Cdc20, inhibit APC/C for the maintenance of MII arrest (Shoji et al. 2006). Sperm mediated Ca^{2+} oscillation activates calcium/calmodulin-dependent protein kinase II (CaMKII) and Emi2 can be phosphorylated by activated CaMKII followed by further phosphorylation by polo-like kinase (Hansen et al. 2006; Madgwick and Jones 2007; Masui and Markert 1971; Shoji et al. 2006). Cdc20 is released from Emi2 and subsequently bind with APC/C that results an active APC/C complex (Liu et al. 2006). Activated APC/C induces the degradation of cyclin B and MPF activity is decreased with an exit of egg from MII arrest by a process of sperm-egg interaction and fusion called fertilization. Another mechanism of MII arrest is by Mos (pp39, serine/threonine kinase), a proto-oncogene product act in the upstream of MEK/MAPK pathway that ultimately activates ribosomal protein S6 kinase (p90^{Rsk}). p90^{Rsk} induces SAC protein activation and thereby inhibition of APC/C (Madgwick and Jones 2007; Maller et al. 2001) to maintain MII arrest. At fertilization Mos is degraded while the MEK/MAPK/p90^{Rsk} is shortly inactivated and release from MII arrest. To the end of this process the sister chromatids are segregated, second polar body is extruded and the first cleavage starts. Postovulatory oocytes mimic the action of egg activation due to aging, increases cytoplasmic Ca^{2+}, and induces exit from MII arrest but they do not progress further and get arrest again in a new metaphase-like stage called MIII in few vertebrate species though the mechanisms for MIII arrest is not well understood (Chaube et al. 2007; Galat et al. 2007; Vincent et al. 1992; Zernicka-Goetz 1991). In aged eggs insufficient Ca^{2+} release and sufficient CSF activity is still present to stabilize the residual or newly formed MPF activity results in MIII arrest (Kubiak et al. 1992; Vincent et al. 1992).

Sperm-induced release of MII of an egg is also termed as "egg activation" that is characterized by so many biochemical changes e.g. Ca^{2+} oscillations, cortical granules exocytosis to block polyspermy, the formation of polar body and male and female pronuclei, recruitment of maternal mRNAs, initiation of DNA synthesis for mitotic divisions to unveil the complete developmental program (Ducibella 1996; Ducibella and Fissore 2008; Schultz and Kopf 1995). The wave of Ca^{2+} initiates at the site of sperm binding/fusion and soon after a wave of intracellular Ca^{2+} traverses the entire volume of the egg (Gilkey et al. 1978; Miyazaki and Ito 2006; Runft et al. 2002; Steinhardt and Epel 1974; Stricker 1999; Whitaker 2006). It is interesting to note that the increase in Ca^{2+} was reported in lysates of sea urchin eggs more than quarter century ago (Mazia 1937). Several excellent review articles have been published describing how egg becomes active in fertilization dependent manner and unite with sperm nuclei to form a zygote (Ajduk et al. 2008; Ducibella and Fissore 2008; Horner and Wolfner 2008; Miyazaki and Ito 2006; Swann et al. 2006; Townley et al. 2006).

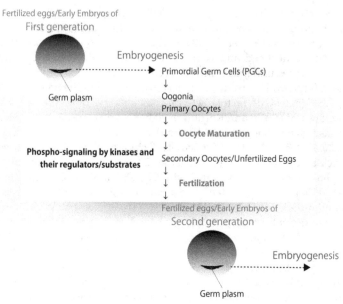

Fig. 1. Germline transmission from one generation to the next generation in sexual reproduction system. In most vertebrate species, the primary oocytes (or immature oocytes) in ovarian tissue pauses their cell cycle at prophase of the first meiosis, resumes the meiosis in response to hormonal signals, re-pauses at metaphase of the second meiotic cell cycle as the secondary oocytes (or mature oocytes), and are subject to ovulation and fertilization. Upon fertilization, eggs undergo a series of extracellular and intracellular reactions/changes, collectively called egg activation that triggers the initiation of development or early embryogenesis. A similar mechanism, although not identical in several species, has been shown to be involved in sexual reproduction system in a diverse array of animal species and maybe in some algae and plants.

Oocyte plasma membrane is surrounded by glycoprotein-rich extracellular matrix called the vitelline envelope (VE) or vitelline membrane (VM) in invertebrates and amphibians, and zona pellucida (ZP) in mammals. Upon fertilization this layer must be modified to prevent additional sperm to bind and fuse to block the polyspermy. Prevention of polyspermy is accomplished in part through Ca^{2+}-dependent cortical granule exocytosis (CGE) (Wessel et al. 2001; Wessel and Wong 2009). Upon egg activation, CGE fuse with the oocyte plasma membrane and release their contents into the perivitelline space that results the biochemical modification of the outer membrane. Ca^{2+}-mediated active CaMKII phosphorylate Emi2 that is further phosphorylated by polo-like kinase and this phosphorylated Emi2 is targeted by APC/C resulting in an active APC/C (Liu and Maller 2005; Rauh et al. 2005). Activated APC/C leads to the degradation of cyclin B that results the inactivation of Cdc2 and might also inactivate the function of Mos (Castro et al. 2001; Madgwick et al. 2006; Madgwick and Jones 2007). Thus, due to the absence of Cdc2/CDK1 activity, meiosis-specific host protein phosphorylations are reduced allowing eggs to exit M-phase. Src family tyrosine kinases (SFKs) are playing important roles in sperm-induced Ca^{2+} oscillation in several species e.g. in starfish (Abassi et al. 2000; Carroll et al. 1999; Giusti et al. 1999a, 1999b), Fyn kinase in sea urchin eggs (Kinsey and Shen 2000) and in rat eggs (Talmor et al. 1998), and Src in frog eggs (Sato et al. 1996; Sato et al. 2006a). In mouse eggs though Src related tyrosine kinase (e.g. Lck, Src) has been reported (Mori et al. 1991) but it is not sufficient or required for fertilization-induced Ca^{2+} oscillation (Kurokawa et al. 2004). In mammals, PLC activity is high enough in sperm that's why even a single sperm equivalent PLC can generate sufficient IP_3 when introduced into the egg cytoplasm (Rice et al. 2000). ζ isoform of PLC present in sperm has been characterized as a soluble sperm factor that evokes Ca^{2+} oscillations in eggs of several mammals e.g. mouse, bovine and human (Malcuit et al. 2005; Rogers et al. 2004; Saunders et al. 2002). Thus upon successful fertilization, the newly formed zygote initiates the developmental program through early stages of embryogenesis until a full different born.

2. Kinases in oocyte maturation, fertilization and activation of development

2.1 Abelson tyrosine kinase (Abl)

Abl has been originally identified as the oncogene product (termed v-Abl) of Abelson murine leukemia virus (Wang et al. 1983). In human, the cellular homolog of v-Abl, c-Abl, is translocated to the Philadelphia chromosome in chronic myelocytic leukemia (so-called Philadelphia syndrome) (de Klein et al. 1982). Gleevec (STI-571), a well-known drug for chronic myeloid leukemia (CML) and some other cancers, has been designed to target the protein product of the CML transforming gene, Bcr (breakpoint cluster region)-Abl (Schindler et al. 2000). In the sea urchin, a 220-kDa Abl-related tyrosine kinase has been identified in the egg cortex. Immunoprecipitation studies demonstrated that it is activated within minutes of fertilization, suggesting a possible role for sperm-induced egg activation, and immunofluorescent studies showed its association with cortical cytoskeleton (Moore and Kinsey 1994; Walker et al. 1996; Wang et al. 1983). However, its mode of activation and physiological substrate has not yet been demonstrated.

2.2 Akt protein kinase (Akt)

The serine/threonine-specific protein kinase **Akt** has been identified first as the oncogene product of murine transforming retrovirus. Because of its structural homology in the catalytic domain to protein kinase A (PKA) and C (PKC), an alternative term "protein kinase B or PKB" is sometimes used. Akt is shown to be involved in several aspects of cellular functions, and most frequently, it is regarded as a kinase that promotes anti-apoptotic growth of cells (Hemmings 1997). Upstream kinases, such as phosphoinositide-dependent protein kinase 1 (PDK1) and mammalian target of rapamycin (mTOR), are responsible for phosphorylation and activation of Akt. Thus, Akt is regulated by metabolism of membrane-bound phosphoinositides as well as extracellular nutrient environments. Akt is shown to be involved in oocyte maturation in some species (e.g. starfish, mouse, and maybe *Xenopus*) (Deng et al. 2011; Feng et al. 2007; Han et al. 2006; Hoshino and Sato 2008; Hoshino et al. 2004; Kalous et al. 2009; Kalous et al. 2006; Mammadova et al. 2009; Okumura et al. 2002; Reddy et al. 2005; Tomek and Smiljakovic 2005; Zhang et al. 2010b). For example, Akt phosphorylation and down-regulation of Myt1, an MPF-inhibitory kinase, and PDE3, a cAMP-antagonizing enzyme, in maturing starfish and mouse oocytes have been demonstrated (Han et al. 2006a). On the other hand, fertilization promotes an activating phosphorylation (on Thr-308) of Akt in *Xenopus* (Mammadova et al. 2009), suggesting its possible role in initiation of development and/or suppression of cell death.

2.3 Adenosine 5'-monophosphate-dependent protein kinase (AMPK)

AMPK is a serine/threonine kinase that is activated in response to high AMP and/or low ATP levels in the cell. AMPK is composed of three subunits; one catalytic subunit and two regulatory subunits, all of which are evolutionary conserved in budding yeast as the SNF1 protein kinase complex (Hardie and Carling 1997). AMPK is phosphorylated and activated by an upstream AMPK kinase. Inhibitory effect of AMP and/or AMPK on oocyte maturation and/or meiotic resumption has been demonstrated in marine worm, starfish, and some mammalian species (Bilodeau-Goeseels et al. 2007; Chen and Downs 2008; Chen et al. 2006; LaRosa and Downs 2006; Stricker 2011; Stricker and Smythe 2006; Stricker et al. 2010b; Tosca et al. 2007). In marine worm, liver kinase B1 (LKB1)-like kinase is likely involved in up-regulation of AMPK (via phosphorylation of Thr-172), and thus suppresses the occurrence of oocyte maturation. On the other hand, MAPK and MPF are shown to simultaneously phosphorylate AMPK on two sites (Ser-485/491), and thereby inactivate the activity of AMPK. Physiological target of AMPK in this species is under investigation.

2.4 Aurora protein kinase (Aurora A/B/C/AIR-2/Eg2/IAK2/Ipl1p)

Aurora is a serine/threonine kinase that has been initially characterized as a protein that regulates proper chromosomal segregation and cytokinesis (Bischoff and Plowman 1999). In *C. elegans*, AIR-2 (homolog of aurora kinase) is involved in the release of chromosomal cohesion. In *Xenopus*, Eg2 (an alternative name of this kinase) has been shown as a component of progesterone-induced maturation of oocytes. H3 histone, a linker histone that regulates the integrity of nucleosome core, has been identified a substrate of aurora in mouse and porcine oocytes. Another substrate known to date includes cytoplasmic polyadenylation element-binding protein (CPEB) that regulates translation of mRNA for Mos, and maskin that regulates the assembly of microtubules. Analyses of cell-free extracts demonstrated that protein phosphatase 2A (PP2A) is responsible for suppression of MPF, which is an upstream activator for aurora kinase. In the meiotic and mitotic exit, aurora undergoes degradation under the control of APC/C interaction with the APC/C recognition domain of aurora kinase. Thus aurora kinase behaves like a component of cytostatic factor (e.g. cyclin, Mos) (Andresson and Ruderman 1998; Detivaud et al. 2003; Ding et al. 2011; Eckerdt et al. 2009; Frank-Vaillant et al. 2000; Hodgman et al. 2001; Jelinkova and Kubelka 2006; Kinoshita et al. 2005; Littlepage and Ruderman 2002; Littlepage et al. 2002; Ma et al. 2003; Maton et al. 2005; Maton et al. 2003; Mendez et al. 2000; Pascreau et al. 2005; Pascreau et al. 2008; Pascreau et al. 2009; Rogers et al. 2002; Roghi et al. 1998; Sardon et al. 2008; Yang et al. 2010b).

2.5 Calmodulin-dependent protein kinase II (CaMKII)

CaMKII is a serine/threonine-specific kinase that is regulated by the intracellular concentration of Ca^{2+} ions. Biochemical analyses demonstrated that the binding of Ca^{2+}/calmodulin to the catalytic core of the kinase as well as the release of an autoinhibitory region from the kinase domain coordinately activates the enzyme (Ishida and Fujisawa 1995). Gene targeting analyses have demonstrated that this kinase is involved in synaptic plasticity such as the long-term potentiation in hippocampus (Silva et al. 1992a; Silva et al. 1992b). In mammalian oocytes and cell-free extracts prepared from *Xenopus* unfertilized eggs, the kinase activity of CaMKII oscillates in response to sperm-induced Ca^{2+} oscillations, well-known phenomenon that is required for the meiotic exit and initiation of embryonic development. The activated CaMKII is believed to be involved in the initiation of signaling cascade involving cyclin/Mos degradation and calcineurin/Rsk activation, both of which leads to the inactivation of Emi2, a suppressor of meiotic exit (Ducibella and Fissore 2008; Hansen et al. 2006; Hudmon et al. 2005; Liu and Maller 2005; Madgwick et al. 2005; Nishiyama et al. 2007b; Nutt et al. 2005).

2.6 Casein kinase II (CKII/CK2)

CKII is a family of serine/threonine-specific kinases that are ubiquitously expressed in eukaryotic organisms including budding yeast (Glover 1998). In the oocyte of *Xenopus*, CKII is shown to localize to the nucleus and transcription factor IIIA has been identified as a substrate of CKII (Leiva et al. 1987; Sanghera et al. 1992; Westmark et al. 2002). CKII can be regulated by PKC and other serine/threonine kinases, and therefore it may be involved in sperm-induced egg activation as well. In this respect, the fact that CKII phosphorylates a serine/threonine residue in the cytoplasmic sequence of uroplakin IIIa (UPIIIa) in uropathological bacteria-infected human urinary bladder cells is interesting. As UPIIIa in *Xenopus* eggs has been suggested to be important for sperm-egg interaction and subsequent phospho-signaling for egg activation (see below) (Mahbub Hasan et al. 2011), CKII may also be an important player in the same system through the phosphorylation of UPIIIa.

2.7 Cyclic AMP-dependent protein kinase (cAPK/PKA)

Inactive **PKA** is a tetrameric protein that is composed of two catalytic subunits and two regulatory (or inhibitory) subunits, latter of which, when cAMP binds, is released from the catalytic subunits. The discovery of AMP as well as PKA as intracellular mediators of several extracellular signal-dependent cellular functions has opened firstly a window of the research field of "phospho-signal transduction" (Robison et al. 1968), followed by discoveries of other important factors such as PKC, receptor/kinase and Src. In vertebrate oocytes, activity of PKA is shown to decrease by phosphodiesterase (PDE)-mediated decrease of intracellular cAMP and then re-increase upon meiotic maturation, and its active state is maintained until fertilization. Upon fertilization, PKA undergoes a rapid decline in its activity. Transition from mitotic phase to interphase in fertilized egg requires MPF-dependent PKA activity. In mammals, maturing oocytes involves PKA phosphorylation of Cdc25B tyrosine phosphatase that leads to up-regulation of MPF activity. In marine worm, AMPK activity has been implicated in oocyte maturation, suggesting that intracellular balance of cAMP and AMP concentrations, as regulated by PDE and adenylate cyclase, is important for oocyte functions (Bornslaeger et al. 1986; Browne et al. 1990; Daar et al. 1993; Faure et al. 1998; Faure et al. 1999; Grieco et al. 1994; Grieco et al. 1996; Matten et al. 1994; Meijer et al. 1989b; Newhall et al. 2006; Pirino et al. 2009; Schmitt and Nebreda 2002a, 2002b; Stricker and Smythe 2006; Stricker et al. 2010b; Wang and Liu 2004; Webb et al. 2008; Yu et al. 2005; Zhang et al. 2008).

2.8 Cyclin-dependent protein kinase (Cdc2/CDK/MPF)

The term "cdc" refers *cell division cycle* and has originally been coined in the study of yeast genetics. While the genetic background as well as biochemical and molecular biological identifications of key regulators for cell division cycle (i.e. several **cdc/CDK** kinases and cyclins) have firstly been demonstrated in the studies of such model organisms as yeast, sea urchin, and clam (Hartwell 1991; Minshull et al. 1989; Nurse 1990), early studies with use of frog oocytes has also contributed to arise a concept of **MPF** (maturation/mitosis-promoting factor) (Masui 1992). It is well known that cdc/CDK kinases are mainly responsible for meiotic cell cycle progression in maturing oocytes, and thereafter acts as an essential component of mitotic cell cycles. Regulatory mechanism of cdc/CDK kinases involves a complex combination of phosphorylation/dephosphorylation on a threonine and tyrosine residues in the ATP-binding pocket (e.g. Wee1, Myt1, Cdc25, PP2A) and a threonine residue in the catalytic domain of cdc/CDK kinase (i.e. CAK kinase), and protein level of activator proteins (e.g. cyclin and RINGO/speedy) and inhibitor proteins (e.g. p16 and p21). Kinase activity of cdc/CDK/MPF has been sometimes regarded as "histone H1 kinase (H1K or HH1K)" because of its in vitro evaluation. Cellular targets of cdc/CDK kinases include aurora kinase and Emi2, which are implicated in chromosomal integrity and meiotic arrest, respectively (Anger et al. 2004; Castilho et al. 2009; Culp and Musci 1999; Eckberg 1997; Edgecombe et al. 1991; Ferrell 1999; Ferrell et al. 1991; Gavin et al. 1999; Grieco et al. 1996; Gutierrez et al. 2006; Karaiskou et al. 1998; Karaiskou et al. 2004; Katsu et al. 1999; Kume et al. 2007; Kuo et al. 2011; Lohka et al. 1988; Masui 2000; Maton et al. 2005; Maton et al. 2003; Meijer et al. 1989a; Meijer et al. 1991; Meijer et al. 1989b; Palmer et al. 1998; Qian et al. 2001; Rime et al. 1994; Ruiz et al. 2008; Sakamoto et al. 1998; Tang et al. 2008; Tokmakov et al. 2005; Wu et al. 2007b; Yu et al. 2005; Yu et al. 2004).

2.9 Dual-specificity tyrosine-regulated kinase 1A/2 (DYRK)/Minibrain-related kinase (Mirk)/MBK-2/Nuclear kinase

DYRK is a dual-specificity protein kinase, whose expression in a wide variety of animal species (e.g. Yak1 in yeast, Mnb in fly, Dyrk1~4 in mammals) has been reported. Tyrosine autophosphorylation in the activation loop is important for enzyme activation of DYRK as a serine/threonine kinase. A similar scheme of kinase regulation has been shown in some other kinases including MAPK, so DYRK is regarded as a member of the MAPK superfamily (Miyata and Nishida 1999). DYRK has been implicated in neurobiological disease such as Down syndrome, cell proliferation and anti-apoptosis in cancer cells, and cell cycle control (Becker 2011; Becker and Sippl 2011). In nematode oocytes, DYRK2/MBK2, a member of DYRK, in cooperation with CDK1 (this kinase catalyses activating phosphorylation of MBK-2 on Ser-68) (Cheng et al. 2009), GSK3, and Kin-19, phosphorylates and promotes degradation of OMA-1 that regulates oocyte-to-embryo transition (Nishi and Lin 2005; Qu et al. 2006; Qu et al. 2007; Stitzel et al. 2007; Stitzel et al. 2006). In *Xenopus* oocytes, Ras-dependent oocyte maturation involves the function of DYRK1A (Qu et al. 2006; Qu et al. 2007).

2.10 Epidermal growth factor receptor (EGFR/HER1)

EGFR is a prototype of the cell surface receptor/kinase that consists of an extracellular ligand -binding domain, a transmembrane hydrophobic sequence, and a cytoplasmic kinase domain that is followed by a non-catalytic sequence, which contains some tyrosine residues to be autophosphorylated in activated molecules. Normally, EGFR is activated by EGF-dependent dimerization (activation as tyrosine kinase) and autophosphorylation (activation as phosphotyrosine-dependent docking protein). Its oncogenic counterpart has been found in avian sarcoma virus that encodes v-erbB, whose protein product lacks entirely the extracellular domain so that the kinase activity is constitutively elevated irrespective of the presence of EGF. While a variety of cellular functions (e.g. normal and malignant growth in several kinds of cells and tissues) have been shown to involve EGF and EGFR, its contribution to oocyte maturation and fertilization remains unclear. In *Xenopus* eggs, ectopically expressed EGFR is capable of inducing egg activation in an EGF-dependent manner (Yim et al. 1994). This could be explained as that active tyrosine kinase can mediate the process of egg activation in this system. In fact, it has been shown that *Xenopus* eggs employ an endogenous tyrosine kinase-dependent egg activation system involving Src and PLCγ.

2.11 ErbB4/HER4

ErbB4 is a member of the EGFR (ErbB1/HER1)/HER family of receptor/tyrosine kinases. Although its involvement in oocyte maturation and fertilization has not yet been shown, implantation of mammalian early embryos involves the actions of ErbB4 and its cognate ligand heparin-binding EGF-like growth factor (HB-EGF) (Chobotova et al. 2002). In this system, metalloproteinase-dependent extracellular shedding of HB-EGF is required for survival of trophoblasts at low oxygen conditions (Armant et al. 2006; Jessmon et al. 2009), one of pro-apoptotic pressures in the embryogenic microenvironment at early stages of pregnancy.

2.12 Focal adhesion kinase (FAK)

FAK is a cytoplasmic tyrosine kinase, whose activity is stimulated by integrin-dependent cell-extracellular matrix (ECM) interactions. Namely, in response to heterodimeric interaction with the ECM-activated integrin α and β subunits, FAK undergoes autophosphorylation and then phosphorylated by Src on tyrosine residues. The activated FAK undergoes a number of molecular interactions with cytoskeletal and signaling proteins, including Src, phosphatidylinositol 3-kinase (PI3K), Grb2, p130[Cas] and paxillin (Cary and Guan 1999). Recent studies also highlight the interaction of FAK with cell cycle control system (e.g. CDK5), pro-apoptotic system (e.g. p53), and cadherin-dependent cell-cell communications (Golubovskaya and Cance 2010; Quadri 2011; Xie et al. 2003). On the other hand, roles of FAK in gamete interaction and gametogenesis have not yet been fully documented. Developmental expression of FAK in porcine oocytes (Okamura et al. 2001) and in *Xenopus* oocytes and early embryos (Hens and DeSimone 1995; Zhang et al. 1995) have only been reported.

2.13 Fer tyrosine kinase (Fer)

DNA microarray analysis demonstrates that Feline encephalitis virus (FES)-related tyrosine kinase protein, named **FER**, is highly expressed in oocytes of the mouse (McGinnis et al. 2011a). It shows a uniform distribution in the ooplasm of small oocytes, but becomes concentrated in the germinal vesicle (GV) during oocyte growth. Association of FER with spindle bodies is seen after GV breakdown (GVBD), suggesting that it is involved in the control of cell cycle and/or chromosomal dynamics (McGinnis et al. 2011b). In support with this, siRNA-mediated knockdown of FER causes the failure of the oocytes to undergo GVBD or during MI (McGinnis et al. 2011b). While upstream and downstream mediators of FER regulation and functions have not yet been shown, other cell systems so far analyzed demonstrate that phospholipase D (PLD)-phosphatidic acid (PA) pathway is capable of stimulating FER activity (Itoh et al. 2009), and that TATA-element modulatory protein is a substrate of nuclear-localized FER (Schwartz et al. 1998). The former fact is of interest because, in amphibian (*Rana pipiens*) oocytes, progesterone-induced oocyte maturation involves a rapid activation of PLD (Kostellow et al. 1996).

2.14 Fibroblast growth factor receptor-1/-2 (FGFR1/2)

To date, 22 members of FGF family of growth factors and 4 members of **FGFR** family of receptor/tyrosine kinase have been identified in human. The FGF-FGFR system is activated in concert with heparin and heparan sulfate proteoglycan on the cell surface, and phosphorylates a number of intracellular substrate to promote a variety of cellular functions (Eswarakumar et al. 2005). In *Xenopus* maturing oocytes, translational activation of FGFR1 has been demonstrated. In the same system, overexpressed FGFR by itself can promote oocyte maturation in response to FGF stimulation, through interaction and/or phosphorylation of the SNT1/FRS2 adaptor protein. In bovine oocytes, FGF10 is shown to enhance the maturation and developmental competence (Zhang et al. 2010a), suggesting that oocytes contain the endogenous and functional FGFR. Developmental expression of FGFR has also been demonstrated in *Xenopus* and zebrafish. However, its involvement in fertilization has not yet been shown (Cailliau et al. 2003; Culp and Musci 1998; Culp and Musci 1999; Mood et al. 2002; Rappolee et al. 1998; Robbie et al. 1995; Tonou-Fujimori et al. 2002).

2.15 Fyn tyrosine kinase (Fyn)

Fyn, 59-kDa protein, is a member of Src family of non-receptor tyrosine kinases (SFKs). Like Src and Yes, another kind of SFK, Fyn is ubiquitously expressed in human tissues and its pleiotropic contribution to cellular functions (e.g. T-lymphocyte activation, spatial learning, and alcohol sensitivity) has been well documented (Palacios and Weiss 2004; Resh 1998; Trepanier et al. 2011). Oocyte-expressing Fyn and Fyn-related protein have been characterized extensively not only in vertebrates (e.g. mammals and fish) but also in sea invertebrates (sea urchin). In sea urchin, sperm-induced tyrosine phosphorylation of oocyte/egg proteins is mainly due to the activated Fyn and Src (and maybe Abl). In this species, tyrosine phosphorylation of phospholipase Cγ (PLCγ) plays an important role in inositol trisphosphate (IP$_3$)-induced Ca^{2+} release. A similar tyrosine kinase-PLCγ pathway also operates in starfish, ascidian, fish, and frog. In mice and rats, Fyn is shown to interact with tubulin and involve in cleavage furrow ingression during meiosis and mitosis. Another report demonstrates that Fyn contributes to establish and maintain polarity of the egg cortex. Further, knockdown of FYN kinase by siRNA resulted in an approximately 50% reduction in progression to metaphase II similar to what was observed in oocytes isolated from Fyn-knockout mice matured in vitro. These results clearly demonstrate that involvement of Fyn in oocyte and egg functions vary among species (Eliyahu et al. 2002; Kierszenbaum et al. 2009; Kinsey 1995; Kinsey 1996; Kinsey and Shen 2000; Kinsey et al. 2003; Levi et al. 2010; Luo et al. 2009; McGinnis et al. 2009;

Rongish and Kinsey 2000; Sette et al. 2002; Sharma and Kinsey 2006; Sharma and Kinsey 2008; Steele et al. 1990; Talmor et al. 1998; Talmor-Cohen et al. 2004b; Wu and Kinsey 2000; Wu and Kinsey 2002; Wu and Kinsey 2004).

2.16 Flagellar protein-tyrosine kinase (Flagellar PTK)

Fertilization in the biflagellated green algae, *Chlamydomonas*, is initiated by flagellar adhesion between gametes of opposite mating types: plus (mt+) and minus (mt-). Flagellar adhesion is followed by an increase in cytoplasmic cAMP concentration that is required for gamete fusion. Pharmacological and biochemical studies have demonstrated that a tyrosine kinase activity, named **flagellar PTK**, which acts upstream of the cAMP elevation, is present in adhering bisexual gametes but not in non-adhering, unisexual gametes. A 105-kDa protein has been identified as a substrate of the flagellar PTK. Analyses of temperature-sensitive mutants have shown that kinesin II is an essential component that connects flagellar adhesion and activation of the tyrosine kinase activity (Kurvari and Snell 1996; Kurvari et al. 1996; Wang and Snell 2003).

2.17 Flagellar p48 protein kinase (SksC)

On the contrary to flagellar PTK, another protein kinase activity is shown to decrease rapidly after gamete adhesion in *Chlamydomonas*. The kinase, named **SksC**, is a 48-kDa protein that is capable of autophosphorylating on serine and tyrosine residues, indicating that it is a dual-specificity kinase. Although it's physiological substrate other than SksC by itself has not yet been identified, adhesion-induced SksC down-regulation and flagellar PTK up-regulation may play important roles simultaneously in gamete fusion and activation of embryogenesis (Pan and Snell 2000; Zhang et al. 1996). In this species, a rapid degradation of two gamete-specific proteins, FUS1 and HAP1, occur upon gamete fusion (Liu et al. 2010). This event is required for polyspermy block. However, its relationship to the aforementioned phospho-signaling is not known.

2.18 Fms-like tyrosine kinase/vascular endothelial growth factor receptor (FLT/Colony-stimulating factor receptor-like/VEGFR)

FLT/VEGFR is a receptor/tyrosine kinase, whose extracellular ligand is VEGF (de Vries et al. 1992). The term FLT is coined because it is structurally related to c-fms/macrophage colony-stimulating factor-1 receptor/kinase. The viral counterpart of c-fms in a feline sarcoma virus (McDonough and HZ-5 strain) arises as a result of alterations in receptor coding sequences that affect its activity as a tyrosine kinase (Sherr et al. 1988). Pleiotropic functions of FLT/VEGFR have been well documented and, most of all, its involvement in angiogenesis in normal as well as cancerous cell conditions has been of clinical interest (Shibuya 1995). In bovine cumulus-oocyte complexes and porcine ovary, expression and physiological impact of VEGF and/or FLT/VEGFR have been investigated. The results so far obtained suggest that VEGF-FLA/VEGFR pathway is involved in viability of oocytes (Einspanier et al. 2002; Okamura et al. 2001).

2.19 Glycogen synthase kinase 3 (GSK3/shaggy/GSK3-B)

GSK3, a serine/threonine protein kinase, have the two isoforms GSK3 (p51) and GSK3 (p47) is known to play roles in many biological processes. Mouse eggs contain centrosomal spindle poles when arrested at meiotic metaphase II. Phosphorylated PKC (p-PKC) and GSK3 are enriched at both centrosomal spindle poles and the kinetochore region (Baluch and Capco 2008). p-PKC phosphorylates GSK3 on the Ser-9 position to inactivate GSK3 and consequently maintaining spindle stability during meiotic metaphase arrest (Baluch and Capco 2008). Similarly, in mouse oocytes, p-GSK3 was increased and phospho-MAPK3/MAPK1 was decreased before GVBD and oocytes were mainly arrested at MI (Uzbekova et al. 2009). GSK3 might be also involved in the local activation of Aurora A kinase that controls MI/MII transition (Uzbekova et al. 2009). GSK3/shaggy along with other downstream components of the Wnt pathway mediate patterning along the primary animal-vegetal axis of the sea urchin embryo (Emily-Fenouil et al. 1998) and along the dorsal-ventral axis in *Xenopus*, suggesting a conserved basis for axial patterning between invertebrate and vertebrate. Double phosphorylation (Thr-239 by DYRK kinase MBK-2 and Thr-339 by GSK-3) on OMA-1 is essential for correctly timed degradation of OMA-1 and ensures a normal oocyte-to-embryo transition in *C. elegans* (Nishi and Lin 2005). Even the conserved function of GSK3 is observed in hydra embryogenesis (Rentzsch et al. 2005), and in zebrafish cardiogenesis (Emily-Fenouil et al. 1998; Lee et al. 2007; Liu et al. 2007; Nishi and Lin 2005; Uzbekova et al. 2009).

2.20 Greatwall kinase (Gwl/GWK)

The balance between Cdc2 kinase/cyclin B also known as M-phase-promoting factor (Arceci et al. 1992), and protein phosphatase 2A (PP2A) is crucial to enable in time mitotic entry and exit. Greatwall (Gwl) kinase (**GWK**) has been identified as a key element in M phase initiation and maintenance in *Drosophila*, *Xenopus* oocytes/eggs, and mammalian cells. GWK is activated by cdk1/cyclin B (Arceci et al. 1992), and promotes the inhibition of protein phosphatase 2A (PP2A) that works on the phosphorylated substrate mediated by CDKs. Activated GWK negatively regulates a crucial phosphatase and thus induce inhibiting phosphorylations of Cdc25 to inhibits M phase induction (Zhao et al. 2008). Thus, mitotic entry and maintenance is not only mediated by the activation of Cdc2 kinase/cyclin B but also by the regulation of PP2A by GWK in *Xenopus* oocytes/eggs (Castilho et al. 2009; Mochida et al. 2010; Vigneron et al. 2009; Yamamoto et al. 2011).

2.21 Histone H1 kinase (HH1K/H1K)

Maturation promoting factor (Arceci et al. 1992) is universally recognized as the biological entity responsible for driving the cell cycle from G2- to M-phase. Histone H1 kinase (**HH1K**) activity is widely accepted as a biochemical indicator of p34Cdc2 protein kinase complex activity and therefore MPF activity. In spontaneously maturing oocytes, HH1K activity increases before GVBD in mouse (Gavin et al. 1994). HH1K activity being higher in the first than in the second cell cycle in mouse embryogenesis that reaches to the basal level (Ciemerych et al. 1998; Fulka et al. 1992). Inhibition of protein phosphatases are correlated with HH1K activity and is sufficient to induce the entry into M phase during the first cell cycle of the mouse parthenogenetic activated oocyte (Rime and Ozon 1990). In fertilized sea urchin eggs the activity of HH1K oscillates during the cell division cycle and there is a striking temporal correlation between HH1K activation and the accumulation of a phosphorylated form of cyclin (Meijer et al. 1989a; Meijer and Pondaven 1988; Tosuji et al. 2003). HH1K activity correlation with the oocyte maturation and after fertilization were carried out in other species e.g. bovine (Collas et al. 1993), cat fish (Balamurugan and Haider 1998), fish (Yamashita et al. 1992), goldfish (Pati et al. 2000), pig (Kikuchi et al. 1995), rabbit (Jelinkova et al. 1994) and sea star (Arion et al. 1988; Pelech et al. 1987).

2.22 Insulin-like growth factor 1 receptor/kinase (IGF-1R/IGFR)

In somatic cell insulin-like growth factor (IGF) receptor (**IGFR**) has the ability to phosphorylate the overall cellular substrates, in particular PLCγ, annexin II and to activate phosphatidylinositol 3-kinase via insulin receptor substrate 1 (Jiang et al. 1996). *Xenopus* oocytes bear both the IGFR-1 and IGFR-2, where IGFR-1, a tyrosine kinase, has the capability of autophosphorylation (Janicot et al. 1991; Nissley et al. 1985). IGF-1-induced oocyte maturation required IGFR-1-mediated endocytosis in *Xenopus* (Taghon and Sadler 1994). IGFR-1 in *Xenopus* ovarian follicle cells somehow supports the IGF-1-stimulated oocyte maturation (Sadler et al. 2010). Expression of IGFR has been shown in the oocytes of rat (Zhao et al. 2001), in bovine (Nuttinck et al. 2004) and in rainbow trout positively correlated with embryonic survival (Aegerter et al. 2004).

2.23 Insulin receptor/kinase (IR)

Insulin/insulin-like growth factor (IGF)-1 receptor (**IR/IGF1R**), a tyrosine kinase, exerts its cellular functions by the phosphorylation of insulin receptor substrate-1 (IRS-1). Tyrosine phosphorylated form of IRS-1 binds to specific Src homology-2 (SH2) domain-containing proteins including the p85 subunit of phosphatidylinositol (PI) 3-kinase and GRB2, a molecule believed to link IRS-1 to the Ras pathway in *Xenopus* oocyte maturation (Chuang et al. 1994; Chuang et al. 1993; El-Etr et al. 1979; Grigorescu et al. 1994). Insulin through IR has influences on oocyte maturation and embryonic development in mouse (Acevedo et al. 2007). Recently, it has been shown that IR and IGF1R are not required for oocyte growth, differentiation, and maturation in mice using genetically ablated mouse (Pitetti et al. 2009). It was shown that IR is the components of sea urchin eggs plasma membrane (Jeanmart et al. 1976) and insulin like peptide 3 acts through mosquito IR in mosquito egg production (Brown et al. 2008).

2.24 c-Jun N-terminal kinase (JNK)

The c-Jun N-terminal kinase (**JNK**) is member of the mitogen-activated protein kinase family that plays critical roles in stress responses and apoptosis. JNK is activated just prior to germinal vesicle breakdown during *Xenopus* oocyte maturation and remains active until the early gastrula stage of embryogenesis (Bagowski et al. 2001). JNK was activated after the microinjection of Mos (Bagowski et al. 2001). Progesterone mediated *Xenopus* oocyte maturation might involve JNK activation both through the raf/MEK (MAPKK)/p42 MAPK-dependent pathway (Bagowski et al. 2001; Chie et al. 2000) and through MEK/p42 MAPK-independent pathways (Bagowski et al. 2001). JNK2 plays an important role in spindle assembly and first polar body extrusion during mouse oocyte meiotic maturation (Huang et al. 2011). JNK mRNA was detected in mouse eggs and pre-implantation embryos (Zhong et al. 2004).

2.25 c-Kit tyrosine kinase (c-Kit)

The proto-oncogene product **c-Kit**, a transmembrane tyrosine kinase, acts as a receptor in mouse oocytes to communicate with the surrounding granulose cells and for its maturation.

Stem cell factor (SCF), a ligand for c-Kit is required for the production of the mature gametes e.g. the growth and maturation of the oocytes in response to gonadotropic hormones (Sette et al. 2000). The level of c-Kit increases during the maturation of mouse oocytes and following fertilization, it decreases rapidly until the early 2-cell stage but it is not detected in the embryos of 4-cell, 8-cell, and morula stages (Arceci et al. 1992; Horie et al. 1991). It is suggested that Kit-PI3K-Akt-GSK-3 pathway might work in the regulation of mouse oocytes growth (Liu et al. 2007).

2.26 Lck tyrosine kinase (Lck)

Lck, a 56-kDa protein, has originally been characterized as a Src-related tyrosine kinase that is specifically expressed in lymphocytes (Lck is named after lymphocyte kinase). In T-cells, Lck associates with CD4/CD8 cell surface receptor for major histocompatibility complex and, upon interaction with antigen-presenting cells, it will be activated by dephosphorylation in the carboxyl-terminal tyrosine residue, as catalyzed by CD45 phosphatase. In murine eggs, it has been reported that CD4-like structures on the vitelline membranes are involved in gamete interaction, and that Lck-like protein could have been detected in association with those CD4-like structures (Mori et al. 2000; Mori et al. 1991). While these studies have been done with the use of specific monoclonal antibodies (e.g. immunofluorescent and immunochemical approaches), biochemical and molecular biological identifications have not yet been demonstrated.

2.27 p38 MAPK/Mipk/Stress-activated protein kinase (SAPK)/Xp38γ

p38/SAPK, which has initially been identified as a stress-activated protein kinase, belongs to the MAPK superfamily (Miyata and Nishida 1999). In the sea star, a p38-related kinase Mipk (meiosis-inhibited protein kinase) has been identified and characterized. Before oocyte maturation, Mipk is highly phosphorylated on tyrosine residues, and during oocyte maturation and some hours after fertilization, it becomes tyrosine-dephosphorylated and enzymatically inactive, suggesting that inhibition of Mipk is related to cell cycle progression during meiosis (Morrison et al. 2000). However, knockdown of Mipk by antisense oligonucleotide is not effective in inducing oocyte maturation. On the other hand, *Xenopus* p38γ/SAPK3 is a major player in G2/M transition of immature oocytes induced by MKK6, a p38 activator. The activated p38γ/SAPK3 is also shown to phosphorylate Ser-205 of and activate Cdc25C phosphatase (Perdiguero et al. 2003). One another interesting feature of p38 in oocyte/egg system is that it may contribute to apoptotic process in starfish eggs left unfertilized for a long time. In this system, inactivation of MAPK is pre-requisite for inducing activation of caspase, a pro-apoptotic protease. p38 has been shown to activate after the MAPK inactivation and seems to be responsible for apoptotic body formation (Morrison et al. 2000; Perdiguero et al. 2003; Sasaki and Chiba 2004).

2.28 Mitogen-activated protein kinase (p42/p44MAPK/ERK)

MAPK is a serine/threonine kinase that has been originally identified as a microtubule-associated protein 2 (MAP2) kinase (this is also termed "MAPK" or "MAP2 kinase") and then well recognized as a mitogen-activated protein kinase (Maller 1990). MAPK is a component of the MAPK kinase, which consists of at least three steps of phospho-dependent activation of kinases that include MAPK (e.g. Erk, p38, JNK), MAPK kinase (MAPKK: e.g. MEK), and MAPKK kinase (MAPKKK: e.g. Mos, Raf). The MAPK cascade is evolutionarily conserved in a variety of unicellular and multicellular organisms and serves as a trigger of multiple cellular functions such as differentiation, nutrition signals, proliferation, and stress responses. In maturing oocytes of several organisms, stoichiometric activation of MAPK will occur (all-or-none signaling of MAPK activation) (Ferrell and Machleder 1998). This MAPK activation seems to be required for maintaining the maturing oocytes to arrest at the metaphase of second meiosis (in mammals and frog), rather than oocyte maturation itself. This is a so-called cytostatic factor's function. MAPK activation can be evaluated by the phosphorylation of a threonine and tyrosine residues in the MAPK molecule, both of which are catalyzed by an upstream dual-specificity kinase, MAPKK. Fertilization promotes Ca^{2+}-dependent degradation and/or inactivation of upstream kinases Mos and MAPKK, and triggers a rapid dephosphorylation/inactivation of all MAPK (inactivation of cytostatic factor). In the actively dividing embryos, a fraction of MAPK will be transiently activated at mitotic phase, and thereafter serves as a component of checkpoint (Chesnel et al. 1997; Chung et al. 1991; Eckberg 1997; Fabian et al. 1993; Ferrell 1999; Ferrell et al. 1991; Gavin et al. 1999; Git et al. 2009; Gross et al. 2000; Huo et al. 2004; Ito et al. 2010; Iwasaki et al. 2008; Katsu et al. 1999; Keady et al. 2007; Kosako et al. 1992; Lee et al. 2006; Lu et al. 2002; Palmer et al. 1998; Philipova and Whitaker 1998; Sackton et al. 2007; Sadler et al. 2004; Sasaki and Chiba 2004; Sato et al. 2001; Sato et al. 2003; Sato et al. 2000; Shibuya et al. 1992; Shibuya et al. 1996; Stricker 2009; Sun et al. 1999; Tokmakov et al. 2005; Verlhac et al. 1996; Zhang et al. 2006).

2.29 MAPK kinase (MAPKK/MEK)

MAPKK is a serine/threonine kinase that will be activated by MAPKKK phosphorylation of its serine residues in the catalytic domain. The activated MAPKK is capable of phosphorylating threonine and tyrosine residues in the catalytic domain of a downstream kinase MAPK, thus MAPKK is a dual-specificity kinase. MAPKK is well known as a mediator of Mos-dependent activation of MAPK cascade in maturing oocytes (Kosako et al. 1992; Xiong et al. 2008).

2.30 Meiosis inhibited protein kinase (MIPK)

p38 type of MAPK is a member of the mitogen-activated protein kinase (MAPK) is usually activated in response to cytokines and various stresses and plays a role in the inhibition of cell proliferation and tumor progression, but its role in oocyte maturation is described recently. In *Xenopus* oocytes, p38MAPK phosphorylated Cdc25C for the meiotic G_2/M progression and this required neither protein synthesis nor activation of p42MAPK-p90[Rsk] pathway (Perdiguero et al. 2003). The function of p38MAPK in accurate chromosome segregation during mouse oocyte meiotic maturation has also been described (Ou et al. 2010). In porcine oocytes, active phosphorylated p38MAPK accumulated in the nucleus before GVBD and remained active through MI to MII (Villa-Diaz and Miyano 2004). A p38MAPK homolog **Mipk** (meiosis-inhibited protein kinase) was highly tyrosine phosphorylated in immature sea star oocytes and subsequently dephosphorylated during the arrest at the G_2/M transition of meiosis I (Morrison et al. 2000). Dephosphorylated Mipk was maintained until the maturation of oocytes and the early mitotic cell divisions but was re-phosphorylated at the time of differentiation and acquisition of G phases in the developing embryos (Morrison et al. 2000).

2.31 Mos protein kinase (Mos)

Mos, a mitogen-activated protein (MAP) kinase kinase kinase that activates the MAPK pathway, is normally expressed only in vertebrate oocytes and take part in their maturation. Cytoplasmic polyadenylation element binding (CPEB) factor is essential for the polyadenylation of c-Mos mRNA and its subsequent translation (Mendez et al. 2000). Early phosphorylation of CPEB is catalyzed by Eg2, a member of the Aurora family of serine/threonine protein kinases (Mendez et al. 2000). Mos in coordination with Cdc2 regulate the translational activation of a maternal FGF receptor-1 (FGFR) mRNA during *Xenopus* oocyte maturation (Culp and Musci 1999). Mos contribute in the first cycle of *Xenopus* embryogenesis (Murakami and Vande Woude 1998) and act like Mos/Raf-1/MAPK pathway (Muslin et al. 1993) or without Raf like Mos/MAPK pathway both in *Xenopus* (Shibuya et al. 1996) and mouse (Verlhac et al. 1996). Mos is also involved in MAPK cascade in the control of microtubule and chromatin organization during meiosis in mouse oocytes (Chesnel et al. 1997; Culp and Musci 1999; Daar et al. 1993; Faure et al. 1998; Mendez et al. 2000; Murakami and Vande Woude 1998; Muslin et al. 1993; Shibuya et al. 1992; Shibuya et al. 1996; Tang et al. 2008; Verlhac et al. 1996; Wu et al. 2007a).

2.32 Myelin basic protein kinase (MBPK)

MBPK is present during maturation and early embryogenesis of the sea star. A meiosis-activated MBP kinase (MBPK) was purified from maturing oocytes of the sea star that rapidly undergo autophosphorylation on serine/threonine residues (Sanghera et al. 1990). MBPK remained highly active until 12 h post-fertilization (Arceci et al. 1992), after which it declined (Lefebvre et al. 1999). During maturation of sea star oocytes, MBPK-II (p110) was fully activated at the time of GVBD, whereas peak activation of MBPK-I (p45) occurred after this event (Pelech et al. 1988). Inhibiting an upstream phosphorylation event in the MBPK activation pathway the sea urchin embryo mitotic cycle at metaphase can be blocked (Pesando et al. 1999). The MBPK activity was at approximately the same high level in all categories (medium, small and tiny) of bovine oocytes after 24 h of culture and remained stable until 40 h (Pavlok et al. 1997; Pelech et al. 1988; Sanghera et al. 1990).

2.33 Myt1 protein kinase (Myt1)

Activation of MPF (composed of cyclin B and Cdc2 kinase) is required to entry into M-phase in all animals. The inhibitory kinase **Myt1**, a member of Wee1 family phosphorylates Cdc2 kinase to keep MPF in an inactive state. During *Xenopus* oocyte maturation MAPK phosphorylates and activates p90[Rsk] and that p90[Rsk] in turn down-regulates Myt1 by phosphorylation, leading to the activation of Cdc2 kinase/cyclin B (Palmer et al. 1998; Ruiz et al. 2010). Alternatively, Mos triggers Myt1 phosphorylation, even in the absence of MAPK activation in a mechanism that directly activates MPF in *Xenopus* oocytes (Peter et al. 2002). Recent model is that up-regulation of cyclin B synthesis causes rapid inactivating phosphorylation of Myt1, mediated by Cdc2 and without any significant contribution of Mos/MAPK or Plx1 (Gaffre et al. 2011). Non-cyclin proteins RINGO/Speedy can phosphorylate Ser residue in the regulatory domain of Myt1 and lead the activation of CDK during G2/M transition in *Xenopus* oocytes (Burrows et al. 2006; Inoue and Sagata 2005; Oh et al. 2010; Palmer et al. 1998; Ruiz et al. 2008).

2.34 Nemo-like kinase 1 (NLK1)

NLK Nemo-like kinase (NLK) is an evolutionary conserved MAPK-like kinase, an atypical MAPK that phosphorylates several transcription factors and is known to function in multiple developmental processes in vertebrates and invertebrates (Ota et al. 2011a; Ota et al. 2011b). Activated NLK directly phosphorylates microtubule-associated protein-1B (MAP1B) and the focal adhesion adaptor protein, paxillin (Ishitani et al. 2009). Inactive NLK1 in immature *Xenopus* oocytes becomes active during maturation depending on Mos protein synthesis but not on p42 MAPK activation (Ota et al. 2011b). NLK1 acts as a kinase downstream of Mos and catalyzes the phosphorylation of Pumilio 1 (Pum1), Pum2, and cytoplasmic polyadenylation element-binding protein (CPEB) to regulate the translation of mRNAs, including cyclin B1 mRNA, stored in oocytes (Ota et al. 2011b). NLK may play a role in neural development together with Sox11 during early *Xenopus* embryogenesis (Hyodo-Miura et al. 2002). NLK appears to function as a positive regulator of Wnt signaling during early zebrafish development (Thorpe and Moon 2004).

2.35 Nerve growth factor receptor (NGFR/TrkA/TrkB)

NGFR, the nerve growth factor (NGF) receptor (NGFR), an integral single membrane protein that is phosphorylated and heavily glycosylated in *Xenopus* oocytes and potentiates the ability of progesterone to induce maturation (Sehgal et al. 1988). NGF treatment on *Xenopus* oocytes results the tyrosine phosphorylation of ectopically expressed human Trk, a proto-oncogene product (p140[proto-Trk]), and meiotic maturation, as determined by germinal vesicle breakdown and the activation of MPF (Nebreda et al. 1991). Thus, the Trk proto-oncogene product can act as a receptor for NGF (Nebreda et al. 1991). In human ovaries, NGF and TrkA (NGF's high-affinity receptor) were detected in granulose cells of preantral and antral follicles and in thecal cells of antral follicles (Salas et al. 2006). NGF/TrkA is present in bovine sperm and might have roles in regulation of sperm physiology relevant to male fertility and infertility (Li et al. 2010).

2.36 Neu tyrosine kinase

p185[Neu], the protein product of the neu gene, is a tyrosine kinase receptor that has the structural similarity to EGFR. The transformed/activated form of p185[Neu] tyrosine kinase (Val664Glu) facilitates the oocyte maturation events reducing the half-life from approximately 9 h to 5 h that are elicited by some steroids (e.g. progesterone) (Narasimhan et al. 1992). However, the activated p185[Neu] tyrosine kinases are not able to mimic the EGF-stimulated EGF receptor tyrosine kinase in triggering oocyte maturation, which suggests that the EGF receptor and the p185[Neu] tyrosine kinase do not work in the same pathways in *Xenopus* oocytes (Narasimhan et al. 1992). But in mouse embryo culture cells, it was shown that mutationally activated Neu protein can substitute the ligand-activated EGF receptor activity to reflect the structural similarity and EGF induced phosphorylation and regulation of p185[Neu] (Kokai et al. 1988; Shirahata et al. 1990).

2.37 p21-activated protein kinase (PAK)

PAK is a serine/threonine kinase that will be activated by its interaction with a small GTP-binding protein (e.g. Rac, Cdc42) and has been implicated in cytoskeletal dynamics and cell motility, transcription through MAPK cascades, death and survival signaling, and cell-cycle progression (Bokoch 2003). In *Xenopus* oocytes, microinjection of catalytically inactive mutant of PAK-related kinase (X-PAK) accelerates cell cycle progression from GV through MII stages. On the other hand, catalytically active mutant of X-PAK is shown to suppress progesterone-induced Mos accumulation and MAPK activation. These data suggest that the endogenous PAK activity is involved in the cell cycle arrest before maturation (Faure et al. 1997; Faure et al. 1999). An inhibitory effect of X-PAK on oocyte maturation seems to be due to a PKA-like mechanism of suppression of the PLK-induced activation of Cdc25 phosphatase, which is a trigger of the activation of Cdc2/cyclin complex (i.e. MPF) (Faure et al. 1999).

2.38 Phosphoinositide-dependent protein kinase 1 (PDK1)

PDK is a serine/threonine kinase that is regulated by a lipid activator phosphatidylinositol 3,4,5-trisphosphate (PIP_3), a product of PI3K phosphorylation of phosphatidylinositol 4,5-bisphosphate (PIP_2) (Toker and Newton 2000). In general, PDK consists of two distinct gene products, PDK1 and PDK2. In 1-methyladenine (1-MA)-dependent maturing starfish oocytes, Akt kinase is responsible for activation of MPF. Akt kinase is, as described above, regulated by upstream kinases such as PDK and mTOR. In fact, starfish PDK1, but not PDK2, is required for 1-MA-induced Akt activation and cell cycle progression (Hiraoka et al. 2004). PI3K-PDK-Akt axis has also been shown in other organisms such as nematode *C. elegans* (Hertweck et al. 2004), however, their involvement in oocyte and egg functions is not yet known.

2.39 Polo-like protein kinase-1 (Plk/1PLK-1/Plx1)

PLK, polo-like kinase, a serine/threonine kinase, is implicated in the regulation of cell cycle progression in all eukaryotes (Sumara et al. 2002). Polo-like kinase type 1 (plk1) is present during meiotic maturation, fertilization, and early embryo cleavage in mouse (Pahlavan et al. 2000; Tong et al. 2002; Xiong et al. 2008), rat (Fan et al. 2003), porcine (Anger et al. 2004; Yao et al. 2003) and parasite trematode oocytes (Long et al. 2010). Though all three *Xenopus* type Plk (Plx); Plx1, Plx2 and Plx3 are observed in oocytes and unfertilized eggs but Plx2 and Plx3 in embryos strongly suggests that individual Plk family members perform distinct functions at later stages of development (Duncan et al. 2001). Plx1 is required for activation of the phosphatase Cdc25C and cyclin B-Cdc2 in *Xenopus* oocytes (Liu and Maller 2005; Qian et al. 2001). The APC/C inhibitor Emi2 or XErp1, a pivotal CSF component, required to maintain metaphase II arrest and rapidly destroyed in response to Ca^{2+} signaling through phosphorylation by Plx1 (Hansen et al. 2006). Interestingly, Plx1 kinase that is required for Cdc25 activation and MPF auto-amplification in fully-grown oocytes is not expressed at the protein level in small stage IV oocytes (Karaiskou et al. 2004). Plx1 acts as a direct inhibitory kinase of Myt1 in the mitotic cell cycles in *Xenopus* (Anger et al. 2004; Eckerdt et al. 2009; Fan et al. 2003; Hansen et al. 2006; Inoue and Sagata 2005; Ito et al. 2008; Karaiskou et al. 2004; Liu and Maller 2005; Pahlavan et al. 2000; Qian et al. 2001; Sumara et al. 2002; Tong et al. 2002; Wianny et al. 1998; Xiong et al. 2008; Yao et al. 2003).

2.40 Protein kinase C (PKC)

PKC is a family of serine/threonine kinase that is primarily regulated by diacylglycerol (DG), a phospholipase C-hydrolyzed product of PIP_2, and intracellular Ca^{2+} (Nishizuka 1984; Nishizuka 1986; Nishizuka 1988). Classical or typical PKCs (α, βI/βII, γ) are also known as an intracellular receptor for phorbol ester, a tumor promoter. Other subfamily members of PKC (atypical or novel types: e.g. δ, ϵ, ζ) have other mechanisms of enzyme regulation such as tyrosine phosphorylation. Live cell imaging studies demonstrated that spatial distribution of PKCs, which differ in both PKC subfamily members and cellular environments, is crucial for PKC activation and its access to substrates. In eggs/oocytes of several organisms (e.g. mammals, marine worms, frog, and sea urchin), activation and PKC(s) and its contribution to oocyte maturation, fertilization, and initiation of development have been well documented. Cellular functions regulated by PKC(s) include the onset of anaphase I, sperm-induced activation of respiratory burst oxidase, MAPK inactivation, reorganization of cytoskeleton, exocytosis of cortical granules, and pronucleus formation (Akabane et al. 2007; Baluch et al. 2004; Capco 2001; Capco et al. 1992; de Barry et al. 1997; Diaz-Meco et al. 1994; Ducibella and LeFevre 1997; Eliyahu et al. 2001; Eliyahu and Shalgi 2002; Eliyahu et al. 2002; Eliyahu et al. 2005; Fan et al. 2002; Gallicano et al. 1995; Gallicano et al. 1997; Haberman et al. 2011; Halet 2004; Heinecke et al. 1990; Kalive et al. 2010; Lu et al. 2002; Luria et al. 2000; Madgwick et al. 2005; Nakaya et al. 2000; Olds et al. 1995; Pauken and Capco 2000; Quan et al. 2003; Sakuma et al. 1997; Sanghera et al. 1992; Shen and Buck 1990; Stricker 2009; Swann et al. 1989; Tatone et al. 2003; Viveiros et al. 2004; Viveiros et al. 2003; Yang et al. 2004; Yu et al. 2004; Yu et al. 2008). In rat eggs, PKC interaction and phosphorylation of RACK (receptor for C-kinase) has been suggested (Haberman et al. 2011). In *Xenopus* oocytes, hormone-induced maturation is accompanied by polarized localization and interaction of atypical PKC(s) and ASIP/PAR-3, a cell polarity regulator, suggesting their involvement in establishing animal-vegetal asymmetry before fertilization (Nakaya et al. 2000).

2.41 Protein kinase M (PKM)

PKM is a catalytic fragment of PKC, produced by a limited proteolysis (probably by calpain, a Ca^{2+}-dependent protease) of the molecule. It has been shown that PKM contributes to the remodeling of cytoskeleton during egg activation in the mouse (Gallicano et al. 1995). Its physiological target (i.e. substrate) is not yet known.

2.42 Proline-rich tyrosine kinase2 (Pyk2)

Pyk2 is a non-receptor tyrosine kinase related to the focal adhesion kinase (FAK; p125) that is rapidly phosphorylated on tyrosine residues in response to various stimuli that elevate the intracellular calcium ion concentration (Lev et al. 1995). Pyk2 is up-regulated in various types of tumors like hepatocellular carcinoma (HCC) (Sun et al. 2011; Sun et al. 2007) and small cell lung cancer (SCLC) (Roelle et al. 2008). Activation of Pyk2 leads to the activation of the MAPK signaling pathways. PYK2 is present in mouse spermatocytes and spermatids (Chieffi et al. 2003). Pyk2 plays a dynamic role during rat oocyte meiotic maturation by regulating the organization of actin filaments (microfilaments) from GV stage to telophase (Meng et al. 2006). Pyk2 has ligand sequences for Src homology 2 and 3 (SH2 and SH3), and has binding sites for paxillin (Li and Earp 1997) and p130[Cas] (Astier et al. 1997).

2.43 Raf protein kinase (A-Raf/B-Raf/Raf-1)

Raf is a serine/threonine kinase that has been originally identified as an oncogene that acts in concert with Myc transcription factor. Raf can be activated by PKC phosphorylation, and the activated Raf acts as a MAPKKK that activates MAPKK-MAPK pathway. Therefore, Raf is a mediator of transmembrane signaling involving hydrolysis of phospholipids and subsequent cytoplasmic kinase cascade. In *Xenopus* oocytes, Raf-1 is shown to act downstream of Mos, another kind of MAPKKK specific to oocyte maturation system, to promote MAPK activation and rearrangement of intracellular pH (from 7.2 to 7.7) in response to progesterone or insulin. The latter phenomenon involves phospho-dependent regulation of Na^+/H^+ exchanger. Whether Raf is responsible for this phosphorylation is unknown (Chesnel et al. 1997; Fabian et al. 1993; Kang et al. 1998; MacNicol et al. 1995; Muslin et al. 1993; Shibuya et al. 1996). Developmental expression of Raf has also been demonstrated in *Xenopus*; however, its role in fertilization has not yet been shown (MacNicol et al. 1995).

2.44 RET tyrosine kinase (Ret)

Receptor tyrosine kinase are rearranged during transfection (**RET**) for activation and about 15 RET gene rearrangement was identified in papillary thyroid carcinoma (PTC) among which RET/PTC1 and RET/PTC3 are the most common type (Marotta et al. 2011). RET was detected in mammalian (human) oocytes (Farhi et al. 2010) and are expressed in embryos throughout the early development with an increase after the early blastocyst stage (Kawamura et al. 2008). Glial cell line-derived neurotrophic factor (GDNF) and both its co-receptors, GDNF family receptor alpha-1 (GFR alpha-1) and RET receptor affect porcine oocyte maturation and pre-implantation embryo developmental competence in a follicular stage-dependent manner (Linher et al. 2007). Receptor tyrosine kinase (RTK1) that is highly similar to RET kinase was not detected in sea urchin unfertilized eggs and was activated after blastula stage (Sakuma et al. 1997).

2.45 S6 kinase (S6K)/ Rsk protein kinase I/II (Rsk)

Several 40S ribosomal protein kinases in vertebrate/frog oocyte stage 6 (**S6K**) are directly phosphorylated and activated by MAPK in order to activate MPF (Barrett et al. 1992; Erikson and Maller 1988). Some S6Ks have been identified and characterized for example in progesterone- and insulin-treated *Xenopus* eggs termed S6K II (S6K II, p92) different from S6K I (Erikson et al. 1987), differential role in *Xenopus* embryogenesis (S6K; p70) (Schwab et al. 1999), in *Rana* oocytes (S6K; p83) (Byun et al. 2002), in porcine oocytes (Sugiura et al. 2002), and G1 phase after completion of meiosis II in starfish unfertilized eggs (Mori et al. 2006) but not in mouse oocytes (Dumont et al. 2005). S6K (p90[Rsk]) inhibits the degradation of cyclin B by anaphase-promoting complex/cyclosome (APC/C) and results the second meiotic metaphase arrest (Maller et al. 2001). S6K phosphorylates and activates the Bub1 protein kinase, which may cause metaphase arrest due to the inhibition of APC (Maller et al. 2001). Mos-dependent phosphorylation of Erp1 by p90[Rsk] at Thr-336, Ser-342 and Ser-344 is crucial for both stabilizing Erp1 that inhibits cyclin B degradation by binding the APC/C and establishing CSF arrest in meiosis II of *Xenopus* oocytes (Nishiyama et al. 2007a).

2.46 Src tyrosine kinase (Src)

Src has been firstly identified as an oncogene of Rous sarcoma virus and thereafter discovered as a normal cellular gene that encodes a 60-kDa protein-tyrosine kinase (Brown and Cooper 1996; Jove and Hanafusa 1987; Thomas and Brugge 1997) that is distributed in a wide range of animal species from a unicellular organism (i.e. *Monosiga ovata*) through multicellular organisms including human (Segawa et al. 2006). In human, Src is ubiquitously expressed in several tissues and seems to be involved in several cellular functions as well (e.g. lymphocyte activation, neuronal signal transduction). In some sea invertebrates (sea urchin, starfish, ascidian, and others) (Abassi et al. 2000; Belton et al. 2001; Dasgupta and Garbers 1983; Giusti et al. 1999a; Giusti et al. 1999b; Giusti et al. 2000a; Giusti et al. 2003; Giusti et al. 2000b; Kamel et al. 1986; O'Neill et al. 2004; Runft et al. 2004; Runft and Jaffe 2000; Runft et al. 2002; Sakuma et al. 1997; Shen et al. 1999; Shilling et al. 1994; Stricker et al. 2010a; Townley et al. 2006; Townley et al. 2009), fish (zebrafish) (in this case, Fyn tyrosine kinase) and frog (African clawed frog) (Glahn et al. 1999; Iwasaki et al. 2008; Iwasaki et al. 2006; Kushima et al. 2011; Mahbub Hasan et al. 2011; Mahbub Hasan et al. 2007; Mahbub Hasan et al. 2005; Mammadova et al. 2009; Sakakibara et al. 2005; Sato et al. 1996; Sato et al. 2006a; Sato et al. 1999; Sato et al. 2004; Sato et al. 2002; Sato et al. 2001; Sato et al. 2003; Sato et al. 2000; Sato et al. 2006b; Steele 1985; Steele et al. 1989b; Tokmakov et al. 2002), the oocyte-expressing Src is suggested to be involved in the initiation of sperm-induced egg activation through the phosphorylation and activation of oocyte proteins such as phospholipase $C\gamma$ (thereby promoting IP_3-dependent Ca^{2+} release). Progesterone-induced oocyte maturation in *Xenopus* also seems to involve the activity of Src (Tokmakov et al. 2005). In mammalian species (i.e. mouse and rat), chromosomal dynamics, rather than sperm-induced Ca^{2+} release (Kurokawa et al. 2004b; McGinnis et al. 2007; Mehlmann and Jaffe 2005; Reut et al. 2007; Tomashov-Matar et al. 2008), seems to be regulated by Src and/or other Src-related kinases (e.g. Fyn) in fertilized oocytes.

2.47 Src64/DSrc

Src64 is a *Drosophila* homolog of the tyrosine kinase Src and is required for ovarian ring canal morphogenesis during oogenesis. Tec29 tyrosine kinase interacts with Src64 and contributes to ring canal development. The Src64-Tec29 axis is also involved in microfilament contraction during cellularization, a *Drosophila*-specific phenomenon. Although the cellular target of Src64 phosphorylation is not yet clearly shown, its upstream regulators such as csk homolog-mediated phosphorylation and phosphoinositide-dependent activation mechanism have been demonstrated (Dodson et al. 1998; Lu et al. 2004; O'Reilly et al. 2006).

2.48 Stigmatic S receptor kinase (SRK)

SRK is a transmembrane receptor/kinase that works as a female determinant for self-incompatibility/self-sterility to prevent inbreeding in *Brassica*, a flowering plant. Upon self-pollination, the pollen-borne ligand S locus protein 11/SCR interacts with SRK expressed in stigma, which in turn autophosphorylates and promotes Ca^{2+}-dependent signal transduction that culminates in self-pollen rejection (Murase et al. 2004). Another protein kinase, named M locus protein kinase, has also been identified as a cytoplasmic mediator of self-incompatibility in this species (Kakita et al. 2007). This is the first example that explains how self-incompatibility, in other words, allogenic authentication, is made possible in sexual reproduction of hermaphrodite organism. More recent studies have demonstrated that a similar system of the allogenic authentication (that utilizes gamete coat/membrane-associated proteins) is also present in animal hermaphrodite organisms (e.g. ascidian) (Harada et al. 2008). Whether such animal system involves protein kinase signaling is unknown.

2.49 T-Cell Origin Protein Kinase/ T-LAK cell-originated protein kinase (TOPK)

TOPK (T-LAK cell-originated protein kinase) is distributed in lymphokine-activated killer T (T-LAK) cell, testis, activated lymphoid cells, and lymphoid tumors, and is related to the dual specific mitogen-activated protein kinase kinase (MAPKK) (Abe et al. 2000). TOPK protein is expressed mainly in the cytosol of spermatocytes and spermatids to support the testicular functions (Fujibuchi et al. 2005). During mitosis, TOPK-Thr-9 was phosphorylated by cdk1/cyclin B and TOPK significantly associates with mitotic spindles (Matsumoto et al. 2004). Insulin-matured *Xenopus* oocytes showed much higher expression of TOPK and nuclear kinase (DYRK1A) but neither of these kinases activates or is activated by MAPK and is therefore unique to insulin-activated wild-type p21[Ras]-induced oocyte maturation via the activation of Raf (Qu et al. 2006; Qu et al. 2007). The functions of insulin-activated wild-type p21[Ras] do not depend on the two classic Raf targets, MEK and MAPK (MAPK or ERK) (Qu et al. 2006; Qu et al. 2007).

2.50 p65[tpr-met], a fused tyrosine kinase (Tpr-Met)

Tpr-met (p65[tpr-met], a fused tyrosine kinase) efficiently induced meiotic maturation in *Xenopus* oocytes and activate MPF through a Mos-dependent pathway (Daar et al. 1991; Park et al. 1986). During *Xenopus* oocyte maturation, receptor tyrosine kinase (RTK) pathway including tpr-met takes part in the activation of MPF that requires activation of Raf and MAPK (Fabian et al. 1993). Aberrant or activated expression of Met receptor (Tpr-Met) in *Xenopus* embryonic system induces ectopic morphogenetic structures during *Xenopus* embryogenesis where recruitment of either the Grb2 or the Shc adaptor protein is sufficient to induce ectopic structures and anterior reduction but the role of PI 3-kinase and PLC recruitment are unclear (Ishimura et al. 2006). Grb2-associated binder 1 (Gab1) when overexpressed in *Xenopus* oocyte is crucial for Tpr-Met-mediated morphological transformation (Mood et al. 2006). Thus, to induce such structure Ras/Raf/MAPK pathway is important.

2.51 Vaccinia-related kinase 1 (VRK1)/*Drosophila* NHK-1

VRK1, a member of the casein kinase I (Minshull et al. 1989) family is a serine/threonine kinase related to vaccinia virus B1R serine/threonine kinase (Klerkx et al. 2009), has been identified as an early response gene required for cyclin D1 expression. VRK1 controls cell survival by phosphorylation of p53, chromatin condensation by phosphorylation of histone, and nuclear envelope assembly by phosphorylation of BANF1 (Valbuena et al. 2011). It is also involved in fragmentation of Golgi apparatus in the G2 phase-cell cycle. In *Drosophila* oocytes, nucleosomal histone kinase-1 (*Drosophila* homolog of VRK1) regulates chromosome-nuclear envelope association via phosphorylation of BAF protein (barrier to auto-integration factor), thereby supports the meiotic progression (karyosome formation) (Lancaster et al. 2007). In the mouse, target disruption of VRK1 causes a delay in meiotic progression and results in the appearance of lagging chromosomes during formation of the metaphase plate (Schober et al. 2011), suggesting that function of VRK1 is evolutionarily conserved, although its substrate has not yet been demonstrated.

2.52 Wee1 protein kinase (Wee1)

Wee1, a protein tyrosine kinase, is the key regulator of cell cycle progression by phosphorylating and inhibiting Cdc2. Wee1, an inhibitor of Cdc2/cyclin B kinase, is decreased for mammalian oocytes meiotic competence (Mitra and Schultz 1996). Wee1 activity is necessary for the control of the first embryonic cell cycle following the fertilization of meiotically mature *Xenopus* oocytes where the protein accumulation is regulated at the level of mRNA translation (Charlesworth et al. 2000). p42 MAPK was found to phosphorylate and activate Wee1 activity towards Cdc2, thus Wee1 might work in the downstream of Mos/MEK/p42 MAPK (Walter et al. 2000). Basically in *Xenopus*, eukaryotic Wee1 homologue, termed Wee1A functions in pre-gastrula embryos with rapid cell cycle and zygotic isoform Wee1B functions post-gastrula embryos where Wee1B inhibits Cdc2 activity and oocyte maturation much more strongly than Wee1A (Okamoto et al. 2002). PKA also involved in the inactive state of Cdc2/cyclin B kinase by regulating Wee1 kinase (Han and Conti 2006).

2.53 Yes tyrosine kinase (Yes/c-Yes)

The egg cortex is known to be rich in cortical structures such as actin cytoskeleton forming microfilaments and cortical vesicles and they are important in many dynamic events in mammalian egg maturation and fertilization, such as sperm incorporation, cortical granule exocytosis, polar body emission, etc. SFKs have been shown to be associated with a wide range of cytoskeletal components and/or to phosphorylate them (Thomas and Brugge 1997). It has demonstrated that Fyn, c-Yes and c-Src are distributed throughout the rat egg cytoplasm, but Fyn and c-Yes are tend to concentrate at the egg cortex whereas only Fyn is localized to the spindle (Talmor-Cohen et al. 2004a). Localization of c-Src, c-Yes and Fyn to different compartments within the egg indicates that these proteins may have different functions within the egg. No change in the subcellular distribution of the three kinases has been observed throughout the stages of the fertilization process, or after parthenogenetic activation (Talmor-Cohen et al. 2004a). Though Yes kinase activity was decreased at fertilization in Zebra fish but it was concentrated in blastoderm cells and maintained the high activity throughout the gastrulation (Tsai et al. 2005). It is possible that the intracellular distribution of c-Src, c-Yes and Fyn imply their association with the cytoskeleton. The involvement of SFKs in reorganization of the cytoskeleton might be involved in egg fertilization. (Steele et al. 1989a; Tsai et al. 2005)

3. Conclusion

For conclusion, please refer to the section 3 of the chapter entitled "Phospho-signaling at Oocyte Maturation and Fertilization: Set Up for Embryogenesis and Beyond Part II. Kinase Regulators and Substrates" by Mahbub Hasan et al.

4. Acknowledgements

We apologize to those whose work was not cited or insufficiently cited. This work is supported by a Grant-in-Aid on Innovative Areas (22112522), and a grant for Private University Strategic Research Foundation Support Program (S0801060) from the Ministry of Education, Culture, Sports, Science and Technology, Japan to K.S.

5. References

Abassi YA, Carroll DJ, Giusti AF, Belton RJ, Jr., Foltz KR. 2000. Evidence that Src-type tyrosine kinase activity is necessary for initiation of calcium release at fertilization in sea urchin eggs. Dev Biol 218(2):206-219.

Abe Y, Matsumoto S, Kito K, Ueda N. 2000. Cloning and expression of a novel MAPKK-like protein kinase, lymphokine-activated killer T-cell-originated protein kinase, specifically expressed in the testis and activated lymphoid cells. J Biol Chem 275(28):21525-21531.

Acevedo N, Ding J, Smith GD. 2007. Insulin signaling in mouse oocytes. Biol Reprod 77(5):872-879.

Aegerter S, Jalabert B, Bobe J. 2004. Messenger RNA stockpile of cyclin B, insulin-like growth factor I, insulin-like growth factor II, insulin-like growth factor receptor Ib, and p53 in the rainbow trout oocyte in relation with developmental competence. Mol Reprod Dev 67(2):127-135.

Ajduk A, Malagocki A, Maleszewski M. 2008. Cytoplasmic maturation of mammalian oocytes: development of a mechanism responsible for sperm-induced Ca2+ oscillations. Reprod Biol 8(1):3-22.

Akabane H, Fan J, Zheng X, Zhu GZ. 2007. Protein kinase C activity in mouse eggs regulates gamete membrane interaction. Mol Reprod Dev 74(11):1465-1472.

Anderson RG. 1998. The caveolae membrane system. Annu Rev Biochem 67:199-225.

Andresson T, Ruderman JV. 1998. The kinase Eg2 is a component of the Xenopus oocyte progesterone-activated signaling pathway. EMBO J 17(19):5627-5637.

Anger M, Klima J, Kubelka M, Prochazka R, Motlik J, Schultz RM. 2004. Timing of Plk1 and MPF activation during porcine oocyte maturation. Mol Reprod Dev 69(1):11-16.

Arceci RJ, Pampfer S, Pollard JW. 1992. Expression of CSF-1/c-fms and SF/c-kit mRNA during preimplantation mouse development. Dev Biol 151(1):1-8.

Arion D, Meijer L, Brizuela L, Beach D. 1988. cdc2 is a component of the M phase-specific histone H1 kinase: evidence for identity with MPF. Cell 55(2):371-378.

Armant DR, Kilburn BA, Petkova A, Edwin SS, Duniec-Dmuchowski ZM, Edwards HJ, Romero R, Leach RE. 2006. Human trophoblast survival at low oxygen concentrations requires metalloproteinase-mediated shedding of heparin-binding EGF-like growth factor. Development 133(4):751-759.

Astier A, Manie SN, Avraham H, Hirai H, Law SF, Zhang Y, Golemis EA, Fu Y, Druker BJ, Haghayeghi N, Freedman AS, Avraham S. 1997. The related adhesion focal tyrosine kinase differentially phosphorylates p130Cas and the Cas-like protein, p105HEF1. J Biol Chem 272(32):19719-19724.

Bagowski CP, Xiong W, Ferrell JE, Jr. 2001. c-Jun N-terminal kinase activation in Xenopus laevis eggs and embryos. A possible non-genomic role for the JNK signaling pathway. J Biol Chem 276(2):1459-1465.

Balamurugan K, Haider S. 1998. Partial purification of maturation-promoting factor from catfish, Clarias batrachus: identification as the histone H1 kinase and its periodic activation. Comp Biochem Physiol C Pharmacol Toxicol Endocrinol 120(3):329-342.

Baluch DP, Capco DG. 2008. GSK3 beta mediates acentromeric spindle stabilization by activated PKC zeta. Dev Biol 317(1):46-58.

Baluch DP, Koeneman BA, Hatch KR, McGaughey RW, Capco DG. 2004. PKC isotypes in post-activated and fertilized mouse eggs: association with the meiotic spindle. Dev Biol 274(1):45-55.

Barrett CB, Erikson E, Maller JL. 1992. A purified S6 kinase kinase from Xenopus eggs activates S6 kinase II and autophosphorylates on serine, threonine, and tyrosine residues. J Biol Chem 267(7):4408-4415.

Becker W. 2011. Recent insights into the function of DYRK1A. FEBS J 278(2):222.

Becker W, Sippl W. 2011. Activation, regulation, and inhibition of DYRK1A. FEBS J 278(2):246-256.

Belton RJ, Jr., Adams NL, Foltz KR. 2001. Isolation and characterization of sea urchin egg lipid rafts and their possible function during fertilization. Mol Reprod Dev 59(3):294-305.

Bilodeau-Goeseels S, Sasseville M, Guillemette C, Richard FJ. 2007. Effects of adenosine monophosphate-activated kinase activators on bovine oocyte nuclear maturation in vitro. Mol Reprod Dev 74(8):1021-1034.

Bischoff JR, Plowman GD. 1999. The Aurora/Ipl1p kinase family: regulators of chromosome segregation and cytokinesis. Trends Cell Biol 9(11):454-459.

Bokoch GM. 2003. Biology of the p21-activated kinases. Annu Rev Biochem 72:743-781.

Bornslaeger EA, Mattei P, Schultz RM. 1986. Involvement of cAMP-dependent protein kinase and protein phosphorylation in regulation of mouse oocyte maturation. Dev Biol 114(2):453-462.

Brown MR, Clark KD, Gulia M, Zhao Z, Garczynski SF, Crim JW, Suderman RJ, Strand MR. 2008. An insulin-like peptide regulates egg maturation and metabolism in the mosquito Aedes aegypti. Proc Natl Acad Sci U S A 105(15):5716-5721.

Brown MT, Cooper JA. 1996. Regulation, substrates and functions of src. Biochim Biophys Acta 1287(2-3):121-149.

Browne CL, Bower WA, Palazzo RE, Rebhun LI. 1990. Inhibition of mitosis in fertilized sea urchin eggs by inhibition of the cyclic AMP-dependent protein kinase. Exp Cell Res 188(1):122-128.

Brunet S, Maro B. 2005. Cytoskeleton and cell cycle control during meiotic maturation of the mouse oocyte: integrating time and space. Reproduction 130(6):801-811.

Burrows AE, Sceurman BK, Kosinski ME, Richie CT, Sadler PL, Schumacher JM, Golden A. 2006. The C. elegans Myt1 ortholog is required for the proper timing of oocyte maturation. Development 133(4):697-709.

Byun HM, Kang SG, Kang HM. 2002. Cloning of ribosomal protein S6 kinase cDNA and its involvement in meiotic maturation in Rana dybowskii oocytes. Mol Cells 14(1):16-23.

Cailliau K, Le Marcis V, Bereziat V, Perdereau D, Cariou B, Vilain JP, Burnol AF, Browaeys-Poly E. 2003. Inhibition of FGF receptor signalling in Xenopus oocytes: differential effect of Grb7, Grb10 and Grb14. FEBS Lett 548(1-3):43-48.

Capco DG. 2001. Molecular and biochemical regulation of early mammalian development. Int Rev Cytol 207:195-235.

Capco DG, Tutnick JM, Bement WM. 1992. The role of protein kinase C in reorganization of the cortical cytoskeleton during the transition from oocyte to fertilization-competent egg. J Exp Zool 264(4):395-405.

Carroll DJ, Albay DT, Terasaki M, Jaffe LA, Foltz KR. 1999. Identification of PLCgamma-dependent and -independent events during fertilization of sea urchin eggs. Dev Biol 206(2):232-247.

Cary LA, Guan JL. 1999. Focal adhesion kinase in integrin-mediated signaling. Front Biosci 4:D102-113.

Castilho PV, Williams BC, Mochida S, Zhao Y, Goldberg ML. 2009. The M phase kinase Greatwall (Gwl) promotes inactivation of PP2A/B55delta, a phosphatase directed against CDK phosphosites. Mol Biol Cell 20(22):4777-4789.

Castro A, Peter M, Magnaghi-Jaulin L, Vigneron S, Galas S, Lorca T, Labbe JC. 2001. Cyclin B/cdc2 induces c-Mos stability by direct phosphorylation in Xenopus oocytes. Mol Biol Cell 12(9):2660-2671.

Charlesworth A, Welk J, MacNicol AM. 2000. The temporal control of Wee1 mRNA translation during Xenopus oocyte maturation is regulated by cytoplasmic polyadenylation elements within the 3'-untranslated region. Dev Biol 227(2):706-719.

Chaube SK, Dubey PK, Mishra SK, Shrivastav TG. 2007. Verapamil reversibly inhibits spontaneous parthenogenetic activation in aged rat eggs cultured in vitro. Cloning Stem Cells 9(4):608-617.

Chen J, Downs SM. 2008. AMP-activated protein kinase is involved in hormone-induced mouse oocyte meiotic maturation in vitro. Dev Biol 313(1):47-57.

Chen J, Hudson E, Chi MM, Chang AS, Moley KH, Hardie DG, Downs SM. 2006. AMPK regulation of mouse oocyte meiotic resumption in vitro. Dev Biol 291(2):227-238.

Cheng KC, Klancer R, Singson A, Seydoux G. 2009. Regulation of MBK-2/DYRK by CDK-1 and the pseudophosphatases EGG-4 and EGG-5 during the oocyte-to-embryo transition. Cell 139(3):560-572.

Chesnel F, Bonnec G, Tardivel A, Boujard D. 1997. Comparative effects of insulin on the activation of the Raf/Mos-dependent MAP kinase cascade in vitellogenic versus postvitellogenic Xenopus oocytes. Dev Biol 188(1):122-133.

Chie L, Amar S, Kung HF, Lin MC, Chen H, Chung DL, Adler V, Ronai Z, Friedman FK, Robinson RC, Kovac C, Brandt-Rauf PW, Yamaizumi Z, Michl J, Pincus MR. 2000. Induction of oocyte maturation by jun-N-terminal kinase (JNK) on the oncogenic ras-p21 pathway is dependent on the raf-MEK signal transduction pathway. Cancer Chemother Pharmacol 45(6):441-449.

Chieffi P, Barchi M, Di Agostino S, Rossi P, Tramontano D, Geremia R. 2003. Prolin-rich tyrosine kinase 2 (PYK2) expression and localization in mouse testis. Mol Reprod Dev 65(3):330-335.

Chobotova K, Spyropoulou I, Carver J, Manek S, Heath JK, Gullick WJ, Barlow DH, Sargent IL, Mardon HJ. 2002. Heparin-binding epidermal growth factor and its receptor ErbB4 mediate implantation of the human blastocyst. Mech Dev 119(2):137-144.

Choi T, Rulong S, Resau J, Fukasawa K, Matten W, Kuriyama R, Mansour S, Ahn N, Vande Woude GF. 1996. Mos/mitogen-activated protein kinase can induce early meiotic phenotypes in the absence of maturation-promoting factor: a novel system for analyzing spindle formation during meiosis I. Proc Natl Acad Sci U S A 93(10):4730-4735.

Chuang LM, Hausdorff SF, Myers MG, Jr., White MF, Birnbaum MJ, Kahn CR. 1994.
    Interactive roles of Ras, insulin receptor substrate-1, and proteins with Src
    homology-2 domains in insulin signaling in Xenopus oocytes. J Biol Chem
    269(44):27645-27649.
Chuang LM, Myers MG, Jr., Backer JM, Shoelson SE, White MF, Birnbaum MJ, Kahn CR.
    1993. Insulin-stimulated oocyte maturation requires insulin receptor substrate 1
    and interaction with the SH2 domains of phosphatidylinositol 3-kinase. Mol Cell
    Biol 13(11):6653-6660.
Chung J, Pelech SL, Blenis J. 1991. Mitogen-activated Swiss mouse 3T3 RSK kinases I and II
    are related to pp44mpk from sea star oocytes and participate in the regulation of
    pp90rsk activity. Proc Natl Acad Sci U S A 88(11):4981-4985.
Ciemerych MA, Tarkowski AK, Kubiak JZ. 1998. Autonomous activation of histone H1
    kinase, cortical activity and microtubule organization in one- and two-cell mouse
    embryos. Biol Cell 90(8):557-564.
Clarke PR, Karsenti E. 1991. Regulation of p34cdc2 protein kinase: new insights into protein
    phosphorylation and the cell cycle. J Cell Sci 100 (Pt 3):409-414.
Collas P, Sullivan EJ, Barnes FL. 1993. Histone H1 kinase activity in bovine oocytes
    following calcium stimulation. Mol Reprod Dev 34(2):224-231.
Culp PA, Musci TJ. 1998. Translational activation and cytoplasmic polyadenylation of FGF
    receptor-1 are independently regulated during Xenopus oocyte maturation. Dev
    Biol 193(1):63-76.
Culp PA, Musci TJ. 1999. c-mos and cdc2 cooperate in the translational activation of
    fibroblast growth factor receptor-1 during Xenopus oocyte maturation. Mol Biol
    Cell 10(11):3567-3581.
Daar I, Yew N, Vande Woude GF. 1993. Inhibition of mos-induced oocyte maturation by
    protein kinase A. J Cell Biol 120(5):1197-1202.
Daar IO, White GA, Schuh SM, Ferris DK, Vande Woude GF. 1991. tpr-met oncogene
    product induces maturation-producing factor activation in Xenopus oocytes. Mol
    Cell Biol 11(12):5985-5991.
Dasgupta JD, Garbers DL. 1983. Tyrosine protein kinase activity during embryogenesis. J
    Biol Chem 258(10):6174-6178.
de Barry J, Kawahara S, Takamura K, Janoshazi A, Kirino Y, Olds JL, Lester DS, Alkon DL,
    Yoshioka T. 1997. Time-resolved imaging of protein kinase C activation during sea
    urchin egg fertilization. Exp Cell Res 234(1):115-124.
de Klein A, van Kessel AG, Grosveld G, Bartram CR, Hagemeijer A, Bootsma D, Spurr NK,
    Heisterkamp N, Groffen J, Stephenson JR. 1982. A cellular oncogene is translocated
    to the Philadelphia chromosome in chronic myelocytic leukaemia. Nature
    300(5894):765-767.
de Vries C, Escobedo JA, Ueno H, Houck K, Ferrara N, Williams LT. 1992. The fms-like
    tyrosine kinase, a receptor for vascular endothelial growth factor. Science
    255(5047):989-991.
Deng X, Feng C, Wang EH, Zhu YQ, Cui C, Zong ZH, Li GS, Liu C, Meng J, Yu BZ. 2011.
    Influence of proline-rich inositol polyphosphate 5-phosphatase, on early
    development of fertilized mouse eggs, via inhibition of phosphorylation of Akt.
    Cell Prolif 44(2):156-165.
Detivaud L, Pascreau G, Karaiskou A, Osborne HB, Kubiak JZ. 2003. Regulation of EDEN-
    dependent deadenylation of Aurora A/Eg2-derived mRNA via phosphorylation

and dephosphorylation in Xenopus laevis egg extracts. J Cell Sci 116(Pt 13):2697-2705.

Diaz-Meco MT, Lozano J, Municio MM, Berra E, Frutos S, Sanz L, Moscat J. 1994. Evidence for the in vitro and in vivo interaction of Ras with protein kinase C zeta. J Biol Chem 269(50):31706-31710.

Ding J, Swain JE, Smith GD. 2011. Aurora kinase-A regulates microtubule organizing center (MTOC) localization, chromosome dynamics, and histone-H3 phosphorylation in mouse oocytes. Mol Reprod Dev 78(2):80-90.

Dodson GS, Guarnieri DJ, Simon MA. 1998. Src64 is required for ovarian ring canal morphogenesis during Drosophila oogenesis. Development 125(15):2883-2892.

Ducibella T. 1996. The cortical reaction and development of activation competence in mammalian oocytes. Hum Reprod Update 2(1):29-42.

Ducibella T, Fissore R. 2008. The roles of Ca2+, downstream protein kinases, and oscillatory signaling in regulating fertilization and the activation of development. Dev Biol 315(2):257-279.

Ducibella T, LeFevre L. 1997. Study of protein kinase C antagonists on cortical granule exocytosis and cell-cycle resumption in fertilized mouse eggs. Mol Reprod Dev 46(2):216-226.

Dumont J, Umbhauer M, Rassinier P, Hanauer A, Verlhac MH. 2005. p90Rsk is not involved in cytostatic factor arrest in mouse oocytes. J Cell Biol 169(2):227-231.

Duncan MJ, Li G, Shin JS, Carson JL, Abraham SN. 2004. Bacterial penetration of bladder epithelium through lipid rafts. J Biol Chem 279(18):18944-18951.

Duncan PI, Pollet N, Niehrs C, Nigg EA. 2001. Cloning and characterization of Plx2 and Plx3, two additional Polo-like kinases from Xenopus laevis. Exp Cell Res 270(1):78-87.

Eckberg WR. 1997. MAP and cdc2 kinase activities at germinal vesicle breakdown in Chaetopterus. Dev Biol 191(2):182-190.

Eckerdt F, Pascreau G, Phistry M, Lewellyn AL, DePaoli-Roach AA, Maller JL. 2009. Phosphorylation of TPX2 by Plx1 enhances activation of Aurora A. Cell Cycle 8(15):2413-2419.

Edgecombe M, Patel R, Whitaker M. 1991. A cyclin-abundance cycle-independent p34cdc2 tyrosine phosphorylation cycle in early sea urchin embryos. EMBO J 10(12):3769-3775.

Einspanier R, Schonfelder M, Muller K, Stojkovic M, Kosmann M, Wolf E, Schams D. 2002. Expression of the vascular endothelial growth factor and its receptors and effects of VEGF during in vitro maturation of bovine cumulus-oocyte complexes (COC). Mol Reprod Dev 62(1):29-36.

El-Etr M, Schorderet-Slatkine S, Baulieu EE. 1979. Meiotic maturation in Xenopus laevis oocytes initiated by insulin. Science 205(4413):1397-1399.

Eliyahu E, Kaplan-Kraicer R, Shalgi R. 2001. PKC in eggs and embryos. Front Biosci 6:D785-791.

Eliyahu E, Shalgi R. 2002. A role for protein kinase C during rat egg activation. Biol Reprod 67(1):189-195.

Eliyahu E, Talmor-Cohen A, Shalgi R. 2002. Signaling through protein kinases during egg activation. J Reprod Immunol 53(1-2):161-169.

Eliyahu E, Tsaadon A, Shtraizent N, Shalgi R. 2005. The involvement of protein kinase C and actin filaments in cortical granule exocytosis in the rat. Reproduction 129(2):161-170.

Emily-Fenouil F, Ghiglione C, Lhomond G, Lepage T, Gache C. 1998. GSK3beta/shaggy
    mediates patterning along the animal-vegetal axis of the sea urchin embryo.
    Development 125(13):2489-2498.
Erikson E, Maller JL. 1988. Substrate specificity of ribosomal protein S6 kinase II from
    Xenopus eggs. Second Messengers Phosphoproteins 12(2-3):135-143.
Erikson E, Stefanovic D, Blenis J, Erikson RL, Maller JL. 1987. Antibodies to Xenopus egg S6
    kinase II recognize S6 kinase from progesterone- and insulin-stimulated Xenopus
    oocytes and from proliferating chicken embryo fibroblasts. Mol Cell Biol 7(9):3147-
    3155.
Eswarakumar VP, Lax I, Schlessinger J. 2005. Cellular signaling by fibroblast growth factor
    receptors. Cytokine Growth Factor Rev 16(2):139-149.
Fabian JR, Morrison DK, Daar IO. 1993. Requirement for Raf and MAP kinase function
    during the meiotic maturation of Xenopus oocytes. J Cell Biol 122(3):645-652.
Fan HY, Tong C, Li MY, Lian L, Chen DY, Schatten H, Sun QY. 2002. Translocation of the
    classic protein kinase C isoforms in porcine oocytes: implications of protein kinase
    C involvement in the regulation of nuclear activity and cortical granule exocytosis.
    Exp Cell Res 277(2):183-191.
Fan HY, Tong C, Teng CB, Lian L, Li SW, Yang ZM, Chen DY, Schatten H, Sun QY. 2003.
    Characterization of Polo-like kinase-1 in rat oocytes and early embryos implies its
    functional roles in the regulation of meiotic maturation, fertilization, and cleavage.
    Mol Reprod Dev 65(3):318-329.
Farhi J, Ao A, Fisch B, Zhang XY, Garor R, Abir R. 2010. Glial cell line-derived neurotrophic
    factor (GDNF) and its receptors in human ovaries from fetuses, girls, and women.
    Fertil Steril 93(8):2565-2571.
Faure S, Morin N, Doree M. 1998. Inactivation of protein kinase A is not required for c-mos
    translation during meiotic maturation of Xenopus oocytes. Oncogene 17(10):1215-
    1221.
Faure S, Vigneron S, Doree M, Morin N. 1997. A member of the Ste20/PAK family of protein
    kinases is involved in both arrest of Xenopus oocytes at G2/prophase of the first
    meiotic cell cycle and in prevention of apoptosis. EMBO J 16(18):5550-5561.
Faure S, Vigneron S, Galas S, Brassac T, Delsert C, Morin N. 1999. Control of G2/M
    transition in Xenopus by a member of the p21-activated kinase (PAK) family: a link
    between protein kinase A and PAK signaling pathways? J Biol Chem 274(6):3573-
    3579.
Feng C, Yu A, Liu Y, Zhang J, Zong Z, Su W, Zhang Z, Yu D, Sun QY, Yu B. 2007.
    Involvement of protein kinase B/AKT in early development of mouse fertilized
    eggs. Biol Reprod 77(3):560-568.
Ferrell JE, Jr. 1999. Xenopus oocyte maturation: new lessons from a good egg. Bioessays
    21(10):833-842.
Ferrell JE, Jr., Machleder EM. 1998. The biochemical basis of an all-or-none cell fate switch in
    Xenopus oocytes. Science 280(5365):895-898.
Ferrell JE, Jr., Wu M, Gerhart JC, Martin GS. 1991. Cell cycle tyrosine phosphorylation of
    p34cdc2 and a microtubule-associated protein kinase homolog in Xenopus oocytes
    and eggs. Mol Cell Biol 11(4):1965-1971.
Frank-Vaillant M, Haccard O, Thibier C, Ozon R, Arlot-Bonnemains Y, Prigent C, Jessus C.
    2000. Progesterone regulates the accumulation and the activation of Eg2 kinase in
    Xenopus oocytes. J Cell Sci 113 (Pt 7):1127-1138.

Fujibuchi T, Abe Y, Takeuchi T, Ueda N, Shigemoto K, Yamamoto H, Kito K. 2005. Expression and phosphorylation of TOPK during spermatogenesis. Dev Growth Differ 47(9):637-644.

Fulka J, Jr., Jung T, Moor RM. 1992. The fall of biological maturation promoting factor (MPF) and histone H1 kinase activity during anaphase and telophase in mouse oocytes. Mol Reprod Dev 32(4):378-382.

Gaffre M, Martoriati A, Belhachemi N, Chambon JP, Houliston E, Jessus C, Karaiskou A. 2011. A critical balance between Cyclin B synthesis and Myt1 activity controls meiosis entry in Xenopus oocytes. Development 138(17):3735-3744.

Galat V, Zhou Y, Taborn G, Garton R, Iannaccone P. 2007. Overcoming MIII arrest from spontaneous activation in cultured rat oocytes. Cloning Stem Cells 9(3):303-314.

Gallicano GI, McGaughey RW, Capco DG. 1995. Protein kinase M, the cytosolic counterpart of protein kinase C, remodels the internal cytoskeleton of the mammalian egg during activation. Dev Biol 167(2):482-501.

Gallicano GI, McGaughey RW, Capco DG. 1997. Activation of protein kinase C after fertilization is required for remodeling the mouse egg into the zygote. Mol Reprod Dev 46(4):587-601.

Gavin AC, Cavadore JC, Schorderet-Slatkine S. 1994. Histone H1 kinase activity, germinal vesicle breakdown and M phase entry in mouse oocytes. J Cell Sci 107 (Pt 1):275-283.

Gavin AC, Ni Ainle A, Chierici E, Jones M, Nebreda AR. 1999. A p90(rsk) mutant constitutively interacting with MAP kinase uncouples MAP kinase from p34(cdc2)/cyclin B activation in Xenopus oocytes. Mol Biol Cell 10(9):2971-2986.

Gilkey JC, Jaffe LF, Ridgway EB, Reynolds GT. 1978. A free calcium wave traverses the activating egg of the medaka, Oryzias latipes. J Cell Biol 76(2):448-466.

Git A, Allison R, Perdiguero E, Nebreda AR, Houliston E, Standart N. 2009. Vg1RBP phosphorylation by Erk2 MAP kinase correlates with the cortical release of Vg1 mRNA during meiotic maturation of Xenopus oocytes. RNA 15(6):1121-1133.

Giusti AF, Carroll DJ, Abassi YA, Foltz KR. 1999a. Evidence that a starfish egg Src family tyrosine kinase associates with PLC-gamma1 SH2 domains at fertilization. Dev Biol 208(1):189-199.

Giusti AF, Carroll DJ, Abassi YA, Terasaki M, Foltz KR, Jaffe LA. 1999b. Requirement of a Src family kinase for initiating calcium release at fertilization in starfish eggs. J Biol Chem 274(41):29318-29322.

Giusti AF, Foltz KR, Jaffe LA. 2000a. The role of Src family kinases in starfish egg fertilisation. Zygote 8 Suppl 1:S16-17.

Giusti AF, O'Neill FJ, Yamasu K, Foltz KR, Jaffe LA. 2003. Function of a sea urchin egg Src family kinase in initiating Ca2+ release at fertilization. Dev Biol 256(2):367-378.

Giusti AF, Xu W, Hinkle B, Terasaki M, Jaffe LA. 2000b. Evidence that fertilization activates starfish eggs by sequential activation of a Src-like kinase and phospholipase cgamma. J Biol Chem 275(22):16788-16794.

Glahn D, Mark SD, Behr RK, Nuccitelli R. 1999. Tyrosine kinase inhibitors block sperm-induced egg activation in Xenopus laevis. Dev Biol 205(1):171-180.

Glover CV, 3rd. 1998. On the physiological role of casein kinase II in Saccharomyces cerevisiae. Prog Nucleic Acid Res Mol Biol 59:95-133.

Golubovskaya VM, Cance W. 2010. Focal adhesion kinase and p53 signal transduction pathways in cancer. Front Biosci 15:901-912.

Grieco D, Avvedimento EV, Gottesman ME. 1994. A role for cAMP-dependent protein
kinase in early embryonic divisions. Proc Natl Acad Sci U S A 91(21):9896-9900.

Grieco D, Porcellini A, Avvedimento EV, Gottesman ME. 1996. Requirement for cAMP-PKA
pathway activation by M phase-promoting factor in the transition from mitosis to
interphase. Science 271(5256):1718-1723.

Grigorescu F, Baccara MT, Rouard M, Renard E. 1994. Insulin and IGF-1 signaling in oocyte
maturation. Horm Res 42(1-2):55-61.

Gross SD, Schwab MS, Taieb FE, Lewellyn AL, Qian YW, Maller JL. 2000. The critical role of
the MAP kinase pathway in meiosis II in Xenopus oocytes is mediated by p90(Rsk).
Curr Biol 10(8):430-438.

Gutierrez GJ, Vogtlin A, Castro A, Ferby I, Salvagiotto G, Ronai Z, Lorca T, Nebreda AR.
2006. Meiotic regulation of the CDK activator RINGO/Speedy by ubiquitin-
proteasome-mediated processing and degradation. Nat Cell Biol 8(10):1084-1094.

Haberman Y, Alon LT, Eliyahu E, Shalgi R. 2011. Receptor for activated C kinase (RACK)
and protein kinase C (PKC) in egg activation. Theriogenology 75(1):80-89.

Halet G. 2004. PKC signaling at fertilization in mammalian eggs. Biochim Biophys Acta
1742(1-3):185-189.

Han SJ, Conti M. 2006. New pathways from PKA to the Cdc2/cyclin B complex in oocytes:
Wee1B as a potential PKA substrate. Cell Cycle 5(3):227-231.

Han SJ, Vaccari S, Nedachi T, Andersen CB, Kovacina KS, Roth RA, Conti M. 2006a. Protein
kinase B/Akt phosphorylation of PDE3A and its role in mammalian oocyte
maturation. EMBO J 25(24):5716-5725.

Hansen DV, Tung JJ, Jackson PK. 2006. CaMKII and polo-like kinase 1 sequentially
phosphorylate the cytostatic factor Emi2/XErp1 to trigger its destruction and
meiotic exit. Proc Natl Acad Sci U S A 103(3):608-613.

Harada Y, Takagaki Y, Sunagawa M, Saito T, Yamada L, Taniguchi H, Shoguchi E, Sawada
H. 2008. Mechanism of self-sterility in a hermaphroditic chordate. Science
320(5875):548-550.

Hardie DG, Carling D. 1997. The AMP-activated protein kinase--fuel gauge of the
mammalian cell? Eur J Biochem 246(2):259-273.

Hartwell LH. 1991. Twenty-five years of cell cycle genetics. Genetics 129(4):975-980.

Heinecke JW, Meier KE, Lorenzen JA, Shapiro BM. 1990. A specific requirement for protein
kinase C in activation of the respiratory burst oxidase of fertilization. J Biol Chem
265(14):7717-7720.

Hemmings BA. 1997. Akt signaling: linking membrane events to life and death decisions.
Science 275(5300):628-630.

Hens MD, DeSimone DW. 1995. Molecular analysis and developmental expression of the
focal adhesion kinase pp125FAK in Xenopus laevis. Dev Biol 170(2):274-288.

Hertweck M, Gobel C, Baumeister R. 2004. C. elegans SGK-1 is the critical component in the
Akt/PKB kinase complex to control stress response and life span. Dev Cell 6(4):577-
588.

Hiraoka D, Hori-Oshima S, Fukuhara T, Tachibana K, Okumura E, Kishimoto T. 2004. PDK1
is required for the hormonal signaling pathway leading to meiotic resumption in
starfish oocytes. Dev Biol 276(2):330-336.

Hodgman R, Tay J, Mendez R, Richter JD. 2001. CPEB phosphorylation and cytoplasmic
polyadenylation are catalyzed by the kinase IAK1/Eg2 in maturing mouse oocytes.
Development 128(14):2815-2822.

Homer H. 2011. New insights into the genetic regulation of homologue disjunction in mammalian oocytes. Cytogenet Genome Res 133(2-4):209-222.

Horie K, Takakura K, Taii S, Narimoto K, Noda Y, Nishikawa S, Nakayama H, Fujita J, Mori T. 1991. The expression of c-kit protein during oogenesis and early embryonic development. Biol Reprod 45(4):547-552.

Horner VL, Wolfner MF. 2008. Transitioning from egg to embryo: triggers and mechanisms of egg activation. Dev Dyn 237(3):527-544.

Hoshino Y, Sato E. 2008. Protein kinase B (PKB/Akt) is required for the completion of meiosis in mouse oocytes. Dev Biol 314(1):215-223.

Hoshino Y, Yokoo M, Yoshida N, Sasada H, Matsumoto H, Sato E. 2004. Phosphatidylinositol 3-kinase and Akt participate in the FSH-induced meiotic maturation of mouse oocytes. Mol Reprod Dev 69(1):77-86.

Huang X, Tong JS, Wang ZB, Yang CR, Qi ST, Guo L, Ouyang YC, Quan S, Sun QY, Qi ZQ, Huang RX, Wang HL. 2011. JNK2 participates in spindle assembly during mouse oocyte meiotic maturation. Microsc Microanal 17(2):197-205.

Hudmon A, Schulman H, Kim J, Maltez JM, Tsien RW, Pitt GS. 2005. CaMKII tethers to L-type Ca2+ channels, establishing a local and dedicated integrator of Ca2+ signals for facilitation. J Cell Biol 171(3):537-547.

Huo LJ, Fan HY, Zhong ZS, Chen DY, Schatten H, Sun QY. 2004. Ubiquitin-proteasome pathway modulates mouse oocyte meiotic maturation and fertilization via regulation of MAPK cascade and cyclin B1 degradation. Mech Dev 121(10):1275-1287.

Hupalowska A, Kalaszczynska I, Hoffmann S, Tsurumi C, Kubiak JZ, Polanski Z, Ciemerych MA. 2008. Metaphase I arrest in LT/Sv mouse oocytes involves the spindle assembly checkpoint. Biol Reprod 79(6):1102-1110.

Hyodo-Miura J, Urushiyama S, Nagai S, Nishita M, Ueno N, Shibuya H. 2002. Involvement of NLK and Sox11 in neural induction in Xenopus development. Genes Cells 7(5):487-496.

Inoue D, Sagata N. 2005. The Polo-like kinase Plx1 interacts with and inhibits Myt1 after fertilization of Xenopus eggs. EMBO J 24(5):1057-1067.

Ishida A, Fujisawa H. 1995. Stabilization of calmodulin-dependent protein kinase II through the autoinhibitory domain. J Biol Chem 270(5):2163-2170.

Ishimura A, Lee HS, Bong YS, Saucier C, Mood K, Park EK, Daar IO. 2006. Oncogenic Met receptor induces ectopic structures in Xenopus embryos. Oncogene 25(31):4286-4299.

Ishitani T, Ishitani S, Matsumoto K, Itoh M. 2009. Nemo-like kinase is involved in NGF-induced neurite outgrowth via phosphorylating MAP1B and paxillin. J Neurochem 111(5):1104-1118.

Ito J, Yoon SY, Lee B, Vanderheyden V, Vermassen E, Wojcikiewicz R, Alfandari D, De Smedt H, Parys JB, Fissore RA. 2008. Inositol 1,4,5-trisphosphate receptor 1, a widespread Ca2+ channel, is a novel substrate of polo-like kinase 1 in eggs. Dev Biol 320(2):402-413.

Ito J, Yoshida T, Kasai Y, Wakai T, Parys JB, Fissore RA, Kashiwazaki N. 2010. Phosphorylation of inositol 1,4,5-triphosphate receptor 1 during in vitro maturation of porcine oocytes. Anim Sci J 81(1):34-41.

Itoh T, Hasegawa J, Tsujita K, Kanaho Y, Takenawa T. 2009. The tyrosine kinase Fer is a downstream target of the PLD-PA pathway that regulates cell migration. Sci Signal 2(87):ra52.

Iwasaki T, Koretomo Y, Fukuda T, Paronetto MP, Sette C, Fukami Y, Sato K. 2008. Expression, phosphorylation, and mRNA-binding of heterogeneous nuclear ribonucleoprotein K in Xenopus oocytes, eggs, and early embryos. Dev Growth Differ 50(1):23-40.

Iwasaki T, Sato K, Yoshino K, Itakura S, Kosuge K, Tokmakov AA, Owada K, Yonezawa K, Fukami Y. 2006. Phylogeny of vertebrate Src tyrosine kinases revealed by the epitope region of mAb327. J Biochem 139(3):347-354.

Janicot M, Flores-Riveros JR, Lane MD. 1991. The insulin-like growth factor 1 (IGF-1) receptor is responsible for mediating the effects of insulin, IGF-1, and IGF-2 in Xenopus laevis oocytes. J Biol Chem 266(15):9382-9391.

Jeanmart J, Uytdenhoef P, De Sutter G, Legros F. 1976. Insulin receptor sites as membrane markers during embryonic development. I. Data obtained with unfertilized and fertilized sea urchin eggs. Differentiation 7(1):23-30.

Jelinkova L, Kubelka M. 2006. Neither Aurora B activity nor histone H3 phosphorylation is essential for chromosome condensation during meiotic maturation of porcine oocytes. Biol Reprod 74(5):905-912.

Jelinkova L, Kubelka M, Motlik J, Guerrier P. 1994. Chromatin condensation and histone H1 kinase activity during growth and maturation of rabbit oocytes. Mol Reprod Dev 37(2):210-215.

Jessmon P, Leach RE, Armant DR. 2009. Diverse functions of HBEGF during pregnancy. Mol Reprod Dev 76(12):1116-1127.

Jiang Y, Chan JL, Zong CS, Wang LH. 1996. Effect of tyrosine mutations on the kinase activity and transforming potential of an oncogenic human insulin-like growth factor I receptor. J Biol Chem 271(1):160-167.

Jove R, Hanafusa H. 1987. Cell transformation by the viral src oncogene. Annu Rev Cell Biol 3:31-56.

Kakita M, Murase K, Iwano M, Matsumoto T, Watanabe M, Shiba H, Isogai A, Takayama S. 2007. Two distinct forms of M-locus protein kinase localize to the plasma membrane and interact directly with S-locus receptor kinase to transduce self-incompatibility signaling in Brassica rapa. Plant Cell 19(12):3961-3973.

Kalive M, Faust JJ, Koeneman BA, Capco DG. 2010. Involvement of the PKC family in regulation of early development. Mol Reprod Dev 77(2):95-104.

Kalous J, Kubelka M, Solc P, Susor A, Motlik J. 2009. AKT (protein kinase B) is implicated in meiotic maturation of porcine oocytes. Reproduction 138(4):645-654.

Kalous J, Solc P, Baran V, Kubelka M, Schultz RM, Motlik J. 2006. PKB/AKT is involved in resumption of meiosis in mouse oocytes. Biol Cell 98(2):111-123.

Kamel C, Veno PA, Kinsey WH. 1986. Quantitation of a src-like tyrosine protein kinase during fertilization of the sea urchin egg. Biochem Biophys Res Commun 138(1):349-355.

Kang MK, Han SJ. 2011. Post-transcriptional and post-translational regulation during mouse oocyte maturation. BMB Rep 44(3):147-157.

Kang MG, Kulisz A, Wasserman WJ. 1998. Raf-1 kinase, a potential regulator of intracellular pH in Xenopus oocytes. Biol Cell 90(6-7):477-485.

Karaiskou A, Cayla X, Haccard O, Jessus C, Ozon R. 1998. MPF amplification in Xenopus oocyte extracts depends on a two-step activation of cdc25 phosphatase. Exp Cell Res 244(2):491-500.

Karaiskou A, Lepretre AC, Pahlavan G, Du Pasquier D, Ozon R, Jessus C. 2004. Polo-like kinase confers MPF autoamplification competence to growing Xenopus oocytes. Development 131(7):1543-1552.

Katsu Y, Minshall N, Nagahama Y, Standart N. 1999. Ca2+ is required for phosphorylation of clam p82/CPEB in vitro: implications for dual and independent roles of MAP and Cdc2 kinases. Dev Biol 209(1):186-199.

Kawamura K, Ye Y, Kawamura N, Jing L, Groenen P, Gelpke MS, Rauch R, Hsueh AJ, Tanaka T. 2008. Completion of Meiosis I of preovulatory oocytes and facilitation of preimplantation embryo development by glial cell line-derived neurotrophic factor. Dev Biol 315(1):189-202.

Keady BT, Kuo P, Martinez SE, Yuan L, Hake LE. 2007. MAPK interacts with XGef and is required for CPEB activation during meiosis in Xenopus oocytes. J Cell Sci 120(Pt 6):1093-1103.

Kierszenbaum AL, Rivkin E, Talmor-Cohen A, Shalgi R, Tres LL. 2009. Expression of full-length and truncated Fyn tyrosine kinase transcripts and encoded proteins during spermatogenesis and localization during acrosome biogenesis and fertilization. Mol Reprod Dev 76(9):832-843.

Kikuchi K, Naito K, Daen FP, Izaike Y, Toyoda Y. 1995. Histone H1 kinase activity during in vitro fertilization of pig follicular oocytes matured in vitro. Theriogenology 43(2):523-532.

Kinoshita K, Noetzel TL, Pelletier L, Mechtler K, Drechsel DN, Schwager A, Lee M, Raff JW, Hyman AA. 2005. Aurora A phosphorylation of TACC3/maskin is required for centrosome-dependent microtubule assembly in mitosis. J Cell Biol 170(7):1047-1055.

Kinsey WH. 1995. Differential phosphorylation of a 57-KDa protein tyrosine kinase during egg activation. Biochem Biophys Res Commun 208(1):204-209.

Kinsey WH. 1996. Biphasic activation of Fyn kinase upon fertilization of the sea urchin egg. Dev Biol 174(2):281-287.

Kinsey WH, Shen SS. 2000. Role of the Fyn kinase in calcium release during fertilization of the sea urchin egg. Dev Biol 225(1):253-264.

Kinsey WH, Wu W, Macgregor E. 2003. Activation of Src-family PTK activity at fertilization: role of the SH2 domain. Dev Biol 264(1):255-262.

Klerkx EP, Lazo PA, Askjaer P. 2009. Emerging biological functions of the vaccinia-related kinase (VRK) family. Histol Histopathol 24(6):749-759.

Kokai Y, Wada T, Myers JN, Brown VI, Dobashi K, Cohen J, Hamuro J, Weiner DB, Greene MI. 1988. The role of the neu oncogene product in cell transformation and normal development. Princess Takamatsu Symp 19:45-57.

Kosako H, Gotoh Y, Matsuda S, Ishikawa M, Nishida E. 1992. Xenopus MAP kinase activator is a serine/threonine/tyrosine kinase activated by threonine phosphorylation. EMBO J 11(8):2903-2908.

Kostellow AB, Ma GY, Morrill GA. 1996. Progesterone triggers the rapid activation of phospholipase D in the amphibian oocyte plasma membrane when initiating the G2/M transition. Biochim Biophys Acta 1304(3):263-271.

Kubiak JZ, Weber M, Geraud G, Maro B. 1992. Cell cycle modification during the transitions between meiotic M-phases in mouse oocytes. J Cell Sci 102 (Pt 3):457-467.

Kume S, Endo T, Nishimura Y, Kano K, Naito K. 2007. Porcine SPDYA2 (RINGO A2) stimulates CDC2 activity and accelerates meiotic maturation of porcine oocytes. Biol Reprod 76(3):440-447.

Kuo P, Runge E, Lu X, Hake LE. 2011. XGef influences XRINGO/CDK1 signaling and CPEB
    activation during Xenopus oocyte maturation. Differentiation 81(2):133-140.
Kurokawa M, Sato K, Smyth J, Wu H, Fukami K, Takenawa T, Fissore RA. 2004. Evidence
    that activation of Src family kinase is not required for fertilization-associated
    [Ca2+]i oscillations in mouse eggs. Reproduction 127(4):441-454.
Kurvari V, Snell WJ. 1996. SksC, a fertilization-related protein kinase in Chlamydomonas, is
    expressed throughout the cell cycle and gametogenesis, and a phosphorylated form
    is present in both flagella and cell bodies. Biochem Biophys Res Commun 228(1):45-
    54.
Kurvari V, Zhang Y, Luo Y, Snell WJ. 1996. Molecular cloning of a protein kinase whose
    phosphorylation is regulated by genetic adhesion during Chlamydomonas
    fertilization. Proc Natl Acad Sci U S A 93(1):39-43.
Kushima S, Mammadova G, Mahbub Hasan AK, Fukami Y, Sato K. 2011. Characterization
    of Lipovitellin 2 as a tyrosine-phosphorylated protein in oocytes, eggs and early
    embryos of Xenopus laevis. Zoolog Sci 28(8):550-559.
Lancaster OM, Cullen CF, Ohkura H. 2007. NHK-1 phosphorylates BAF to allow karyosome
    formation in the Drosophila oocyte nucleus. J Cell Biol 179(5):817-824.
LaRosa C, Downs SM. 2006. Stress stimulates AMP-activated protein kinase and meiotic
    resumption in mouse oocytes. Biol Reprod 74(3):585-592.
Ledan E, Polanski Z, Terret ME, Maro B. 2001. Meiotic maturation of the mouse oocyte
    requires an equilibrium between cyclin B synthesis and degradation. Dev Biol
    232(2):400-413.
Lee B, Vermassen E, Yoon SY, Vanderheyden V, Ito J, Alfandari D, De Smedt H, Parys JB,
    Fissore RA. 2006. Phosphorylation of IP3R1 and the regulation of [Ca2+]i responses
    at fertilization: a role for the MAP kinase pathway. Development 133(21):4355-4365.
Lee HC, Tsai JN, Liao PY, Tsai WY, Lin KY, Chuang CC, Sun CK, Chang WC, Tsai HJ. 2007.
    Glycogen synthase kinase 3 alpha and 3 beta have distinct functions during
    cardiogenesis of zebrafish embryo. BMC Dev Biol 7:93.
Lefebvre DL, Charest DL, Yee A, Crawford BJ, Pelech SL. 1999. Characterization of
    fertilization-modulated myelin basic protein kinases from sea star: regulation of
    Mapk. J Cell Biochem 75(2):272-287.
Leiva L, Carrasco D, Taylor A, Veliz M, Gonzalez C, Allende CC, Allende JE. 1987. Casein
    kinase II is a major protein phosphorylating activity in the nuclei of Xenopus laevis
    oocytes. Biochem Int 14(4):707-717.
Lev S, Moreno H, Martinez R, Canoll P, Peles E, Musacchio JM, Plowman GD, Rudy B,
    Schlessinger J. 1995. Protein tyrosine kinase PYK2 involved in Ca(2+)-induced
    regulation of ion channel and MAP kinase functions. Nature 376(6543):737-745.
Levi M, Maro B, Shalgi R. 2010. Fyn kinase is involved in cleavage furrow ingression during
    meiosis and mitosis. Reproduction 140(6):827-834.
Li M, Li S, Yuan J, Wang ZB, Sun SC, Schatten H, Sun QY. 2009. Bub3 is a spindle assembly
    checkpoint protein regulating chromosome segregation during mouse oocyte
    meiosis. PLoS One 4(11):e7701.
Li C, Sun Y, Yi K, Ma Y, Zhang W, Zhou X. 2010. Detection of nerve growth factor (NGF)
    and its specific receptor (TrkA) in ejaculated bovine sperm, and the effects of NGF
    on sperm function. Theriogenology 74(9):1615-1622.
Li X, Earp HS. 1997. Paxillin is tyrosine-phosphorylated by and preferentially associates
    with the calcium-dependent tyrosine kinase in rat liver epithelial cells. J Biol Chem
    272(22):14341-14348.

Liang CG, Su YQ, Fan HY, Schatten H, Sun QY. 2007. Mechanisms regulating oocyte meiotic resumption: roles of mitogen-activated protein kinase. Mol Endocrinol 21(9):2037-2055.

Linher K, Wu D, Li J. 2007. Glial cell line-derived neurotrophic factor: an intraovarian factor that enhances oocyte developmental competence in vitro. Endocrinology 148(9):4292-4301.

Littlepage LE, Ruderman JV. 2002. Identification of a new APC/C recognition domain, the A box, which is required for the Cdh1-dependent destruction of the kinase Aurora-A during mitotic exit. Genes Dev 16(17):2274-2285.

Littlepage LE, Wu H, Andresson T, Deanehan JK, Amundadottir LT, Ruderman JV. 2002. Identification of phosphorylated residues that affect the activity of the mitotic kinase Aurora-A. Proc Natl Acad Sci U S A 99(24):15440-15445.

Liu J, Grimison B, Lewellyn AL, Maller JL. 2006. The anaphase-promoting complex/cyclosome inhibitor Emi2 is essential for meiotic but not mitotic cell cycles. J Biol Chem 281(46):34736-34741.

Liu J, Maller JL. 2005. Calcium elevation at fertilization coordinates phosphorylation of XErp1/Emi2 by Plx1 and CaMK II to release metaphase arrest by cytostatic factor. Curr Biol 15(16):1458-1468.

Liu L, Rajareddy S, Reddy P, Jagarlamudi K, Du C, Shen Y, Guo Y, Boman K, Lundin E, Ottander U, Selstam G, Liu K. 2007. Phosphorylation and inactivation of glycogen synthase kinase-3 by soluble kit ligand in mouse oocytes during early follicular development. J Mol Endocrinol 38(1-2):137-146.

Liu Y, Misamore MJ, Snell WJ. 2010. Membrane fusion triggers rapid degradation of two gamete-specific, fusion-essential proteins in a membrane block to polygamy in Chlamydomonas. Development 137(9):1473-1481.

Lohka MJ, Hayes MK, Maller JL. 1988. Purification of maturation-promoting factor, an intracellular regulator of early mitotic events. Proc Natl Acad Sci U S A 85(9):3009-3013.

Long T, Cailliau K, Beckmann S, Browaeys E, Trolet J, Grevelding CG, Dissous C. 2010. Schistosoma mansoni Polo-like kinase 1: A mitotic kinase with key functions in parasite reproduction. Int J Parasitol 40(9):1075-1086.

Lu N, Guarnieri DJ, Simon MA. 2004. Localization of Tec29 to ring canals is mediated by Src64 and PtdIns(3,4,5)P3-dependent mechanisms. EMBO J 23(5):1089-1100.

Lu Q, Smith GD, Chen DY, Han ZM, Sun QY. 2002. Activation of protein kinase C induces mitogen-activated protein kinase dephosphorylation and pronucleus formation in rat oocytes. Biol Reprod 67(1):64-69.

Luo J, McGinnis LK, Kinsey WH. 2009. Fyn kinase activity is required for normal organization and functional polarity of the mouse oocyte cortex. Mol Reprod Dev 76(9):819-831.

Luria A, Tennenbaum T, Sun QY, Rubinstein S, Breitbart H. 2000. Differential localization of conventional protein kinase C isoforms during mouse oocyte development. Biol Reprod 62(6):1564-1570.

Ma C, Cummings C, Liu XJ. 2003. Biphasic activation of Aurora-A kinase during the meiosis I- meiosis II transition in Xenopus oocytes. Mol Cell Biol 23(5):1703-1716.

Machaca K. 2007. Ca2+ signaling differentiation during oocyte maturation. J Cell Physiol 213(2):331-340.

MacNicol AM, Muslin AJ, Howard EL, Kikuchi A, MacNicol MC, Williams LT. 1995.
    Regulation of Raf-1-dependent signaling during early Xenopus development. Mol
    Cell Biol 15(12):6686-6693.
Madgwick S, Hansen DV, Levasseur M, Jackson PK, Jones KT. 2006. Mouse Emi2 is required
    to enter meiosis II by reestablishing cyclin B1 during interkinesis. J Cell Biol
    174(6):791-801.
Madgwick S, Jones KT. 2007. How eggs arrest at metaphase II: MPF stabilisation plus
    APC/C inhibition equals Cytostatic Factor. Cell Div 2:4.
Madgwick S, Nixon VL, Chang HY, Herbert M, Levasseur M, Jones KT. 2004. Maintenance
    of sister chromatid attachment in mouse eggs through maturation-promoting factor
    activity. Dev Biol 275(1):68-81.
Madgwick S, Levasseur M, Jones KT. 2005. Calmodulin-dependent protein kinase II, and not
    protein kinase C, is sufficient for triggering cell-cycle resumption in mammalian
    eggs. J Cell Sci 118(Pt 17):3849-3859.
Mahbub Hasan AK, Fukami Y, Sato KI. 2011. Gamete membrane microdomains and their
    associated molecules in fertilization signaling. Mol Reprod Dev. 78(10-11):814-830.
Mahbub Hasan AK, Ou Z, Sakakibara K, Hirahara S, Iwasaki T, Sato K, Fukami Y. 2007.
    Characterization of Xenopus egg membrane microdomains containing uroplakin
    Ib/III complex: roles of their molecular interactions for subcellular localization and
    signal transduction. Genes Cells 12(2):251-267.
Mahbub Hasan AK, Sato K, Sakakibara K, Ou Z, Iwasaki T, Ueda Y, Fukami Y. 2005.
    Uroplakin III, a novel Src substrate in Xenopus egg rafts, is a target for sperm
    protease essential for fertilization. Dev Biol 286(2):483-492.
Malcuit C, Knott JG, He C, Wainwright T, Parys JB, Robl JM, Fissore RA. 2005. Fertilization
    and inositol 1,4,5-trisphosphate (IP3)-induced calcium release in type-1 inositol
    1,4,5-trisphosphate receptor down-regulated bovine eggs. Biol Reprod 73(1):2-13.
Maller JL. 1990. Xenopus oocytes and the biochemistry of cell division. Biochemistry
    29(13):3157-3166.
Maller JL, Schwab MS, Roberts BT, Gross SD, Taieb FE, Tunquist BJ. 2001. The pathway of
    MAP kinase mediation of CSF arrest in Xenopus oocytes. Biol Cell 93(1-2):27-33.
Mammadova G, Iwasaki T, Tokmakov AA, Fukami Y, Sato K. 2009. Evidence that
    phosphatidylinositol 3-kinase is involved in sperm-induced tyrosine kinase
    signaling in Xenopus egg fertilization. BMC Dev Biol 9:68.
Marangos P, Verschuren EW, Chen R, Jackson PK, Carroll J. 2007. Prophase I arrest and
    progression to metaphase I in mouse oocytes are controlled by Emi1-dependent
    regulation of APC(Cdh1). J Cell Biol 176(1):65-75.
Marotta V, Guerra A, Sapio MR, Vitale M. 2011. RET/PTC rearrangement in benign and
    malignant thyroid diseases: a clinical standpoint. Eur J Endocrinol 165(4):499-507.
Masui Y, Markert CL. 1971. Cytoplasmic control of nuclear behavior during meiotic
    maturation of frog oocytes. J Exp Zool 177(2):129-145.
Masui Y. 1992. Towards understanding the control of the division cycle in animal cells.
    Biochem Cell Biol 70(10-11):920-945.
Masui Y. 2000. The elusive cytostatic factor in the animal egg. Nat Rev Mol Cell Biol
    1(3):228-232.
Maton G, Lorca T, Girault JA, Ozon R, Jessus C. 2005. Differential regulation of Cdc2 and
    Aurora-A in Xenopus oocytes: a crucial role of phosphatase 2A. J Cell Sci 118(Pt
    11):2485-2494.

Maton G, Thibier C, Castro A, Lorca T, Prigent C, Jessus C. 2003. Cdc2-cyclin B triggers H3 kinase activation of Aurora-A in Xenopus oocytes. J Biol Chem 278(24):21439-21449.

Matsumoto S, Abe Y, Fujibuchi T, Takeuchi T, Kito K, Ueda N, Shigemoto K, Gyo K. 2004. Characterization of a MAPKK-like protein kinase TOPK. Biochem Biophys Res Commun 325(3):997-1004.

Matten W, Daar I, Vande Woude GF. 1994. Protein kinase A acts at multiple points to inhibit Xenopus oocyte maturation. Mol Cell Biol 14(7):4419-4426.

Mayes MA, Sirard MA. 2002. Effect of type 3 and type 4 phosphodiesterase inhibitors on the maintenance of bovine oocytes in meiotic arrest. Biol Reprod 66(1):180-184.

Mazia D. 1937. The release of calcium in *Arbacia* eggs upon fertilization. J Cell and Comp Phys 10(3):291-304.

McGinnis LK, Albertini DF, Kinsey WH. 2007. Localized activation of Src-family protein kinases in the mouse egg. Dev Biol 306(1):241-254.

McGinnis LK, Carroll DJ, Kinsey WH. 2011a. Protein tyrosine kinase signaling during oocyte maturation and fertilization. Mol Reprod Dev. 78(10-11):831-845.

McGinnis LK, Hong X, Christenson LK, Kinsey WH. 2011b. Fer tyrosine kinase is required for germinal vesicle breakdown and meiosis-I in mouse oocytes. Mol Reprod Dev 78(1):33-47.

McGinnis LK, Kinsey WH, Albertini DF. 2009. Functions of Fyn kinase in the completion of meiosis in mouse oocytes. Dev Biol 327(2):280-287.

Mehlmann LM. 2005. Stops and starts in mammalian oocytes: recent advances in understanding the regulation of meiotic arrest and oocyte maturation. Reproduction 130(6):791-799.

Mehlmann LM, Jaffe LA. 2005. SH2 domain-mediated activation of an SRC family kinase is not required to initiate Ca2+ release at fertilization in mouse eggs. Reproduction 129(5):557-564.

Mehlmann LM, Jones TL, Jaffe LA. 2002. Meiotic arrest in the mouse follicle maintained by a Gs protein in the oocyte. Science 297(5585):1343-1345.

Meijer L, Arion D, Golsteyn R, Pines J, Brizuela L, Hunt T, Beach D. 1989a. Cyclin is a component of the sea urchin egg M-phase specific histone H1 kinase. EMBO J 8(8):2275-2282.

Meijer L, Azzi L, Wang JY. 1991. Cyclin B targets p34cdc2 for tyrosine phosphorylation. EMBO J 10(6):1545-1554.

Meijer L, Dostmann W, Genieser HG, Butt E, Jastorff B. 1989b. Starfish oocyte maturation: evidence for a cyclic AMP-dependent inhibitory pathway. Dev Biol 133(1):58-66.

Meijer L, Pondaven P. 1988. Cyclic activation of histone H1 kinase during sea urchin egg mitotic divisions. Exp Cell Res 174(1):116-129.

Mendez R, Hake LE, Andresson T, Littlepage LE, Ruderman JV, Richter JD. 2000. Phosphorylation of CPE binding factor by Eg2 regulates translation of c-mos mRNA. Nature 404(6775):302-307.

Meng XQ, Zheng KG, Yang Y, Jiang MX, Zhang YL, Sun QY, Li YL. 2006. Proline-rich tyrosine kinase2 is involved in F-actin organization during in vitro maturation of rat oocyte. Reproduction 132(6):859-867.

Minshull J, Pines J, Golsteyn R, Standart N, Mackie S, Colman A, Blow J, Ruderman JV, Wu M, Hunt T. 1989. The role of cyclin synthesis, modification and destruction in the control of cell division. J Cell Sci Suppl 12:77-97.

Mitra J, Schultz RM. 1996. Regulation of the acquisition of meiotic competence in the mouse: changes in the subcellular localization of cdc2, cyclin B1, cdc25C and wee1, and in

the concentration of these proteins and their transcripts. J Cell Sci 109 (Pt 9):2407-
2415.

Miyata Y, Nishida E. 1999. Distantly related cousins of MAP kinase: biochemical properties
and possible physiological functions. Biochem Biophys Res Commun 266(2):291-
295.

Miyazaki S, Ito M. 2006. Calcium signals for egg activation in mammals. J Pharmacol Sci
100(5):545-552.

Mochida S, Maslen SL, Skehel M, Hunt T. 2010. Greatwall phosphorylates an inhibitor of
protein phosphatase 2A that is essential for mitosis. Science 330(6011):1670-1673.

Mood K, Friesel R, Daar IO. 2002. SNT1/FRS2 mediates germinal vesicle breakdown
induced by an activated FGF receptor1 in Xenopus oocytes. J Biol Chem
277(36):33196-33204.

Mood K, Saucier C, Bong YS, Lee HS, Park M, Daar IO. 2006. Gab1 is required for cell cycle
transition, cell proliferation, and transformation induced by an oncogenic met
receptor. Mol Biol Cell 17(9):3717-3728.

Moore KL, Kinsey WH. 1994. Identification of an abl-related protein tyrosine kinase in the
cortex of the sea urchin egg: possible role at fertilization. Dev Biol 164(2):444-455.

Mori M, Hara M, Tachibana K, Kishimoto T. 2006. p90Rsk is required for G1 phase arrest in
unfertilized starfish eggs. Development 133(9):1823-1830.

Mori T, Guo MW, Sato E, Baba T, Takasaki S, Mori E. 2000. Molecular and immunological
approaches to mammalian fertilization. J Reprod Immunol 47(2):139-158.

Mori T, Wu GM, Mori E. 1991. Expression of CD4-like structure on murine egg vitelline
membrane and its signal transductive roles through p56lck in fertilization. Am J
Reprod Immunol 26(3):97-103.

Morrison DL, Yee A, Paddon HB, Vilimek D, Aebersold R, Pelech SL. 2000. Regulation of the
meiosis-inhibited protein kinase, a p38(MAPK) isoform, during meiosis and
following fertilization of seastar oocytes. J Biol Chem 275(44):34236-34244.

Murakami MS, Vande Woude GF. 1998. Analysis of the early embryonic cell cycles of
Xenopus; regulation of cell cycle length by Xe-wee1 and Mos. Development
125(2):237-248.

Murase K, Shiba H, Iwano M, Che FS, Watanabe M, Isogai A, Takayama S. 2004. A
membrane-anchored protein kinase involved in Brassica self-incompatibility
signaling. Science 303(5663):1516-1519.

Muslin AJ, MacNicol AM, Williams LT. 1993. Raf-1 protein kinase is important for
progesterone-induced Xenopus oocyte maturation and acts downstream of mos.
Mol Cell Biol 13(7):4197-4202.

Nakaya M, Fukui A, Izumi Y, Akimoto K, Asashima M, Ohno S. 2000. Meiotic maturation
induces animal-vegetal asymmetric distribution of aPKC and ASIP/PAR-3 in
Xenopus oocytes. Development 127(23):5021-5031.

Narasimhan V, Hamill O, Ccrione RA. 1992. The effects of the normal and oncogenic forms
of the neu tyrosine kinase, and the corresponding forms of an immunoglobulin E
receptor/neu tyrosine kinase fusion protein, on Xenopus oocyte maturation. FEBS
Lett 303(2-3):164-168.

Nebreda AR, Martin-Zanca D, Kaplan DR, Parada LF, Santos E. 1991. Induction by NGF of
meiotic maturation of Xenopus oocytes expressing the trk proto-oncogene product.
Science 252(5005):558-561.

Newhall KJ, Criniti AR, Cheah CS, Smith KC, Kafer KE, Burkart AD, McKnight GS. 2006. Dynamic anchoring of PKA is essential during oocyte maturation. Curr Biol 16(3):321-327.

Niault T, Hached K, Sotillo R, Sorger PK, Maro B, Benezra R, Wassmann K. 2007. Changing Mad2 levels affects chromosome segregation and spindle assembly checkpoint control in female mouse meiosis I. PLoS One 2(11):e1165.

Nishi Y, Lin R. 2005. DYRK2 and GSK-3 phosphorylate and promote the timely degradation of OMA-1, a key regulator of the oocyte-to-embryo transition in C. elegans. Dev Biol 288(1):139-149.

Nishiyama T, Ohsumi K, Kishimoto T. 2007a. Phosphorylation of Erp1 by p90rsk is required for cytostatic factor arrest in Xenopus laevis eggs. Nature 446(7139):1096-1099.

Nishiyama T, Yoshizaki N, Kishimoto T, Ohsumi K. 2007b. Transient activation of calcineurin is essential to initiate embryonic development in Xenopus laevis. Nature 449(7160):341-345.

Nishizuka Y. 1984. The role of protein kinase C in cell surface signal transduction and tumour promotion. Nature 308(5961):693-698.

Nishizuka Y. 1986. Studies and perspectives of protein kinase C. Science 233(4761):305-312.

Nishizuka Y. 1988. The molecular heterogeneity of protein kinase C and its implications for cellular regulation. Nature 334(6184):661-665.

Nissley SP, Haskell JF, Sasaki N, De Vroede MA, Rechler MM. 1985. Insulin-like growth factor receptors. J Cell Sci Suppl 3:39-51.

Norris RP, Ratzan WJ, Freudzon M, Mehlmann LM, Krall J, Movsesian MA, Wang H, Ke H, Nikolaev VO, Jaffe LA. 2009. Cyclic GMP from the surrounding somatic cells regulates cyclic AMP and meiosis in the mouse oocyte. Development 136(11):1869-1878.

Nurse P. 1990. Universal control mechanism regulating onset of M-phase. Nature 344(6266):503-508.

Nutt LK, Margolis SS, Jensen M, Herman CE, Dunphy WG, Rathmell JC, Kornbluth S. 2005. Metabolic regulation of oocyte cell death through the CaMKII-mediated phosphorylation of caspase-2. Cell 123(1):89-103.

Nuttinck F, Charpigny G, Mermillod P, Loosfelt H, Meduri G, Freret S, Grimard B, Heyman Y. 2004. Expression of components of the insulin-like growth factor system and gonadotropin receptors in bovine cumulus-oocyte complexes during oocyte maturation. Domest Anim Endocrinol 27(2):179-195.

O'Neill FJ, Gillett J, Foltz KR. 2004. Distinct roles for multiple Src family kinases at fertilization. J Cell Sci 117(Pt 25):6227-6238.

O'Reilly AM, Ballew AC, Miyazawa B, Stocker H, Hafen E, Simon MA. 2006. Csk differentially regulates Src64 during distinct morphological events in Drosophila germ cells. Development 133(14):2627-2638.

Oh JS, Han SJ, Conti M. 2010. Wee1B, Myt1, and Cdc25 function in distinct compartments of the mouse oocyte to control meiotic resumption. J Cell Biol 188(2):199-207.

Okamoto K, Nakajo N, Sagata N. 2002. The existence of two distinct Wee1 isoforms in Xenopus: implications for the developmental regulation of the cell cycle. EMBO J 21(10):2472-2484.

Okamura Y, Myoumoto A, Manabe N, Tanaka N, Okamura H, Fukumoto M. 2001. Protein tyrosine kinase expression in the porcine ovary. Mol Hum Reprod 7(8):723-729.

Okumura E, Fukuhara T, Yoshida H, Hanada Si S, Kozutsumi R, Mori M, Tachibana K, Kishimoto T. 2002. Akt inhibits Myt1 in the signalling pathway that leads to meiotic G2/M-phase transition. Nat Cell Biol 4(2):111-116.

Olds JL, Favit A, Nelson T, Ascoli G, Gerstein A, Cameron M, Cameron L, Lester DS, Rakow T, De Barry J, et al. 1995. Imaging protein kinase C activation in living sea urchin eggs after fertilization. Dev Biol 172(2):675-682.

Ota R, Kotani T, Yamashita M. 2011a. Biochemical characterization of Pumilio1 and Pumilio2 in Xenopus oocytes. J Biol Chem 286(4):2853-2863.

Ota R, Kotani T, Yamashita M. 2011b. Possible involvement of Nemo-like kinase 1 in Xenopus oocyte maturation as a kinase responsible for Pumilio1, Pumilio2, and CPEB phosphorylation. Biochemistry 50(25):5648-5659.

Ou XH, Li S, Xu BZ, Wang ZB, Quan S, Li M, Zhang QH, Ouyang YC, Schatten H, Xing FQ, Sun QY. 2010. p38alpha MAPK is a MTOC-associated protein regulating spindle assembly, spindle length and accurate chromosome segregation during mouse oocyte meiotic maturation. Cell Cycle 9(20):4130-4143.

Pahlavan G, Polanski Z, Kalab P, Golsteyn R, Nigg EA, Maro B. 2000. Characterization of polo-like kinase 1 during meiotic maturation of the mouse oocyte. Dev Biol 220(2):392-400.

Palacios EH, Weiss A. 2004. Function of the Src-family kinases, Lck and Fyn, in T-cell development and activation. Oncogene 23(48):7990-8000.

Palmer A, Gavin AC, Nebreda AR. 1998. A link between MAP kinase and p34(cdc2)/cyclin B during oocyte maturation: p90(rsk) phosphorylates and inactivates the p34(cdc2) inhibitory kinase Myt1. EMBO J 17(17):5037-5047.

Pan J, Snell WJ. 2000. Signal transduction during fertilization in the unicellular green alga, Chlamydomonas. Curr Opin Microbiol 3(6):596-602.

Park M, Dean M, Cooper CS, Schmidt M, O'Brien SJ, Blair DG, Vande Woude GF. 1986. Mechanism of met oncogene activation. Cell 45(6):895-904.

Pascreau G, Delcros JG, Cremet JY, Prigent C, Arlot-Bonnemains Y. 2005. Phosphorylation of maskin by Aurora-A participates in the control of sequential protein synthesis during Xenopus laevis oocyte maturation. J Biol Chem 280(14):13415-13423.

Pascreau G, Delcros JG, Morin N, Prigent C, Arlot-Bonnemains Y. 2008. Aurora-A kinase Ser349 phosphorylation is required during Xenopus laevis oocyte maturation. Dev Biol 317(2):523-530.

Pascreau G, Eckerdt F, Lewellyn AL, Prigent C, Maller JL. 2009. Phosphorylation of p53 is regulated by TPX2-Aurora A in xenopus oocytes. J Biol Chem 284(9):5497-5505.

Pati D, Lohka MJ, Habibi HR. 2000. Time-related effect of GnRH on histone H1 kinase activity in the goldfish follicle-enclosed oocyte. Can J Physiol Pharmacol 78(12):1067-1071.

Pauken CM, Capco DG. 2000. The expression and stage-specific localization of protein kinase C isotypes during mouse preimplantation development. Dev Biol 223(2):411-421.

Pavlok A, Kalab P, Bobak P. 1997. Fertilisation competence of bovine normally matured or aged oocytes derived from different antral follicles: morphology, protein synthesis, H1 and MBP kinase activity. Zygote 5(3):235-246.

Pelech SL, Meijer L, Krebs EG. 1987. Characterization of maturation-activated histone H1 and ribosomal S6 kinases in sea star oocytes. Biochemistry 26(24):7960-7968.

Pelech SL, Tombes RM, Meijer L, Krebs EG. 1988. Activation of myelin basic protein kinases during echinoderm oocyte maturation and egg fertilization. Dev Biol 130(1):28-36.

Perdiguero E, Pillaire MJ, Bodart JF, Hennersdorf F, Frodin M, Duesbery NS, Alonso G, Nebreda AR. 2003. Xp38gamma/SAPK3 promotes meiotic G(2)/M transition in Xenopus oocytes and activates Cdc25C. EMBO J 22(21):5746-5756.

Perry AC, Verlhac MH. 2008. Second meiotic arrest and exit in frogs and mice. EMBO Rep 9(3):246-251.

Pesando D, Pesci-Bardon C, Huitorel P, Girard JP. 1999. Caulerpenyne blocks MBP kinase activation controlling mitosis in sea urchin eggs. Eur J Cell Biol 78(12):903-910.

Peter M, Labbe JC, Doree M, Mandart E. 2002. A new role for Mos in Xenopus oocyte maturation: targeting Myt1 independently of MAPK. Development 129(9):2129-2139.

Philipova R, Whitaker M. 1998. MAP kinase activity increases during mitosis in early sea urchin embryos. J Cell Sci 111 (Pt 17):2497-2505.

Pirino G, Wescott MP, Donovan PJ. 2009. Protein kinase A regulates resumption of meiosis by phosphorylation of Cdc25B in mammalian oocytes. Cell Cycle 8(4):665-670.

Pitetti JL, Torre D, Conne B, Papaioannou MD, Cederroth CR, Xuan S, Kahn R, Parada LF, Vassalli JD, Efstratiadis A, Nef S. 2009. Insulin receptor and IGF1R are not required for oocyte growth, differentiation, and maturation in mice. Sex Dev 3(5):264-272.

Potapova TA, Daum JR, Byrd KS, Gorbsky GJ. 2009. Fine tuning the cell cycle: activation of the Cdk1 inhibitory phosphorylation pathway during mitotic exit. Mol Biol Cell 20(6):1737-1748.

Qian YW, Erikson E, Taieb FE, Maller JL. 2001. The polo-like kinase Plx1 is required for activation of the phosphatase Cdc25C and cyclin B-Cdc2 in Xenopus oocytes. Mol Biol Cell 12(6):1791-1799.

Qu Y, Adler V, Chu T, Platica O, Michl J, Pestka S, Izotova L, Boutjdir M, Pincus MR. 2006. Two dual specificity kinases are preferentially induced by wild-type rather than by oncogenic RAS-P21 in Xenopus oocytes. Front Biosci 11:2420-2427.

Qu Y, Adler V, Izotova L, Pestka S, Bowne W, Michl J, Boutjdir M, Friedman FK, Pincus MR. 2007. The dual-specificity kinases, TOPK and DYRK1A, are critical for oocyte maturation induced by wild-type--but not by oncogenic--ras-p21 protein. Front Biosci 12:5089-5097.

Quadri SK. 2011. Cross talk between focal adhesion kinase and cadherins: Role in regulating endothelial barrier function. Microvasc Res.

Quan HM, Fan HY, Meng XQ, Huo LJ, Chen DY, Schatten H, Yang PM, Sun QY. 2003. Effects of PKC activation on the meiotic maturation, fertilization and early embryonic development of mouse oocytes. Zygote 11(4):329-337.

Rappolee DA, Patel Y, Jacobson K. 1998. Expression of fibroblast growth factor receptors in peri-implantation mouse embryos. Mol Reprod Dev 51(3):254-264.

Rauh NR, Schmidt A, Bormann J, Nigg EA, Mayer TU. 2005. Calcium triggers exit from meiosis II by targeting the APC/C inhibitor XErp1 for degradation. Nature 437(7061):1048-1052.

Reddy P, Shen L, Ren C, Boman K, Lundin E, Ottander U, Lindgren P, Liu YX, Sun QY, Liu K. 2005. Activation of Akt (PKB) and suppression of FKHRL1 in mouse and rat oocytes by stem cell factor during follicular activation and development. Dev Biol 281(2):160-170.

Rentzsch F, Hobmayer B, Holstein TW. 2005. Glycogen synthase kinase 3 has a proapoptotic function in Hydra gametogenesis. Dev Biol 278(1):1-12.

Resh MD. 1998. Fyn, a Src family tyrosine kinase. Int J Biochem Cell Biol 30(11):1159-1162.

Reut TM, Mattan L, Dafna T, Ruth KK, Ruth S. 2007. The role of Src family kinases in egg activation. Dev Biol 312(1):77-89.

Rice A, Parrington J, Jones KT, Swann K. 2000. Mammalian sperm contain a Ca(2+)-sensitive phospholipase C activity that can generate InsP(3) from PIP(2) associated with intracellular organelles. Dev Biol 228(1):125-135.

Rime H, Huchon D, De Smedt V, Thibier C, Galaktionov K, Jessus C, Ozon R. 1994. Microinjection of Cdc25 protein phosphatase into Xenopus prophase oocyte activates MPF and arrests meiosis at metaphase I. Biol Cell 82(1):11-22.

Rime H, Ozon R. 1990. Protein phosphatases are involved in the in vivo activation of histone H1 kinase in mouse oocyte. Dev Biol 141(1):115-122.

Robbie EP, Peterson M, Amaya E, Musci TJ. 1995. Temporal regulation of the Xenopus FGF receptor in development: a translation inhibitory element in the 3' untranslated region. Development 121(6):1775-1785.

Robison GA, Butcher RW, Sutherland EW. 1968. Cyclic AMP. Annu Rev Biochem 37:149-174.

Roelle S, Grosse R, Buech T, Chubanov V, Gudermann T. 2008. Essential role of Pyk2 and Src kinase activation in neuropeptide-induced proliferation of small cell lung cancer cells. Oncogene 27(12):1737-1748.

Rogers E, Bishop JD, Waddle JA, Schumacher JM, Lin R. 2002. The aurora kinase AIR-2 functions in the release of chromosome cohesion in Caenorhabditis elegans meiosis. J Cell Biol 157(2):219-229.

Rogers NT, Hobson E, Pickering S, Lai FA, Braude P, Swann K. 2004. Phospholipase Czeta causes Ca2+ oscillations and parthenogenetic activation of human oocytes. Reproduction 128(6):697-702.

Roghi C, Giet R, Uzbekov R, Morin N, Chartrain I, Le Guellec R, Couturier A, Doree M, Philippe M, Prigent C. 1998. The Xenopus protein kinase pEg2 associates with the centrosome in a cell cycle-dependent manner, binds to the spindle microtubules and is involved in bipolar mitotic spindle assembly. J Cell Sci 111 (Pt 5):557-572.

Rongish BJ, Kinsey WH. 2000. Transient nuclear localization of Fyn kinase during development in zebrafish. Anat Rec 260(2):115-123.

Ruiz EJ, Hunt T, Nebreda AR. 2008. Meiotic inactivation of Xenopus Myt1 by CDK/XRINGO, but not CDK/cyclin, via site-specific phosphorylation. Mol Cell 32(2):210-220.

Ruiz EJ, Vilar M, Nebreda AR. 2010. A two-step inactivation mechanism of Myt1 ensures CDK1/cyclin B activation and meiosis I entry. Curr Biol 20(8):717-723.

Runft LL, Carroll DJ, Gillett J, Giusti AF, O'Neill FJ, Foltz KR. 2004. Identification of a starfish egg PLC-gamma that regulates Ca2+ release at fertilization. Dev Biol 269(1):220-236.

Runft LL, Jaffe LA. 2000. Sperm extract injection into ascidian eggs signals Ca(2+) release by the same pathway as fertilization. Development 127(15):3227-3236.

Runft LL, Jaffe LA, Mehlmann LM. 2002. Egg activation at fertilization: where it all begins. Dev Biol 245(2):237-254.

Sackton KL, Buehner NA, Wolfner MF. 2007. Modulation of MAPK activities during egg activation in Drosophila. Fly (Austin) 1(4):222-227.

Sadler KC, Yuce O, Hamaratoglu F, Verge V, Peaucellier G, Picard A. 2004. MAP kinases regulate unfertilized egg apoptosis and fertilization suppresses death via Ca2+ signaling. Mol Reprod Dev 67(3):366-383.

Sadler SE, Angleson JK, Dsouza M. 2010. IGF-1 receptors in Xenopus laevis ovarian follicle cells support the oocyte maturation response. Biol Reprod 82(3):591-598.

Sakakibara K, Sato K, Yoshino K, Oshiro N, Hirahara S, Mahbub Hasan AK, Iwasaki T, Ueda Y, Iwao Y, Yonezawa K, Fukami Y. 2005. Molecular identification and characterization of Xenopus egg uroplakin III, an egg raft-associated transmembrane protein that is tyrosine-phosphorylated upon fertilization. J Biol Chem 280(15):15029-15037.

Sakamoto I, Takahara K, Yamashita M, Iwao Y. 1998. Changes in cyclin B during oocyte maturation and early embryonic cell cycle in the newt, Cynops pyrrhogaster: requirement of germinal vesicle for MPF activation. Dev Biol 195(1):60-69.

Sakuma M, Onodera H, Suyemitsu T, Yamasu K. 1997. The protein tyrosine kinases of the sea urchin Anthocidaris crassispina. Zoolog Sci 14(6):941-946.

Salas C, Julio-Pieper M, Valladares M, Pommer R, Vega M, Mastronardi C, Kerr B, Ojeda SR, Lara HE, Romero C. 2006. Nerve growth factor-dependent activation of trkA receptors in the human ovary results in synthesis of follicle-stimulating hormone receptors and estrogen secretion. J Clin Endocrinol Metab 91(6):2396-2403.

Sam MR, Elliott BE, Mueller CR. 2007. A novel activating role of SRC and STAT3 on HGF transcription in human breast cancer cells. Mol Cancer 6:69.

Sanghera JS, Charlton LA, Paddon HB, Pelech SL. 1992. Purification and characterization of echinoderm casein kinase II. Regulation by protein kinase C. Biochem J 283 (Pt 3):829-837.

Sanghera JS, Paddon HB, Bader SA, Pelech SL. 1990. Purification and characterization of a maturation-activated myelin basic protein kinase from sea star oocytes. J Biol Chem 265(1):52-57.

Sardon T, Peset I, Petrova B, Vernos I. 2008. Dissecting the role of Aurora A during spindle assembly. EMBO J 27(19):2567-2579.

Sasaki K, Chiba K. 2004. Induction of apoptosis in starfish eggs requires spontaneous inactivation of MAPK (extracellular signal-regulated kinase) followed by activation of p38MAPK. Mol Biol Cell 15(3):1387-1396.

Sato K, Aoto M, Mori K, Akasofu S, Tokmakov AA, Sahara S, Fukami Y. 1996. Purification and characterization of a Src-related p57 protein-tyrosine kinase from Xenopus oocytes. Isolation of an inactive form of the enzyme and its activation and translocation upon fertilization. J Biol Chem 271(22):13250-13257.

Sato K, Fukami Y, Stith BJ. 2006a. Signal transduction pathways leading to Ca2+ release in a vertebrate model system: lessons from Xenopus eggs. Semin Cell Dev Biol 17(2):285-292.

Sato K, Iwao Y, Fujimura T, Tamaki I, Ogawa K, Iwasaki T, Tokmakov AA, Hatano O, Fukami Y. 1999. Evidence for the involvement of a Src-related tyrosine kinase in Xenopus egg activation. Dev Biol 209(2):308-320.

Sato K, Iwasaki T, Hirahara S, Nishihira Y, Fukami Y. 2004. Molecular dissection of egg fertilization signaling with the aid of tyrosine kinase-specific inhibitor and activator strategies. Biochim Biophys Acta 1697(1-2):103-121.

Sato K, Iwasaki T, Ogawa K, Konishi M, Tokmakov AA, Fukami Y. 2002. Low density detergent-insoluble membrane of Xenopus eggs: subcellular microdomain for tyrosine kinase signaling in fertilization. Development 129(4):885-896.

Sato K, Ogawa K, Tokmakov AA, Iwasaki T, Fukami Y. 2001. Hydrogen peroxide induces Src family tyrosine kinase-dependent activation of Xenopus eggs. Dev Growth Differ 43(1):55-72.

Sato K, Tokmakov AA, He CL, Kurokawa M, Iwasaki T, Shirouzu M, Fissore RA, Yokoyama
    S, Fukami Y. 2003. Reconstitution of Src-dependent phospholipase Cgamma
    phosphorylation and transient calcium release by using membrane rafts and cell-
    free extracts from Xenopus eggs. J Biol Chem 278(40):38413-38420.
Sato K, Tokmakov AA, Iwasaki T, Fukami Y. 2000. Tyrosine kinase-dependent activation of
    phospholipase Cgamma is required for calcium transient in Xenopus egg
    fertilization. Dev Biol 224(2):453-469.
Sato K, Yoshino K, Tokmakov AA, Iwasaki T, Yonezawa K, Fukami Y. 2006b. Studying
    fertilization in cell-free extracts: focusing on membrane/lipid raft functions and
    proteomics. Methods Mol Biol 322:395-411.
Saunders CM, Larman MG, Parrington J, Cox LJ, Royse J, Blayney LM, Swann K, Lai FA.
    2002. PLC zeta: a sperm-specific trigger of Ca(2+) oscillations in eggs and embryo
    development. Development 129(15):3533-3544.
Schindler T, Bornmann W, Pellicena P, Miller WT, Clarkson B, Kuriyan J. 2000. Structural
    mechanism for STI-571 inhibition of abelson tyrosine kinase. Science
    289(5486):1938-1942.
Schmitt A, Nebreda AR. 2002b. Inhibition of Xenopus oocyte meiotic maturation by
    catalytically inactive protein kinase A. Proc Natl Acad Sci U S A 99(7):4361-4366.
Schmidt A, Rauh NR, Nigg EA, Mayer TU. 2006. Cytostatic factor: an activity that puts the
    cell cycle on hold. J Cell Sci 119(Pt 7):1213-1218.
Schmitt A, Nebreda AR. 2002a. Signalling pathways in oocyte meiotic maturation. J Cell Sci
    115(Pt 12):2457-2459.
Schober CS, Aydiner F, Booth CJ, Seli E, Reinke V. 2011. The kinase VRK1 is required for
    normal meiotic progression in mammalian oogenesis. Mech Dev 128(3-4):178-190.
Schultz RM, Kopf GS. 1995. Molecular basis of mammalian egg activation. Curr Top Dev
    Biol 30:21-62.
Schwab MS, Kim SH, Terada N, Edfjall C, Kozma SC, Thomas G, Maller JL. 1999. p70(S6K)
    controls selective mRNA translation during oocyte maturation and early
    embryogenesis in Xenopus laevis. Mol Cell Biol 19(4):2485-2494.
Schwartz Y, Ben-Dor I, Navon A, Motro B, Nir U. 1998. Tyrosine phosphorylation of the
    TATA element modulatory factor by the FER nuclear tyrosine kinases. FEBS Lett
    434(3):339-345.
Segawa Y, Suga H, Iwabe N, Oneyama C, Akagi T, Miyata T, Okada M. 2006. Functional
    development of Src tyrosine kinases during evolution from a unicellular ancestor to
    multicellular animals. Proc Natl Acad Sci U S A 103(32):12021-12026.
Sehgal A, Wall DA, Chao MV. 1988. Efficient processing and expression of human nerve
    growth factor receptors in Xenopus laevis oocytes: effects on maturation. Mol Cell
    Biol 8(5):2242-2246.
Sette C, Dolci S, Geremia R, Rossi P. 2000. The role of stem cell factor and of alternative c-kit
    gene products in the establishment, maintenance and function of germ cells. Int J
    Dev Biol 44(6):599-608.
Sette C, Paronetto MP, Barchi M, Bevilacqua A, Geremia R, Rossi P. 2002. Tr-kit-induced
    resumption of the cell cycle in mouse eggs requires activation of a Src-like kinase.
    EMBO J 21(20):5386-5395.
Sharma D, Kinsey WH. 2006. Fertilization triggers localized activation of Src-family protein
    kinases in the zebrafish egg. Dev Biol 295(2):604-614.
Sharma D, Kinsey WH. 2008. Regionalized calcium signaling in zebrafish fertilization. Int J
    Dev Biol 52(5-6):561-570.

Shen SS, Buck WR. 1990. A synthetic peptide of the pseudosubstrate domain of protein kinase C blocks cytoplasmic alkalinization during activation of the sea urchin egg. Dev Biol 140(2):272-280.

Shen SS, Kinsey WH, Lee SJ. 1999. Protein tyrosine kinase-dependent release of intracellular calcium in the sea urchin egg. Dev Growth Differ 41(3):345-355.

Sherr CJ, Roussel MF, Rettenmier CW. 1988. Colony-stimulating factor-1 receptor (c-fms). J Cell Biochem 38(3):179-187.

Shibuya EK, Boulton TG, Cobb MH, Ruderman JV. 1992. Activation of p42 MAP kinase and the release of oocytes from cell cycle arrest. EMBO J 11(11):3963-3975.

Shibuya EK, Morris J, Rapp UR, Ruderman JV. 1996. Activation of the Xenopus oocyte mitogen-activated protein kinase pathway by Mos is independent of Raf. Cell Growth Differ 7(2):235-241.

Shibuya M. 1995. Role of VEGF-flt receptor system in normal and tumor angiogenesis. Adv Cancer Res 67:281-316.

Shilling FM, Carroll DJ, Muslin AJ, Escobedo JA, Williams LT, Jaffe LA. 1994. Evidence for both tyrosine kinase and G-protein-coupled pathways leading to starfish egg activation. Dev Biol 162(2):590-599.

Shirahata S, Rawson C, Loo D, Chang YJ, Barnes D. 1990. ras and neu oncogenes reverse serum inhibition and epidermal growth factor dependence of serum-free mouse embryo cells. J Cell Physiol 144(1):69-76.

Shoji S, Yoshida N, Amanai M, Ohgishi M, Fukui T, Fujimoto S, Nakano Y, Kajikawa E, Perry AC. 2006. Mammalian Emi2 mediates cytostatic arrest and transduces the signal for meiotic exit via Cdc20. EMBO J 25(4):834-845.

Silva AJ, Paylor R, Wehner JM, Tonegawa S. 1992a. Impaired spatial learning in alpha-calcium-calmodulin kinase II mutant mice. Science 257(5067):206-211.

Silva AJ, Stevens CF, Tonegawa S, Wang Y. 1992b. Deficient hippocampal long-term potentiation in alpha-calcium-calmodulin kinase II mutant mice. Science 257(5067):201-206.

Sirard MA. 2001. Resumption of meiosis: mechanism involved in meiotic progression and its relation with developmental competence. Theriogenology 55(6):1241-1254.

Sirard MA, Bilodeau S. 1990a. Effects of granulosa cell co-culture on in-vitro meiotic resumption of bovine oocytes. J Reprod Fertil 89(2):459-465.

Sirard MA, Bilodeau S. 1990b. Granulosa cells inhibit the resumption of meiosis in bovine oocytes in vitro. Biol Reprod 43(5):777-783.

Solc P, Schultz RM, Motlik J. 2010. Prophase I arrest and progression to metaphase I in mouse oocytes: comparison of resumption of meiosis and recovery from G2-arrest in somatic cells. Mol Hum Reprod 16(9):654-664.

Stanford JS, Ruderman JV. 2005. Changes in regulatory phosphorylation of Cdc25C Ser287 and Wee1 Ser549 during normal cell cycle progression and checkpoint arrests. Mol Biol Cell 16(12):5749-5760.

Steele RE. 1985. Two divergent cellular src genes are expressed in Xenopus laevis. Nucleic Acids Res 13(5):1747-1761.

Steele RE, Deng JC, Ghosn CR, Fero JB. 1990. Structure and expression of fyn genes in Xenopus laevis. Oncogene 5(3):369-376.

Steele RE, Irwin MY, Knudsen CL, Collett JW, Fero JB. 1989a. The yes proto-oncogene is present in amphibians and contributes to the maternal RNA pool in the oocyte. Oncogene Res 4(3):223-233.

Steele RE, Unger TF, Mardis MJ, Fero JB. 1989b. The two Xenopus laevis SRC genes are co-expressed and each produces functional pp60src. J Biol Chem 264(18):10649-10653.

Steinhardt RA, Epel D. 1974. Activation of sea-urchin eggs by a calcium ionophore. Proc Natl Acad Sci U S A 71(5):1915-1919.

Stitzel ML, Cheng KC, Seydoux G. 2007. Regulation of MBK-2/Dyrk kinase by dynamic cortical anchoring during the oocyte-to-zygote transition. Curr Biol 17(18):1545-1554.

Stitzel ML, Pellettieri J, Seydoux G. 2006. The C. elegans DYRK Kinase MBK-2 Marks Oocyte Proteins for Degradation in Response to Meiotic Maturation. Curr Biol 16(1):56-62.

Stricker SA. 1999. Comparative biology of calcium signaling during fertilization and egg activation in animals. Dev Biol 211(2):157-176.

Stricker SA. 2009. Interactions between mitogen-activated protein kinase and protein kinase C signaling during oocyte maturation and fertilization in a marine worm. Mol Reprod Dev 76(8):708-721.

Stricker SA. 2011. Potential upstream regulators and downstream targets of AMP-activated kinase signaling during oocyte maturation in a marine worm. Reproduction 142(1):29-39.

Stricker SA, Carroll DJ, Tsui WL. 2010a. Roles of Src family kinase signaling during fertilization and the first cell cycle in the marine protostome worm Cerebratulus. Int J Dev Biol 54(5):787-793.

Stricker SA, Smythe TL. 2006. Differing mechanisms of cAMP- versus seawater-induced oocyte maturation in marine nemertean worms I. The roles of serine/threonine kinases and phosphatases. Mol Reprod Dev 73(12):1578-1590.

Stricker SA, Swiderek L, Nguyen T. 2010b. Stimulators of AMP-activated kinase (AMPK) inhibit seawater- but not cAMP-induced oocyte maturation in a marine worm: Implications for interactions between cAMP and AMPK signaling. Mol Reprod Dev 77(6):497-510.

Sugiura K, Naito K, Iwamori N, Kagii H, Goto S, Ohashi S, Naruoka H, Yada E, Yamanouchi K, Tojo H. 2002. Activation of ribosomal S6 kinase (RSK) during porcine oocyte maturation. Zygote 10(1):31-36.

Sumara I, Vorlaufer E, Stukenberg PT, Kelm O, Redemann N, Nigg EA, Peters JM. 2002. The dissociation of cohesin from chromosomes in prophase is regulated by Polo-like kinase. Mol Cell 9(3):515-525.

Sun CK, Ng KT, Lim ZX, Cheng Q, Lo CM, Poon RT, Man K, Wong N, Fan ST. 2011. Proline-rich tyrosine kinase 2 (Pyk2) promotes cell motility of hepatocellular carcinoma through induction of epithelial to mesenchymal transition. PLoS One 6(4):e18878.

Sun CK, Ng KT, Sun BS, Ho JW, Lee TK, Ng I, Poon RT, Lo CM, Liu CL, Man K, Fan ST. 2007. The significance of proline-rich tyrosine kinase2 (Pyk2) on hepatocellular carcinoma progression and recurrence. Br J Cancer 97(1):50-57.

Sun QY, Rubinstein S, Breitbart H. 1999. MAP kinase activity is downregulated by phorbol ester during mouse oocyte maturation and egg activation in vitro. Mol Reprod Dev 52(3):310-318.

Sun QY, Miao YL, Schatten H. 2009. Towards a new understanding on the regulation of mammalian oocyte meiosis resumption. Cell Cycle 8(17):2741-2747.

Swann K, Igusa Y, Miyazaki S. 1989. Evidence for an inhibitory effect of protein kinase C on G-protein-mediated repetitive calcium transients in hamster eggs. EMBO J 8(12):3711-3718.

Swann K, Saunders CM, Rogers NT, Lai FA. 2006. PLCzeta(zeta): a sperm protein that triggers Ca2+ oscillations and egg activation in mammals. Semin Cell Dev Biol 17(2):264-273.

Taghon MS, Sadler SE. 1994. Insulin-like growth factor 1 receptor-mediated endocytosis in Xenopus laevis oocytes. A role for receptor tyrosine kinase activity. Dev Biol 163(1):66-74.

Talmor A, Kinsey WH, Shalgi R. 1998. Expression and immunolocalization of p59c-fyn tyrosine kinase in rat eggs. Dev Biol 194(1):38-46.

Talmor-Cohen A, Tomashov-Matar R, Eliyahu E, Shapiro R, Shalgi R. 2004a. Are Src family kinases involved in cell cycle resumption in rat eggs? Reproduction 127(4):455-463.

Talmor-Cohen A, Tomashov-Matar R, Tsai WB, Kinsey WH, Shalgi R. 2004b. Fyn kinase-tubulin interaction during meiosis of rat eggs. Reproduction 128(4):387-393.

Tang W, Wu JQ, Guo Y, Hansen DV, Perry JA, Freel CD, Nutt L, Jackson PK, Kornbluth S. 2008. Cdc2 and Mos regulate Emi2 stability to promote the meiosis I-meiosis II transition. Mol Biol Cell 19(8):3536-3543.

Tatone C, Delle Monache S, Francione A, Gioia L, Barboni B, Colonna R. 2003. Ca2+-independent protein kinase C signalling in mouse eggs during the early phases of fertilization. Int J Dev Biol 47(5):327-333.

Thomas SM, Brugge JS. 1997. Cellular functions regulated by Src family kinases. Annu Rev Cell Dev Biol 13:513-609.

Thorpe CJ, Moon RT. 2004. nemo-like kinase is an essential co-activator of Wnt signaling during early zebrafish development. Development 131(12):2899-2909.

Toker A, Newton AC. 2000. Cellular signaling: pivoting around PDK-1. Cell 103(2):185-188.

Tokmakov A, Iwasaki T, Itakura S, Sato K, Shirouzu M, Fukami Y, Yokoyama S. 2005. Regulation of Src kinase activity during Xenopus oocyte maturation. Dev Biol 278(2):289-300.

Tokmakov AA, Sato KI, Iwasaki T, Fukami Y. 2002. Src kinase induces calcium release in Xenopus egg extracts via PLCgamma and IP3-dependent mechanism. Cell Calcium 32(1):11-20.

Tomashov-Matar R, Levi M, Shalgi R. 2008. The involvement of Src family kinases (SFKs) in the events leading to resumption of meiosis. Mol Cell Endocrinol 282(1-2):56-62.

Tomek W, Smiljakovic T. 2005. Activation of Akt (protein kinase B) stimulates metaphase I to metaphase II transition in bovine oocytes. Reproduction 130(4):423-430.

Tong C, Fan HY, Lian L, Li SW, Chen DY, Schatten H, Sun QY. 2002. Polo-like kinase-1 is a pivotal regulator of microtubule assembly during mouse oocyte meiotic maturation, fertilization, and early embryonic mitosis. Biol Reprod 67(2):546-554.

Tonou-Fujimori N, Takahashi M, Onodera H, Kikuta H, Koshida S, Takeda H, Yamasu K. 2002. Expression of the FGF receptor 2 gene (fgfr2) during embryogenesis in the zebrafish Danio rerio. Mech Dev 119 Suppl 1:S173-178.

Tornell J, Billig H, Hillensjo T. 1991. Regulation of oocyte maturation by changes in ovarian levels of cyclic nucleotides. Hum Reprod 6(3):411-422.

Tosca L, Uzbekova S, Chabrolle C, Dupont J. 2007. Possible role of 5'AMP-activated protein kinase in the metformin-mediated arrest of bovine oocytes at the germinal vesicle stage during in vitro maturation. Biol Reprod 77(3):452-465.

Tosuji H, Fusetani N, Seki Y. 2003. Calyculin A causes the activation of histone H1 kinase and condensation of chromosomes in unfertilized sea urchin eggs independently of the maturation-promoting factor. Comp Biochem Physiol C Toxicol Pharmacol 135(4):415-424.

Townley IK, Roux MM, Foltz KR. 2006. Signal transduction at fertilization: the Ca2+ release
pathway in echinoderms and other invertebrate deuterostomes. Semin Cell Dev
Biol 17(2):293-302.

Townley IK, Schuyler E, Parker-Gur M, Foltz KR. 2009. Expression of multiple Src family
kinases in sea urchin eggs and their function in Ca2+ release at fertilization. Dev
Biol 327(2):465-477.

Trepanier CH, Jackson MF, Macdonald JF. 2011. Regulation of NMDA receptors by the
tyrosine kinase Fyn. FEBS J.

Tripathi A, Kumar KV, Chaube SK. 2010. Meiotic cell cycle arrest in mammalian oocytes. J
Cell Physiol 223(3):592-600.

Trounson A, Anderiesz C, Jones G. 2001. Maturation of human oocytes in vitro and their
developmental competence. Reproduction 121(1):51-75.

Truong LD, Shen SS. 2011. Immunohistochemical diagnosis of renal neoplasms. Arch Pathol
Lab Med 135(1):92-109.

Tsafriri A, Chun SY, Zhang R, Hsueh AJ, Conti M. 1996. Oocyte maturation involves
compartmentalization and opposing changes of cAMP levels in follicular somatic
and germ cells: studies using selective phosphodiesterase inhibitors. Dev Biol
178(2):393-402.

Tsai WB, Zhang X, Sharma D, Wu W, Kinsey WH. 2005. Role of Yes kinase during early
zebrafish development. Dev Biol 277(1):129-141.

Uzbekova S, Salhab M, Perreau C, Mermillod P, Dupont J. 2009. Glycogen synthase kinase
3B in bovine oocytes and granulosa cells: possible involvement in meiosis during in
vitro maturation. Reproduction 138(2):235-246.

Vaccari S, Horner K, Mehlmann LM, Conti M. 2008. Generation of mouse oocytes defective
in cAMP synthesis and degradation: endogenous cyclic AMP is essential for
meiotic arrest. Dev Biol 316(1):124-134.

Valbuena A, Sanz-Garcia M, Lopez-Sanchez I, Vega FM, Lazo PA. 2011. Roles of VRK1 as a
new player in the control of biological processes required for cell division. Cell
Signal 23(8):1267-1272.

Vigneron S, Brioudes E, Burgess A, Labbe JC, Lorca T, Castro A. 2009. Greatwall maintains
mitosis through regulation of PP2A. EMBO J 28(18):2786-2793.

Villa-Diaz LG, Miyano T. 2004. Activation of p38 MAPK during porcine oocyte maturation.
Biol Reprod 71(2):691-696.

Vincent C, Cheek TR, Johnson MH. 1992. Cell cycle progression of parthenogenetically
activated mouse oocytes to interphase is dependent on the level of internal calcium.
J Cell Sci 103 (Pt 2):389-396.

Viveiros MM, O'Brien M, Eppig JJ. 2004. Protein kinase C activity regulates the onset of
anaphase I in mouse oocytes. Biol Reprod 71(5):1525-1532.

Viveiros MM, O'Brien M, Wigglesworth K, Eppig JJ. 2003. Characterization of protein kinase
C-delta in mouse oocytes throughout meiotic maturation and following egg
activation. Biol Reprod 69(5):1494-1499.

Walker G, Burgess D, Kinsey WH. 1996. Fertilization promotes selective association of the
Abl [correction of AbI] kinase with the egg cytoskeleton. Eur J Cell Biol 70(2):165-
171.

Walter SA, Guadagno SN, Ferrell JE, Jr. 2000. Activation of Wee1 by p42 MAPK in vitro and
in cycling xenopus egg extracts. Mol Biol Cell 11(3):887-896.

Wang JY, Prywes R, Baltimore D. 1983. Structure and function of the Abelson murine
leukemia virus transforming gene. Prog Clin Biol Res 119:57-63.

Wang Q, Snell WJ. 2003. Flagellar adhesion between mating type plus and mating type minus gametes activates a flagellar protein-tyrosine kinase during fertilization in Chlamydomonas. J Biol Chem 278(35):32936-32942.

Wassmann K, Niault T, Maro B. 2003. Metaphase I arrest upon activation of the Mad2-dependent spindle checkpoint in mouse oocytes. Curr Biol 13(18):1596-1608.

Westmark CJ, Ghose R, Huber PW. 2002. Phosphorylation of Xenopus transcription factor IIIA by an oocyte protein kinase CK2. Biochem J 362(Pt 2):375-382.

Wessel GM, Brooks JM, Green E, Haley S, Voronina E, Wong J, Zaydfudim V, Conner S. 2001. The biology of cortical granules. Int Rev Cytol 209:117-206.

Wessel GM, Wong JL. 2009. Cell surface changes in the egg at fertilization. Mol Reprod Dev 76(10):942-953.

Whitaker M. 2006. Calcium at fertilization and in early development. Physiol Rev 86(1):25-88.

Wianny F, Tavares A, Evans MJ, Glover DM, Zernicka-Goetz M. 1998. Mouse polo-like kinase 1 associates with the acentriolar spindle poles, meiotic chromosomes and spindle midzone during oocyte maturation. Chromosoma 107(6-7):430-439.

Wojcik EJ, Sharifpoor S, Miller NA, Wright TG, Watering R, Tremblay EA, Swan K, Mueller CR, Elliott BE. 2006. A novel activating function of c-Src and Stat3 on HGF transcription in mammary carcinoma cells. Oncogene 25(19):2773-2784.

Wu JQ, Hansen DV, Guo Y, Wang MZ, Tang W, Freel CD, Tung JJ, Jackson PK, Kornbluth S. 2007a. Control of Emi2 activity and stability through Mos-mediated recruitment of PP2A. Proc Natl Acad Sci U S A 104(42):16564-16569.

Wu JQ, Kornbluth S. 2008. Across the meiotic divide - CSF activity in the post-Emi2/XErp1 era. J Cell Sci 121(Pt 21):3509-3514.

Wu Q, Guo Y, Yamada A, Perry JA, Wang MZ, Araki M, Freel CD, Tung JJ, Tang W, Margolis SS, Jackson PK, Yamano H, Asano M, Kornbluth S. 2007b. A role for Cdc2- and PP2A-mediated regulation of Emi2 in the maintenance of CSF arrest. Curr Biol 17(3):213-224.

Wu W, Kinsey WH. 2000. Fertilization triggers activation of Fyn kinase in the zebrafish egg. Int J Dev Biol 44(8):837-841.

Wu W, Kinsey WH. 2002. Role of PTPase(s) in regulating Fyn kinase at fertilization of the zebrafish egg. Dev Biol 247(2):286-294.

Wu W, Kinsey WH. 2004. Detection and measurement of membrane-bound protein tyrosine kinases in the zebrafish egg. Methods Mol Biol 253:273-283.

Wu XR, Kong XP, Pellicer A, Kreibich G, Sun TT. 2009. Uroplakins in urothelial biology, function, and disease. Kidney Int 75(11):1153-1165.

Wu XR, Lin JH, Walz T, Haner M, Yu J, Aebi U, Sun TT. 1994. Mammalian uroplakins. A group of highly conserved urothelial differentiation-related membrane proteins. J Biol Chem 269(18):13716-13724.

Wu XR, Manabe M, Yu J, Sun TT. 1990. Large scale purification and immunolocalization of bovine uroplakins I, II, and III. Molecular markers of urothelial differentiation. J Biol Chem 265(31):19170-19179.

Xie Z, Sanada K, Samuels BA, Shih H, Tsai LH. 2003. Serine 732 phosphorylation of FAK by Cdk5 is important for microtubule organization, nuclear movement, and neuronal migration. Cell 114(4):469-482.

Xiong B, Sun SC, Lin SL, Li M, Xu BZ, OuYang YC, Hou Y, Chen DY, Sun QY. 2008. Involvement of Polo-like kinase 1 in MEK1/2-regulated spindle formation during mouse oocyte meiosis. Cell Cycle 7(12):1804-1809.

Xu X, Sun TT, Gupta PK, Zhang P, Nasuti JF. 2001. Uroplakin as a marker for typing
metastatic transitional cell carcinoma on fine-needle aspiration specimens. Cancer
93(3):216-221.

Yamamoto TM, Blake-Hodek K, Williams BC, Lewellyn AL, Goldberg ML, Maller JL. 2011.
Regulation of Greatwall kinase during Xenopus oocyte maturation. Mol Biol Cell
22(13):2157-2164.

Yamashita M, Fukada S, Yoshikuni M, Bulet P, Hirai T, Yamaguchi A, Yasuda H, Ohba Y,
Nagahama Y. 1992. M-phase-specific histone H1 kinase in fish oocytes. Purification,
components and biochemical properties. Eur J Biochem 205(2):537-543.

Yang D, Hinton SD, Eckberg WR. 2004. Regulation of cleavage by protein kinase C in
Chaetopterus. Mol Reprod Dev 69(3):308-315.

Yang KT, Li SK, Chang CC, Tang CJ, Lin YN, Lee SC, Tang TK. 2010b. Aurora-C kinase
deficiency causes cytokinesis failure in meiosis I and production of large polyploid
oocytes in mice. Mol Biol Cell 21(14):2371-2383.

Yao LJ, Fan HY, Tong C, Chen DY, Schatten H, Sun QY. 2003. Polo-like kinase-1 in porcine
oocyte meiotic maturation, fertilization and early embryonic mitosis. Cell Mol Biol
(Noisy-le-grand) 49(3):399-405.

Yim DL, Opresko LK, Wiley HS, Nuccitelli R. 1994. Highly polarized EGF receptor tyrosine
kinase activity initiates egg activation in Xenopus. Dev Biol 162(1):41-55.

Yu B, Wang Y, Liu Y, Li X, Wu D, Zong Z, Zhang J, Yu D. 2005. Protein kinase A regulates
cell cycle progression of mouse fertilized eggs by means of MPF. Dev Dyn
232(1):98-105.

Yu BZ, Zheng J, Yu AM, Shi XY, Liu Y, Wu DD, Fu W, Yang J. 2004. Effects of protein kinase
C on M-phase promoting factor in early development of fertilized mouse eggs. Cell
Biochem Funct 22(5):291-298.

Yu Y, Halet G, Lai FA, Swann K. 2008. Regulation of diacylglycerol production and protein
kinase C stimulation during sperm- and PLCzeta-mediated mouse egg activation.
Biol Cell 100(11):633-643.

Zernicka-Goetz M. 1991. Spontaneous and induced activation of rat oocytes. Mol Reprod
Dev 28(2):169-176.

Zhang K, Hansen PJ, Ealy AD. 2010a. Fibroblast growth factor 10 enhances bovine oocyte
maturation and developmental competence in vitro. Reproduction 140(6):815-826.

Zhang L, Liang Y, Liu Y, Xiong CL. 2010b. The role of brain-derived neurotrophic factor in
mouse oocyte maturation in vitro involves activation of protein kinase B.
Theriogenology 73(8):1096-1103.

Zhang WL, Huitorel P, Geneviere AM, Chiri S, Ciapa B. 2006. Inactivation of MAPK in
mature oocytes triggers progression into mitosis via a Ca2+ -dependent pathway
but without completion of S phase. J Cell Sci 119(Pt 17):3491-3501.

Zhang M, Ouyang H, Xia G. 2009. The signal pathway of gonadotrophins-induced
mammalian oocyte meiotic resumption. Mol Hum Reprod 15(7):399-409.

Zhang X, Wright CV, Hanks SK. 1995. Cloning of a Xenopus laevis cDNA encoding focal
adhesion kinase (FAK) and expression during early development. Gene 160(2):219-
222.

Zhang Y, Luo Y, Emmett K, Snell WJ. 1996. Cell adhesion-dependent inactivation of a
soluble protein kinase during fertilization in Chlamydomonas. Mol Biol Cell
7(4):515-527.

Zhang Y, Zhang Z, Xu XY, Li XS, Yu M, Yu AM, Zong ZH, Yu BZ. 2008. Protein kinase A modulates Cdc25B activity during meiotic resumption of mouse oocytes. Dev Dyn 237(12):3777-3786.

Zhao J, Taverne MA, Van Der Weijden GC, Bevers MM, Van Den Hurk R. 2001. Insulin-like growth factor-I (IGF-I) stimulates the development of cultured rat pre-antral follicles. Mol Reprod Dev 58(3):287-296.

Zhao Y, Haccard O, Wang R, Yu J, Kuang J, Jessus C, Goldberg ML. 2008. Roles of Greatwall kinase in the regulation of cdc25 phosphatase. Mol Biol Cell 19(4):1317-1327.

Zhong W, Sun T, Wang QT, Wang Y, Xie Y, Johnson A, Leach R, Puscheck EE, Rappolee DA. 2004. SAPKgamma/JNK1 and SAPKalpha/JNK2 mRNA transcripts are expressed in early gestation human placenta and mouse eggs, preimplantation embryos, and trophoblast stem cells. Fertil Steril 82 Suppl 3:1140-1148.

Zhou RP, Oskarsson M, Paules RS, Schulz N, Cleveland D, Vande Woude GF. 1991. Ability of the c-mos product to associate with and phosphorylate tubulin. Science 251(4994):671-675.

8

Morphometry as a Method of Studying Adaptive Regulation of Embryogenesis in Polluted Environments

Elena A. Severtsova, David R. Aguillón Gutiérrez
and Aleksey S. Severtsov
Lomonosov Moscow State University
Russian Federation

1. Introduction

Amphibians- one of the most convenient natural objects to study the effect of various pollutants (Dawson et al, 1985). They are very sensitive to changes in the environment. Larval and adult specimens have a high permeable skin and their life cycle takes place in water and in terrestrial ecosystems, depending on the stage of the life cycle; they have different types of food; their special reproduction allows monitoring of development and the quantity to conduct representative research (Cooke, 1974; Greenhouse, 1976; Sparling et al, 2001).

In addition, some amphibian species are laboratory animals with a well-studied biology and genetics, for example the clawed frog (*Xenopus laevis*), which is the biological model in the FETAX project (Fort et al, 2004; Morgan et al, 1996) and *Xenopus* metamorphosis assay (XEMA) (Opitz et al, 2005) which conducted test of various chemical pollutants. Applied nature of these programs involves testing the impact on the embryogenesis of each substance alone (FETAX, 1991), although in nature, in the embryo affects a complex of pollutants of the lake. Therefore, in recent years, within the confines of FETAX system, are conducted analyzes of the synergistic complex of pollutants (Orton et al., 2006; Ettler et al., 2008), which allows to evaluate the natural biotoxicity and anthropogenic pollutants.

However, being just test models, laboratory animals because of its unsuitability to the full range of factors in nature, can not be regarded as indicators of ecosystem pollution. Thus, the experimental exposure to pesticides (particularly DDT [dichlorodiphenyltrichloroethane] and HCH [Hexachlorocyclohexane]) on eggs of Clawed and Moor frogs, showed significantly more sensitive reaction in *Xenopus laevis* than natural populations of *Rana arvalis*, making an incorrect comparative analysis (Voronova et al, 1983).

Therefore, to study the effects of pollution in natural habitats are more informative the representatives of natural populations (Stroganov, 1971). Field studies are conducted on widely distributed amphibian species: *Rana pipens* (Allran & Karasov, 2000, 2001), *Rana arvalis* (Andren et al, 1989), *Rana temporaria* (Leontieva & Semenov, 1997; Dunson et al, 1992; Johansson et al, 2001), *Bufo americanus* (Hecnar, 1995) and other species which by 1992 had already been tested 211 types of pollutants in 45 amphibian species (Hall & Henry, 1992). Among these works dominates the studies on tadpoles or adult animals (Vershin, 1997; Freda, 1986; Horne & Dunson, 1995). Typically, field studies are limited to the proportion of those who died and/or are abnormal embryos (Beattie et al, 1991). This is certainly an important indicator of overall population status, but, in our view, insufficient because they do not reflect the actual adaptation processes in natural populations under the action of a new environmental evolutionary factor (Vershinin, 1997; Severtsova, 2002), and is an indicator of only a fraction of genotypes who dropped out of population diversity. More informative is the method of morphometric evaluation of the nature of variability (Severtsova & Severtsov, 2005).

2. Morphometric analysis of early development

The object of developmental studies is the development of the individual, whereas the subject of population studies - the aggregate of individuals united in a population. These two lines of research have traditionally developed independently. In the analysis of specimens is estimated the condition of the body at a certain stage of development and its dynamics in time and space. The study of ontogeny itself, which often is limited to embryonic development and identify differences between individuals.

In the study of population is estimated the populations status in a determinate moment and also its dynamics in time and space, it means the assessment of population dynamics and detection of inter and intrapopulation differences. This is usually not taken into account that each individual is ontogeny, and the population estimate is a slice of the trajectories of individual development. In those cases where this is taken into account the possible ontogenetic change as something that interferes with population estimate, and the problem usually comes down to their elimination by the analysis of individuals of the same age. At the same time as a special analysis of individual ontogeny is not only necessary for the correct assessment of populations and their dynamics in time and space, but also for understanding the mechanisms of the stability of population processes.

Stabilized development is one of the most common characteristics of the developing organism. It was shown that high stability is maintained on the basis of genetic coadaptation under optimal conditions of development (Zakharov, 1989; Moller & Swaddle, 1997). A growing number of studies of developmental stability and incessant debate about the significance of such studies to characterize the state of the population determines the need to assess the possibility of using developmental stability as a measure of environmental stress for monitoring populations. In this case, the most convenient is the morphometric method based on an evaluation of morphological variability. Themselves morphometric methods for assessing variability are used frequently, including to assess the variability of synanthropic populations of animals and plants (see "Intraspecific variation ..." 1980; "Animals in ...", 1990, "Structure and functional role ...", 2001).

Assessment of variability in the earliest stages of development in this aspect was not carried out, although the method of study of their variability exists. Apparently, the quantitative study of the variability of morphogenetic traits was a Gurwitsch's idea (Gurwitsch, 1922: in Cherdantsev, 2003). The idea is that by studying the relation between the average and the variance of quantitative traits of the embryo or its parts, you can get a fairly accurate idea of the nature of interactions corresponding to the morphogenetic field. Specific studies on the Gurwitsch's planned program are very few (Cherdantsev & Scobeyeva, 1994; Glukhova & Cherdantsev, 1999; Severtsova, 2002), although this method can be used to study the variability of embryogenesis during the development in different contexts, including the anthropogenic pollution condition.

The model of some-like research is the most convenient anuran eggs at the stage of mid-late gastrula (Cherdantsev & Scobeyeva, 1994), since the earlier stages are not enough informative, and the analysis of later stages must take into account a large number of complex processes, including those related to organogenesis. Gastrula stage is best studied in terms of morphogenetic processes and the genetic basis of regulation of these processes (Beetschen, 2001). In particular, revealed that in the analysis of morphological variability of gastrulation, in some amphibian species (*Rana temporaria, Rana lessonae* and *Pelobates fuscus*) there is huge value of the coefficients of variation of characters in the embryos that are genetically homogeneous (from the same clutch of eggs), developing in quite the same conditions, and the most importantly, developing normally in terms of the final result of embryonic development. It may be more than 50%, ie almost an order of magnitude higher than normal variability of quantitative traits in definitive stages of development of the phenotype (see Falconer, 1981). In addition, is shown that this is the stage in early development of amphibians most sensitive to environmental effects (Saber & Dunson, 1978; Beattie et al., 1992; Severtsova, 2005), that allows us to estimate the variability in the development stages, when the abnormality of the morphogenetic processes have not yet discernible to the naked eye.

For morphometric analysis, we selected the most popular anuran species in Moscow: Common frog (*Rana temporaria* L.), Moor frog (*Rana arvalis* Nills.) and the Marsh frog (*Pelophylax ridibundus* Pallas). In fixed eggs of these species were removed all the membranes, including the vitelline, made by standard embryological methods, divided sagittal and measured under a binocular microscope with eyepiece micrometer (accurate to 1 division of ocular micrometer; 20 divisions of ocular micrometer range – 1mm) following features (Fig. 1A): *D1* - total diameter of the gastrula, *D2* - diameter of yolk plug, *ArthG* - roof height of gastrocele, *vh* - the maximum height of the yolk duct, *LbalD* - the depth of the dorsal lip of blastopore screwing, *LbalV* - the depth of the ventral lip of blastopore screwing, *G* - the distance between the deepest dorsal screwing and ventral blastopore lip, *preg* - the distance between the cavities of gastrocele and blastocele, *ArthB* - blastocoel roof height. The choice of these features provides a fairly complete description of gastrulation - one of the most important stages of embryogenesis (Slack et al., 1992; Gilbert, 1993; Cherdantsev, 2003). Such signs as *LbalD, LbalV* and *G* are key indicators of advanced gastrulation processes, as they reflect the extent and nature of the blastopore lip formation. *ArthG* - roof height of gastrocele - describes the process of forming chordomesoderm in the investigates development stage. Signs *preg* and *ArthB* are closely related to the previous stage of embryogenesis - blastulation, as they mark the location of the reducing blastocoel, and thus may serve, in conjunction with *ArthG* an index of "looseness" of the embryo. "Loose", commonly referred to the gastrula with non-dense intercellular contacts, or with abnormally large cells that do not allow because of the physical features of morphogenesis to form cavities or even to continue the further development (Cherdantsev, 2003).

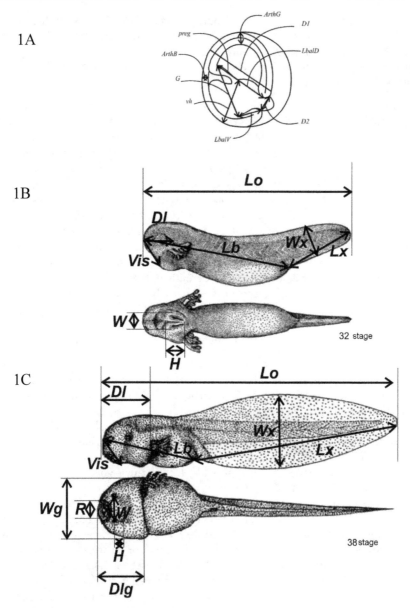

Fig. 1. Research of morphometrical signs. See notation in text.

Assessment of variability is possible at later stages of embryogenesis, for example at the stage of hatching larvae from the eggs (Fig. 1B). This stage of development is longer in comparison with gastrulation, but also allows to evaluate representatively the intra-population and intra-egg mass variability. Signs for the morphometric analysis are indicators such as: *Lo*- tadpole total length (from the outermost point of the snout to the tip of the tail plate); Lx - the length of the tail plate (from the cloaca to the tip of the tail plate); Dl - the length of the muzzle (from the base of the gill filaments to the outermost point of the muzzle); Vis - the height of the muzzle (from oral sucker to the outermost point of the muzzle); W - width of oral sucker, H - height of the oral sucker; G - the number of gill filaments, summarized on the left and right sides of the embryo. In the analysis of development assessment is possible also include indices that reflect the proportional development of larvae. Correlation Lx / Lo - index of proportion of the tail plate's size relative to the total body length. Correlation Dl / Vis is a measure of proportionality of head structures development. We considered the area of the oral sucker $(W * H)$ / 2, as an indicator of the reliability of attachment when the tadpole had just hatched from the egg, where it is some time before moving on to the stage of free swimming. An indicator of the symmetry of the gill filaments development was like a asymmetry coefficient . At tadpole stages of development, including at the stage prior to the commencement of metamorphosis, the number of features can be extended (Fig. 1C), introducing measures such as the maximum width of the tail plate (Wx), head width (Wg) - the distance measured along the line of gill slits in the tadpole; the length of the head (Dlg) – the distance measured from the ventral surface of the tadpole through gills slits to the outermost point of the snout; sucker width (W) – the distance between the sucker rollers; sucker heigh (H) in the sagittal direction; mouth opening width (R) - the distance between the corners of the mouth opening; the distance between the pupils of the eyes (eye). Evaluation of these indicators reflect the two most important processes: growth and differentiation. The same measurements can be made even in vivo, without damaging the tadpoles and without exerting a strong influence on the course of development. Last is the most valuable in studies aimed to studying the dynamics of development, including passing under the influence of various pollutants.

3. Variability and correlation of early development

One of the main parameters, with which operates the morphometric analysis - is the concept of "variability". From a biological point of view, it implies the diversity of individuals in the study group. Expressed mathematically in such quantities as the variance, standard deviation or coefficient of variation. The latter parameter is used in our studies because, as a dimensionless quantity allows the comparison of mixed-signs. No less important is the concept of "correlation", i.e. consistency of the emerging structures of the developing organism. Mathematically, the nature and strength of this mutual influence is expressed through the correlation coefficient calculated for the signs in the role of characteristics of structures.

The results of morphometric research of gastrula stage in three species of anurans: Common frog (Rana temporaria L.), Moor frog (Rana arvalis Nills.) and Marsh frog (Pelophylax ridibundus Pallas, «Rana ridibunda in fig. 2»), inhabiting the territory of Moscow (Russia) show the ratio between the variability and the correlation of the processes of morphogenesis, as a response to environmental degradation (Severtsova, Severtsov, 2005, 2007). At the same time such a change occurs in different areas in different ways. In the district of Ramenki, in the Moor frog is observed a high variability and low correlation of morphogenetic processes (Fig. 2).

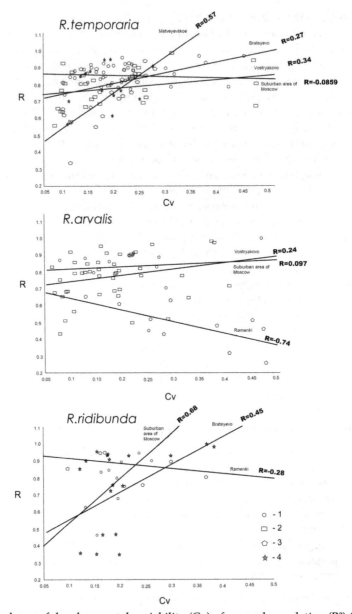

Fig. 2. Dependence of developmental variability (Cv) of general correlation (R^2) for clutches from: 1 Suburban areas of Moscow; 2 – Vostryakovo area; 3 – Ramenky area or Matveyevskoe area; 4 – Brateyevo area.

In the first place at the end of gastrulation comes the preparation for the neurulation processes, up to heterochronies, when the yolk plug has not disappeared yet, but it's not just the formation of the neural plate, but the rise of the neural crest (Fig. 3).

 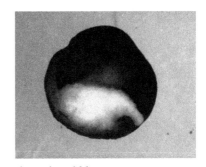

Fig. 3. Embryos of Common frog in 22nd stage with unclosed blastopore.

However, this accelerated development of high variability and miscorrelation leads to high mortality (about 5 times more in Moscow, compared with the population living outside the city of Moscow, which is characterized by low levels of anthropogenic pollution). In the Marsh frog in the district of Ramenki, the low overall correlation of development, compensates the large number of weakly inter-dependent processes. In the district Brateevo, in the Marsh frog is observed a combination of low variability and a high correlation (see Fig. 2), which allows embryos to grow more consistently and leads to low mortality. A somewhat different picture from Moor frog, living in the same area: with very low overall correlation of development, the key role is played by the ventral lip of blastopore screwing. A similar pattern was observed in embryos of the Common frog, which clutches were fixed from the pond, located 5 km downstream of the Setun river - in the district of Matveevskoe. In the Common frog gastrula, with high variability of the gastrulation processes, is observed a significant increase of the correlation among morphogenetic processes, ie with a correlation coefficient of less than 0.6 in the relationship between developing structures. This ensures the integrity of the eggs with the predominant role of the process of screwing in the ventral lip of blastopore.

The development of Common and moor frogs in the district of Vostryakovo varied. In the Common frog with a relatively low correlation of the general development, as well as in the district of Brateevo, an important role played the processes of changes in the diameter of eggs associated with the beginning of the neural elongation of the embryo. In the moor frog with a very high proportion of significance, but not high correlation coefficients between the studied traits is observed as a characteristic of gastrulation the predominance of the role on the ventral lip of blastopore invagination. In the eggs in gastrulation from nests near Moscow also significantly affects the overall high correlation of development in conjunction with their low variability (see Fig. 2). However, for the eggs of the Common frog populations near Moscow, the dimensional characteristics are closely related to the amount of yolk in the eggs. To Moor and Marsh frogs, fixed at 17th stage, the predominant role is played by processes of the screwing of the ventral lip of blastopore.

As can be seen from the above analysis the changing nature of morphogenetic processes during early development in anurans is due to changes in the correlation of variability and correlation of the emerging structures. This is manifested by increasing the total number of interrelated features that allows to keep and maintain the integrity of the embryo and thus provide a clearer differentiation of developing structures, despite their high variability. The second possible way of regulating early ontogenesis is realized in the amplification of correlative linked structures formed precisely at the stage of investigation, coupled with the increasing variability of morphogenetic processes. As a result, the development of this structure is accelerated. An extreme version of such correlation of variability and correlation of development is the emergence of heterochronies, i.e. situation where the structures of the embryo is formed, advancing the general process of development. Our data show that the occurrence of heterochrony changes may occur as a result of a sharp drop of the general correlation of development, against the backdrop of significantly increased variability in the emerging structures. This leads to a significant increase in mortality and in the case of the population of the Ramenki district to its extinction, aside from the small number of this population.

No less important is the conclusion that, in spite of interspecific ecological and morphological differences between Common and Marsh frogs, the mechanisms of regulation of early development occur in a general scheme. This indicates a non-specific reactions of early embryogenesis. The differences in overall mortality between them, apparently due to a later spawning, characteristic of the Marsh frog. Its embryogenesis is under more favorable terms than those with Common frogs against a background of lower concentrations of pollutants.

4. Adaptation of embryogenesis to a new environmental evolutionary factor – Is it possible?

Modern concepts are based on the division of any sign of variability into three components: genetic, paratypic (environmental) and epigenetic. As the name implies, genetic variability is formed due to the work of the genotype, paratypic - is the result of the impact of external environmental factors on the ontogeny, and epigenetic variability appears as a result of interactions among cells, tissues, organs developing, so changes in one element leads to changes of many interacting structures and functions (Cherdantsev & Scobeyeva, 1994; Horder, 2006, 2008). As shown by the few experimental data, the proportion of the genetic components of the total variability in the early stages of embryogenesis is not large and varies considerably in the investigated characteristics, depending on experimental conditions (Surova, 1988a; Travis, 1980, 1981; Berven, 1982; Berven & Gill, 1983).

For example, in tadpoles of *Rana silvatica* of populations of the plains of Maryland, the mountains of Virginia and tundra of northern Canada, the heritability of growth rate is 0.08, 0.58 and 0.27, respectively (Berven & Gill, 1983). Our own calculation of the coefficient of heritability (in the broadest sense of the term) of a number of morphometric characters of Common frog embryos at the stage of hatching, showed that the average proportion of the genetic component is about 10%. This allows us to consider the overall variability of the early embryonic stages as a combination of paratypic and epigenetic components of variability and to assess their contribution to the variability in development. To do this, we performed an experiment using the cross-combinations of two environmental factors: water chemistry and density of embryos per unit volume of water. At the same time as the other parameters are aligned (temperature, light, etc.). Eggs, collected in the natural ponds of Moscow, has evolved in the water from their home pond and water from the ponds, which are located outside the city of Moscow in an area with low anthropogenic load. Eggs from the ponds that are located outside Moscow has evolved in the water of the native ponds and water from the city's ponds. Because in the experiment were used clutches of Common frogs from Moscow ponds, placing these eggs in ponds from outside Moscow, for the chemical composition that is different from that in which live and breed frogs of urban populations for many generations, should not be considered as a control or optimal conditions. The second environmental factor - tadpoles density - is no less important. Of course, its impact is more significant in the later stages of development, when begins to operate the so-called "group effect" (Schwartz et al, 1976; Severtsov, 1996). However, even at stages of development before hatching, the cluster of eggs is also important because is a regulator of oxygen inside the cluster (Surova & Severtsov, 1985). The results of factor analysis performed on the results of this experiment, allow us to conclude that the effect of the chemical composition of water and different densities of tadpoles have little impact on the overall morphogenesis. Plays a fundamental role the tempo of growth characteristics: an increase in the overall size of the larvae, depending on the stage of development that allow us to demonstrate that in the early development plays an important role the epigenetic component of variability.

The idea of epigenetic effect is to shape the developing embryo belongs to Huxley (1942) and developed by Waddington (1956). All stages of the development are potentially open to evolutionary changes, but as epigenetic interactions is so fundamental, early evolutionary stages are stable (Horder, 2006). These stages are called critical (Svetlov, 1978), nodal (Cherdantsev 2003), or even Phylotypic (Sander, 1983), as this stages may coincide with the conservative Haeckel's stages (Hall, 1997; Richardson, 1998). But is possible find a name such as "Korpergrundestalt" (Seidel, 1960), "Phyletic stage" (Cohen, 1977) "Zootip / Phylotype" (Slack et al., 1993), "Phylotypic period" (Richardson, 1995) "Hounglass model" (Duboule, 1994). However, the critical stages (or even critical periods of development) do not always coincide with the phylogenetic stages. Allocation of such stages in the development, usually is based on the idea of laying new structures of the embryo at this stage and determination of the fact that it is at these stages when is observed a high mortality of embryos.

Morphometric analysis of several successive stages of development showed that the duration of the critical stages may be limited to one or several consecutive stages - a critical period in development (Severtsova & Severtsov, 2011). During these stages (period) value of the estimated coefficients of variation of traits in embryos from different nests are not significantly different. Between the critical periods, the variability can have high or low significance, characteristic of the critical period. This is clearly seen in the analysis of changes in the values of the coefficients of variation in egg, died during the experiment. To them were characteristic the values of the coefficients of variation above or below the critical value. Among the survivors clutches in the critical period of development was observed an increase in mortality and the occurrence of anomalies incompatible with life, but the proportion of those embryos was low and the clutch continues the development. The earliest nodal stage can be seen from 18 to 20 stage of development.

The second critical period in early embryogenesis involves stage after hatching, i.e. 32 - 33rd stage of development. This period is critical for the formation of the overall length of the embryo (*Lo*) and the length of the tail plate (*Lx*). At the same time, critical for the formation of the caudal plate width (*Wx*) is the stage 34, where there is a transition of tadpoles from attached to embryonic jelly coat (gallert) to free-swimming lifestyle. 36th stage of development is critical for the formation of signs of "length of the body» (*Lb*) and "long tail plate» (*Lx*), but not for the sign " total length of the embryo» (*Lo*). For the features that characterize the differentiation of the embryo were selected the following nodal stages: " length of the muzzle " (*Dl*) and "width of the head" (*Wg*) – 36th stage, for the feature "length of the head" (*Dlg*) – 34th stage, and for the feature "width of the sucker" (*W*) – 33th stage of development. For other characters, *Vis*, *H*, *R*, and *eye*, - the allocation of nodal stages in the investigated interval of development is difficult. Perhaps, all the investigated features of the critical period are present even in the 39th stage of development, but to confirm this fact, studies should be undertaken at more advanced stages of development. Thus, in the development the critical periods can be distinguished, but these periods are critical not only for the development of the embryo as a whole, but also for the process of formation of its individual structures. In some cases, these periods can be the same for different structures and then a group is formed ("modules") signs, that changes the correlation between "variability - correlation" which is similar. Thus, the signs of the first group, *Lo*, *Lb*, *Lx* and *Wx*, describing the growth processes exhibit insignificance, with the development, reduced the general variability with a significant increase of the correlation (Fig. 4).

At the same correlation of growth processes sharply increases from 29 to 32 stage and 34 to 36th, i.e. just to pre-nodal and nodal stages. Signs that do not have the segment of the nodal stage (*Vis*, *H*, *R*, and *eye*) are characterized by an increase in the variability of development and low coherence. In signs (*Dl*, *Dlg*, *Wg*, *W*), for which was isolated nodal stage, with low variability increases, anyway there is an increase in the coherence of development. Our findings are confirmed by other researchers, which also showed that the so-called Phylotypic stages of development increases the level of morphological interactions by a decline in variability (Irmler et al., 2004), and such stage is also characterized by modularity (Raff & Sly, 2000; Galis & Metz, 2001; Galis & Sinervo, 2002).

It is shown that modularity acts as a buffer mechanism for deviant development in extreme environments (Schmidt & Starck, 2010). Thus, influencing on the signs themselves, the environment does not change the modularity of its structure. Only when the impact causes an increase in the variability of development, there is either independent development units (up to the emergence of heterochronies (Richardson, 1995), or the emergence of new correlation coefficients between the modules and, consequently, the formation of a unified system of non-rigid correlation interactions. (Severtsova & Severtsov, 2011).

In a variety of environmental conditions the extent to which you can change a sign, is limited by its norm of reaction. This term was proposed by I. I. Schmalhausen (1969) and is special an individual characteristic, reflecting the breadth of variability in body shape in response to exposure to the environment and the ongoing without changing its genotype (Severtsov, 2004).

In the english-language evolutionary studies, the concept of norm of reaction criterion is close to the concept of "phenotypic plasticity" (Gordon, 1992; Via et al., 1995; Pigliucci, 2005; Garland & Kelly, 2006). As wider is the norm of reaction of genotype, wider is the variation of sign limits and is wider the range of environmental conditions in which this feature ensures the survival of the individual. It is in unstable conditions where the individual with the most wide norm of reaction will get a selective advantage (Severtsov, 1985; Severtsov & Surova, 1981).

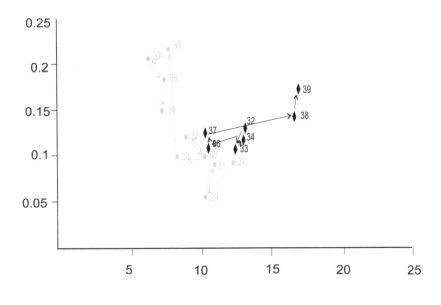

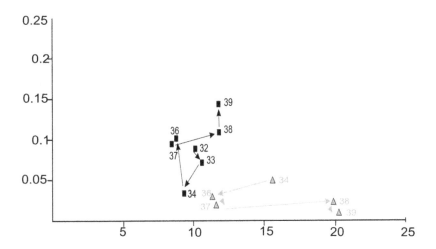

Fig. 4. Stepwise analysis of change in values of a generalized coefficient of determination ($Cv_{обш}$) (axis X) from determination coefficient (R^2) (axis Y). Numbers means the stage number. For signs: • – Lo, Lb, Lx, Wx and ♦ – Dl, Vis, Wg, Dlg, W, H, R, eye. For signs: ■ – for signs Dl, Wg, W, Dlg and ▲ –for signs Vis, H, R, eye.

Mechanisms of "retention" of the formation of character in this framework for early embryogenesis practically have not been studied. In our works we have considered as a possible regulatory mechanism the changing of the correlation of the variability and consistency (correlation) of the formed structures. This is manifested by increasing the total number of interrelated features that allows to keep and maintain the integrity of the embryo's development and thus provide a clearer differentiation of developing structures, despite their high variability. The second possible way of regulating early ontogenesis is realized by the amplification of correlative linked structures, formed precisely at the stage of investigation, coupled with the increasing variability of morphogenetic processes. As a result, the developmental process of this structure is accelerated up to the emergence of heterochronies. Perhaps this way of regulation can be illustrated by a moving ball in the trench chreod in the model of epigenetic landscape proposed by Waddington (1947) (Shishkin, 1984).

If we consider the ball as a separate character state, not as a body (in the original model) then the path traveled by the ball-sign would be the way of development for this sign. In a stable environment occurs the moving of the ball on the bottom of the "trench" and we, analyzing the variability of this path, will see that it is not high. In the case of "interference" occurs "swing" of the ball in the trench chreod and thus an increase in the variability of the characteristic (Fig. 5). In this case we need constant adjustment the path of development that holds "the ball" in the trench, which will be implemented by changing the consistency of development of a specific sign with other signs of the developing embryo. But we should not forget that epigenomics has no additive genetic component, so the variability in developing systems is characterized by stability and equifinality (Schmalhausen 1942; Shishkin, 1984; Cherdantsev, 2003).

In this form of organization of morphogenetic variability the direct action of natural selection on early signs of morphogenesis is hampered. By itself, the variability of development can not be used as material for selection, because it´s basically epigenomic and also is lost on the nodal stages of development. However, this does not mean that the adaptive evolution of the early stages of ontogeny to the effects of pollutants is not possible (Holloway et al., 1990; Forbes & Calow, 1997), although in many cases, the rate of environmental change might be too quick to create adaptations (Lynch & Lande, 1993; Burger & Lynch, 1997). However, published data show an increase in resistance to pollutants (Hesnar, 1995; Forbes & Calow, 1997; Gu et al., 2000; Johanson et al., 2001) and even the direct effects of environment on gene expression (Morozova et al., 2006; Wittkopp, 2007).

In some cases, is shown the stability of embryogenesis to the action of pollutants in amphibians from populations living in conditions of water pollution by organic substances (Hecnar, 1995; Johansson et al., 2001; Severtsova, 2002) or strong acidification (Andren et al., 1989). Comparison of resistance to pollution by nitrates of two populations of *Bufo americanus* showed a higher tolerance for this type of pollution in populations that spends a long time under the action of nitrates (Hecnar, 1995). Studies in Sweden have shown that the more resistant to prolonged exposure to a solution of ammonium nitrate are frog eggs from the southern regions, where the concentration of nitrate in nature is higher than in the northern (Johansson et al., 2001).

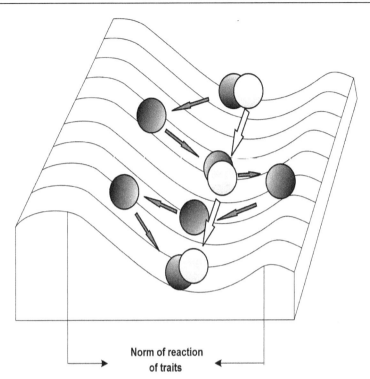

Norm of reaction
of traits

Fig. 5. Model of epigenetic landscape of Waddington (Shishkin, 1984)with changes. Plot of epigenetic landscape of the valley (chreod) which moves the ball, which symbolizes the path of the trait development. The chreoda's width is determinate by the norm of reaction of a trait, the depth, by the genetic determinism. Moving the light ball in the chreoda symbolizes the development of a trait in the right conditions, the dark ball, under adverse conditions that requiring regulation of development.

All these data indicate the possibility of adaptive evolution. Certainly, the formation of such adaptations are in all stages of the life cycle. However, as shown by the results, the factor analysis carried by us, according on the results of experiments with cross-coupling conditions for the development of tadpoles, response to exposure to the environment occurs individually for each clutch in each year of analysis (Severtsova, 2009). Consequently, the evolutionary transformation will occur at the level of change in the way of individual development of each clutch, driving changes in the way of development of each trait (West-Eberhard, 2003).

5. Conclusion

In conclusion, we emphasize that our approach using morphometric analysis of early embryogenesis allow us to evaluate the general condition of the population, living in conditions of anthropogenic pollution of the environment and to answer the important question of the possible adaptations of early embryogenesis in this new evolutionary environmental factor. Variability- an inherent property of life. Variability is the material for natural selection and, therefore, the methodological approaches to the study of variability in understanding the degree of diversity and mechanisms of evolutionary changes have been well worked. Embryological studies dealing with the investigation of the features of ontogeny or evolutionary change of early development use morphometric approach for the study of variability in a lower grade.

6. References

Animals in conditions of anthropogenic landscape. (1990). Sverdlovsk ASUSSR UrO pp. 125. (Russ.).

Intraspecific variation in animal ontogenesis. (1980) M.: Nauka. pp 226 (Russ.).

Structure and functional role of animal populations in natural and transformed ecosystems. (2001) Abstract: I International Conference 17-20 September. Dnepropetrovsk. pp. 247. (Russ).

Allran, J. & Karasov, W. (2000). Effects of atrazine and nitrate on northern leopard frog (*Rana pipience*) larvae exposed in the laboratory from posthatch through metamorphosis. *Environ. Toxicol. Chem.* V. 19, №.11, pp. 2850-2855.

Allran, J. & Karasov, W. (2001) Effects of atrazine on embryos, larvae and adults of anuran amphibians. *Environ. Toxicol. Chem.* V. 20, №. 4, pp. 769-775.

Andren C., Marden M. & Nilson G. (1989). Tolerance to low pH in a population of moor frogs, *Rana arvalis*, from an acid and a neutral environment: a possible case of rapid evolutionary response to acidification. *Oikos.* V. 56, pp. 215-223.

Beattie R.C., Aston R.J. & Milner A.G .P. (1991)A Field-Study of Fertilization and Embryonic-Development in the Common Frog (*Rana temporaria*) with Particular Reference to Acidity and Temperature. *J. of Applied ecology.* V. 28, № 1, pp. 346-357.

Beattie R.S., Tyler-Jones R. & Baxter M.J. (1992). Influence of pH, aluminum, temperature on embryonic development of *Rana temporaria.* *J. Zool.* V. 228. № 4. pp.557-570.

Beetschen, J.C. (2001). Amphibian gastrulation: history and evolution of a 125 year-old concept. *Int J Dev Biol.* V. 45, pp. 771-795.

Berven K.A. (1982). The genetic basis of altitudinal variation in the wood frog, *Rana sylvatica.* 1. An experimental analysis of larval development. *Oecologia.* v. 52. № 3. pp. 360-369.

Berven K.A. & Gill D.E. (1983). Interpreting geographic variation in life-history traits. *Amer. Zool.* v. 23. № 1. pp. 85-97.

Burger R. & Lynch M. (1997). Adaptation and extinction in changing environments. In: Environmental stress, Adaptation and Evolution. Bijlsma R.; Loeschcke V. Springer Verlag, pp. 209-240.

Cherdantsev, V.G. & Scobeyeva, V.A. (1994). The morphological basis of self-organization. Developmental and evolutionary aspects. *RiVista di Biologyca- Biology forum.* v. 87. № 1. pp. 57-85.

Cherdantsev, V. G. (2003). *Morphogenesis and evolution.* M. KMK. pp. 359 (Russ.).

Cohen, J. (1977) *Reproduction.* London: Butterworths.

Cooke A.S. (1974). The effect of pp'-DDT on adult frog (*Rana temporaria*). *Brit. J. Herp.* №4. pp. 647-652.

Dawson, D.A., McCormick, C.A. & Bantle, J.A. (1985). Detection of teratogenic substances in acidic mine water samples using the frog embryo teratogenesis assay - *Xenopus* (FETAX). *J. Appl. Toxicol.* V. 5. № 4. pp. 234-44.

Duboule, D. (1994) Temporal colinearity and the phylotipic progression: a basis for the stability of a vertebrate Bauplan and the evolution of morphologies through heterochrony. *Dev Suppl* pp. 135-142.

Dunson, W.A., Wyman, R.L. & Corbett, E.S. (1992) A symposium of amphibians declines and habitat acidification. *J. Herpetol.* V.26. pp. 349-352.

Ettler, V., Mihaljevič, M., Matura, M., Skalová, M., Šebek, O. & Bezdička P. (2008). Temporal Variation of Trace Elements in Waters Polluted by Municipal Solid Waste Landfill Leachate. *Bulletin of Environmental Contamination and Toxicology.* V. 80, №3, pp. 274-279.

Falconer, D. (1981) *Introduction to qualitative genetics.* M.: Agropromizdat. pp. 486 (Russ).

Fetax. (1991). Standard guide for conduction the frog embryo teratogenesis assays- Xenopus (Fetax). *Amer. Soc.for test. and mat.* pp. 1-11.

Forbes, V.E. & Calow, P. (1997). Responses of aquatic organisms to pollutant stress: theoretical and practical implications. In: Environmental stress, Adaptation and Evolution. Bijlsma R. & Loeschcke V. B: Springer Verlag. pp. 25-42.

Fort, D.J., Guiney P., Weeks, J., Thomas J., Rogers, R., Noll, A. & Spaulding, C. (2004). Effect of Methoxychlor on Various Life Stages of Xenopus laevis. Toxicological sciences. V. 81, pp. 454–466.

Freda, J. (1986). The influence of acidic ponds water on amphibians: A review. *Water Air Soil Poll.* V. 20. pp. 439-450.

Galis, F. & Metz, J.A. (2001). Testing the vulnerability of the phylotipic stage: on modularity and evolutionary conservation. *J exp Zool (Mol Dev Evol).* V. 291, pp. 195-204.

Galis, F., & Sinervo, B. (2002). Divergence and convergence in early embryonic stages of metazoans. *Contrib Zool.* V. 71, pp. 101–113.

Garland, T. & Kelly, S. A. (2006). Phenotypic plasticity and experimental evolution. J. Exp. Biol. V. 209. pp. 2344-2361.

Gilbert, S. (1993).*Developmental Biology.* T. 1. M.:Mir. pp. 110-144. (Russ.).

Glukhova, E.V. & Cherdantsev V.G. (1999) Microfolds in suprablastopore zone o frog´s gastrula and its relation to mechanics and geometrics of the gastrula. *Ontogenez.* V. 30. № 6. pp. 369-379. (Russ.).

Gordon, D.M. (1992). *Phenotypic plasticity.* In keywords in evolutionary biology. Cambridge, MA: Harvard University Press. pp. 255-262.

Greenhouse, G. (1976). The evaluation of toxic effects of chemicals in fresh water by using frog embryos and larval. *Environ. Poll.* V.11. №4. pp. 303-315.

Gu, Y.Z., Hogenesch, J.B. & Bradfield C.A. (2000). The PAS superfamily: sensors of environmental and developmental signals. *Annu. Rev. Pharmacol. Toxicol.* V.40. P.519–561.

Hall, R.J. & Henry P.F.P. (1992). Assessing effects of pesticides on amphibians and reptiles: status and needs. *Herpetol. J.* V. 2. pp. 65-71.

Hall, B.K. (1997). Phylotypic stage or phantom: is there a highly conserved embryonic stage in vertebrates? *Trends Ecol Evol.* V. 12. pp. 461-463.

Hecnar, S.J. (1995). Acute and chronic toxicity of ammonium nitrate fertilizer to amphibians from southern Ontario. *Environ. Tox. and Chem.* V.14. pp. 2131-2137.

Holloway, G.J., Sibly, R.M. & Pover, S.R. (1990). Evolution in toxic stress environments. *Funct. Ecology.* V. 4. pp. 289-294.

Horder, T.J. (2006). Gavin Rylands de Beer: how embryology foreshadowed the dilemmas of the genome. *Nat Rev Genet.* V. 7. pp. 892-898.

Horder, T. (2008). A history of evo-devo in Britain. Theoretical ideals confront with biological complexity. *Ann Hist Phil Biol.* V. 13 pp. 101-174.

Horne, M.T. & Dunson W.A. (1995). Toxicity of metals and low pH to embryos and larvae of the Jefferson salamander, *Ambystoma jeffersonianum. Arch. Environ. Contam. Toxicol.* V. 29. pp. 110-114.

Huxley, J.S. (1942). *Evolution: The modern synthesis.* London, pp. 74.

Irmler, I., Schmidt, K. & Starck, J.M. (2004). Developmental variability during early embryonic development of zebra fish, *Danio rerio. J exp Zool B: Mol Dev Evol.* V. 302 pp. 446-457.

Johansson, M., Rasanen, K. & Merila J. (2001). Comparison of nitrate tolerance between different populations of the common frog, *Rana temporaria. Aquatic Toxicol.* V. 54. pp. 1-14.

Leontieva, O. A. & Semenov D. V. (1997). Amphibians as bioindicators of anthropogenic change in environment. *Uspeji Sobrem. Biologii.* T. 117. No. .6. pp. 726-736. (Russ.).

Lynch, M. & Lande, R. (1993). Evolution and extinction in response to environmental change In *Biotic Interactions and Global Change.* Eds. Kareiva P.; Kingsolver J. & Huey R.Sinauer Associates, Inc. Sunderland. MA. pp. 234-250.

Moller, A.P. & Swaddle, J.P. *Asymmetry, developmental stability, and evolution.* Oxford: University Press, pp. 1997. 291.

Morgan, M. K., Scheuerman, P.R., Bishop, C.S. & Pyles, R.A. (1996) Teratogenic potential of atrazine and 2,4-D using FETAX. *J. Toxicol Environ Health.* V. 48(2). pp. 151-168.

Morozova, T. V., Anholt, R. R. & Mackay, T. F. (2006). Transcriptional response to alcohol exposure in *Drosophila melanogaster. Genome Biol.* V. 7. pp. 95.

Opitz, R., Braunbeck, T., Bögi, C., Pickford, D.B., Nentwig, G., Oehlmann, J., Tooi, O., Lutz, I. & Kloas, W. (2005). Description and initial evaluation of a Xenopus metamorphosis assay for detection of thyroid system-disrupting activities of environmental compounds. *Environ Toxicol Chem.* V. 24(3) pp.653-64.

Orton, F., Carr, J.A. & Handy, R.D. (2006). Effects of nitrate and atrazine on larval development and sexual differentiation in the northern leopard frog *Rana pipiens. Environ Toxicol Chem.* V. 25(1) pp. 65-71.

Pigliucci, M. (2005). Evolution of phenotypic plasticity: where are we going now? *Trends Ecol. Evol.* V. 20. pp. 481-486.

Raff, R.A. & Sly, B.J. (2000). Modularity and dissociation in the evolution of gene expression territories in development. *Evol Dev.* V. 2 pp. 102-113.

Richardson, M.K. (1995). Heterochrony and the phylotypic period. *Dev Biol.* V. 172, pp. 412-421.

Richardson, M.K., Minelli, A., Coates, M. & Hanken J. (1998). Phylotypic stage theory. *Tree.* V. 13. № 4. pp. 158.

Saber, D.A. & Dunson W.A. (1978). Toxicity of bog water to embryonic and larval anuran amphibians. *J. Exp. Zool.* V. 204. № 1. pp. 33-34.

Sander, K. (1983) The evolution of patterning mechanism: gleaning from insect embryogenesis and spermatogenesis. In: *Development and Evolution.* Goodwin, B.C.; Holder, N. & Wylie, C.C. editors. Cambridge University Press, Cambridge. pp. 123-159

Schmalhausen I. I. (1942) Organism as a whole in individual and hitorical development. M.; L. AS USSR. pp. 212. (Russ.).

Schmalhausen, I. I. (1969) Problems of Darwinism.L. Nauka. pp. 492. (Russ.).

Schmidt, K. & Starck, J.M.. (2010). Developmental plasticity, modularity, and heterochrony during the phylotypic stage of the zebra fish, *Danio rerio. J Exp Zool B Mol Dev Evol.* V. 15 № 314(2), pp. 166-78.

Schwartz, S.S., Pyastolova, O.A., Dobrininskaya, A.A. & Runova, G.G. (1976). *Group effect in populations of aquatic animals and chemical ecology.* M.:Nauka. pp. 1-152 (Russ.).

Seidel, F. (1960). Korpergrundgestalt und Keimstruktur. Eine erorterung uber die Grundlagen der vergleichenden und experimentellen Embryologie und deren Gultigkeit bei phylogenetischen Uberlegungen. *Zool Anz.* V. 164 pp. 245-305.

Severtsov, A.S. (1985). Emergence of plastic features in phenotype under selection. *Zhurnal obshei biologii.* T. 46. № 5. pp. 579-589. (Russ.).

Severtsov, A.S. (1996). Group selection as a cause of emergence group adaptations. *Zoologichesky zhurnal.* T 75, No. 10, pp. 1525-1540. (Russ.).

Severtsov, A.C. (2004) *Theory of evolution.* Vlados. pp. 380. (Russ.).

Severtsov, A.S. & Surova, G.S. (1981). Individual variability o norm of reaction and adaptation of population. *Zhurnal obshei biologii* T XLII, №2, pp. 181-192. (Russ.).

Severtsova, E.A. (2002) *Adaptive processes and variability of embryogenesis of anuran amphibians in urban populations.* Synopsis PhD thesis M.: MSU. 2002. 24 p russ.

Severtsova, E.A. (2005). Adaptation of Amphibian gastrulating embryos to an influence of anthropogenic pollution. *10th congress European Society for Evolutionary Biology.* Poland. 15-20.08.2005. pp.514.

Severtsova, E.A. (2009). Experimental work on the correlation of variability components of development of Common frog tadpoles (*Rana temporaria*, Anura, Amphibia) in hatching stage. *Zoologichesky zhurnal* T 88, №1, pp. 1-10. (Russ.).

Severtsova, E.A. & Severtsov, A.C. (2005). Interpopulation comparison of variability o gastrulation of Common frog embryos living in conditions of anthropogenic polluted ponds. *Ontogenez.* T. 36. №2. p. 110-122. (Russ.).

Severtsova, E.A. & Severtsov, A.C. (2007). Interpopulation comparison of gastrula of Moor and Marsh frogs living in conditions of anthropogenic polluted ponds. *Ontogenez.* T. 38. № 1. pp. 1-14. (Russ.).

Severtsova, E.A. & Severtsov A.C. (2011). Crucial stages of embryogenesis of *R.arvalis* Part 1: Linear Measurements of Embryonic structures. *Russian Journal of Developmental Biology.* Vol. 42, № 5, pp. 331-341.

Shishkin. (1984). Individual development and natural selectionp. *Ontogenez.* T. 15. № 2. pp. 115-136. (Russ.).

Slack, J.M.., Isaacs, H.V., Johnson, G.E., Lettice, L.A., Tannahill, D. & Thompson J. (1992). Specification of the body plan during Xenopus gastrulation: dorsoventral and anteroposterior of the mesoderm. *Dev. Suppl.* V. 1. pp.143-149.

Slack, J.M.., Holland, P.W. & Graham C.F. (1993). The zootype and the phylotypic stage. *Nature,* V. 361, pp. 490–492.

Sparling, D.W., Linder, G. & Bishop C.A. (2001). *Ecotoxicology of amphibian and reptiles.* SETAC press. Pensacola. FL.

Stroganov, N.N. (1971). Methods for determinate the toxicity in aquatic environments . In *Methods for biological research in aquatic toxicology.* M.: Nauka.. pp. 14-60.

Svetlov. (1978). *Physiology (mechanics) of development.* L.: Nauka, T.1. Processes of morphogenesis in cellular and organismic levels. 279 p. T.2. Internal and external actors of development. 263 p.

Surova, G. (1988). Interaction of tadpoles of Common frog in natural conditions. *Ekologiya* № 4. pp. 49-54.

Surova, G. (1988a). Environmental and hereditary components of periods of ontogenesis of Common (*Rana temporaria*) and Moor (*R. arvalis*) frog tadpoles. *Zool Zhur* T. 67.No. 3. pp. 396-405. (Russ.).

Surova, G. & Severtsov A.C. (1985). Death in Common frog (*Rana temporaria* L.) in early ontogenesis and factors of it cause. *Zool. Zhur.* T. 44. No. 1, pp. 61-71. (Russ.).

Travis, J. (1980). Genetic variation for larval specific growth rate in the *Hyla gratiosa. Growth.* V. 44. № 3. pp. 167-181.

Travis, J. (1981). Control of larval growth variation in a population of *Pseudacris triseriata* (Amura:Hylidae). *Evolution.* V. 35. № 3. pp. 424-432.

Vershin, V.L. (1997) *Ecological features in amphibian populations of urban territories.* Synopsis: Doctoral degree thesis. Ekaterinburg: Institute of plant and animal ecology UrORAS pp. 47.

Via, S., Gomulkiewicz, R., De Jong, G., Scheiner, S. M., Schlichting, C. D. & Van Tienderen, P. H. (1995). Adaptive phenotypic plasticity: consensus and controversy. *Trends Ecol. Evol.* V. 10. pp. 212-217.

Voronova, L.D., Golichenkov, V.A., Popov, D.V., Kalistratova, E.N. & Sokolova Z.A. (1983). Reaction of pigmentary system of amphibian tadpoles in lower concentrations of some pesticides. In *Problems of ecological monitoring and ecosystem model.* L. Gidmeteoizdat . T. VI. pp. 77-90. (Russ.).

Waddington, C.H. (1947). *Organizers and genes.* M.: Gos. Izdat. Inostran. Lit. pp. 240. (Russ).

Waddington, C.H. (1956). *Principles of Embryology.* Allen & Unwin, London.

West-Eberhard. (2003). *Developmental Plasticity and Evolution.* N. Y.: Oxford University Press. pp. 794.

Wittkopp. (2007) Variable gene expression in eukaryotes: a network perspective. *J. Exp. Biol.* V. 210. pp. 1567-1575.

Zakharov, V.M. (1989). Future prospects for population phenogenetics. *Sov. Sci. Rev. Physiol. Gen. Biol.* V.4. pp.1-79.

9

Microspore Embryogenesis

Tara D. Silva
University of Colombo
Sri Lanka

1. Introduction

Microspores are the precursors of the male gametes of plants and are equivalent to spermatids of animals. Microspores develop into pollen grains within the anthers of a flower in angiosperms. Mature pollen grains are the male gametophytes. The function of the male gametophyte is to participate in the sexual reproduction of plants.

Quite separate from this intended pathway of gametogenesis, a microspore can also be induced to assume sporophytic development. "Totipotency", which is unique to plant cells, allows a microspore that is already destined to develop into a male gametophyte, to re-direct its development pathway so that a haploid sporophyte is regenerated under specified conditions. The cellular totipotency displayed by the microspore is considered an adaptive mechanism for survival that is brought about under stressful environmental conditions (Bonet et al. 1998). The process that leads to the development of a sporophyte from a microspore is referred to as microspore embryogenesis (or pollen embryogenesis). It is also commonly termed androgenesis.

Androgenesis rarely occurs in nature, but is relatively easily induced in several plant species under in vitro conditions. The first report on regeneration of androgenic plants was in the Solanaceous species, *Datura innoxia,* where Guha and Maheshwari (1964) demonstrated that the anthers cultured in vitro yielded haploid plants. Following this initial discovery, many attempts have been made to repeat the success in other plant species. Early studies were mostly empirical and were directed at identifying suitable culture media, pre-treatment conditions that are required for inducing sporophytic development in microspores, and other such practical considerations. Since then, much progress has been made, particularly toward the understanding of the basic processes that occur at cellular level with the switching of developmental pathway of the microspore, from gametogenesis to microspore embryogenesis. Concerted efforts on many different plants have resulted in the documentation of the microspore embryogenesis process in over 250 species (Maluszynski et al. 2003), even though achieving success with androgenesis is still restricted to annuals and herbaceous or non-woody plants mostly. The two main techniques that are employed to generate androgenic plants in vitro are the anther culture and the microspore (pollen) culture methods. Anther culture is the more widely applied, where the excised immature anthers dissected out from flower buds are cultured whole on suitable growth media under appropriate in vitro conditions. The microspores inside the anther develop into plants through a process by which their normal gametophytic development is stalled and sporophytic development is initiated. As the microspores are endowed with only one set of chromosomes instead of the two sets present in the somatic cells of the diploid plant, the pollen-derived plants are haploid. In the anther culture method, even though the goal is to produce plants of microspore origin, there is a danger of plants regenerating from somatic tissue of the anther wall rather than from the haploid microspore cells. In the microspore culture technique, individual pollen cells isolated from anther tissue are placed directly in culture. It is possible to isolate microspores by mechanical means such as crushing of anthers and release into medium by magnetic stirring. They can also be isolated as naturally shed microspores in the culture medium after pre-culture of anthers. Although microspore culture is technically more demanding and is less efficient in some plant species than the culture of whole anthers, it has several advantages over anther culture. Mainly it eliminates the danger of plant development from anther wall tissue. As individual plants develop from separate pollen cells, the chance of chimera production is also low. Further, the culture of individual microspores makes possible the tracing of events that occur from microspore initiation through to embryogenic development, by cell tracking studies allowing for the greater understanding of the process of androgenesis in plants.

Haploid plants are weak and sterile and have no regular means to produce sexual progeny. Therefore, haploid plants by themselves serve no useful purpose. However, haploid plants regenerated from cultured anthers or microspores can afterwards be brought back to the normal diploid state by duplicating their chromosome number through application of chemicals such as colchicine, or other in vitro techniques. The resulting plants will have two identical chromosome sets and be perfectly homozygous, and therefore will give rise to fertile homogenous progenies. Thus, the usefulness of the technique is not in deriving haploid plants but in producing doubled haploids or dihaploids. The great interest in androgenesis among the scientific community, particularly among those involved in practical plant breeding, is this perceived potential of the technique to rapidly develop homozygous lines in the breeding material, which otherwise would require several generations of inbreeding through conventional procedures. The effectiveness of the technique depends on the efficiency of haploid plant regeneration from microspores contained within the anthers, and the conversion of these haploids to doubled haploid plants either spontaneously during the tissue culture phase or induced thereafter. With this method, true-breeding lines are produced in the immediately succeeding generation, and thus the technique has immense utility for developing homozygous breeding lines in a relatively short period.

The objective of this Chapter is to detail out the process of microspore embryogenesis in plants and discuss the factors that influence its induction, particularly with reference to the recent progress made in the understanding of molecular events that occur during the re-programming of the microspore development pathway.

2. Gametogenesis versus embryogenesis

In normal gametogenesis, the diploid microsporocytes or pollen mother cells in the anther undergo meiotic division to yield haploid microspores. Each uni-nulceate microspore then divides mitotically to produce a cell with two haploid nuclei, which is the bi-nucleate state of pollen. Cytokinesis that follows produces two cells that are of unequal size. The larger cell is termed the vegetative cell and the smaller one is known as the generative cell. The generative nucleus divides again by mitosis to produce two haploid sperm nuclei. The vegetative nucleus remains without further division. Thus a mature male gametophyte at the time of affecting fertilization is often bi-cellular (and sometimes tri-cellular) with three haploid nuclei of which two nuclei of the generative cell origin participate in the "double fertilization" that is characteristic of angiosperm reproduction, producing the diploid zygote and the triploid endosperm of the seed.

In contrast, the embryogenic development of a uni-nucleate microspore that is induced in vitro, is usually initiated by a symmetric mitotic division that results in two equal-sized cells. Further mitotic divisions result in a group of undifferentiated cells, and their rapid proliferation gives rise to a multi-cellular mass. Depending on the in vitro conditions provided, the multi-cellular structures may become embryogenic or remain meristematic. Embryogenic cells, following further division and pattern-formation, are able to develop into structures referred to as somatic embryos. The microspore-derived somatic embryos are initiated as globular structures, and become heart-shaped, and with further morphogenesis become torpedo-shaped and finally develop into cotyledonary embryos with bipolar embryo axes, mimicking closely the zygotic embryo development process. Under some in vitro conditions, regeneration of shoots and roots may occur directly from proliferating callus through a process of organogenesis without producing somatic embryos. The regenerated plants may possess a haploid chromosome complement, similar to the microspore cell of origin, or the chromosome number may have doubled during cycles of cell division, to give rise to doubled haploid plants that are completely homozygous.

3. Conditions under which microspore embryogenesis can be induced

To deviate from the normal pathway of gametogenesis and for re-direction towards a pathway of embryogenesis, a microspore has to be subjected to specific conditions. The conditions that are required to induce such developmental reprogramming will obviously vary with the species, and often within a species among its genotypes or varieties also. Nevertheless, several key factors have been recognized that positively influence the microspore embryogenesis induction in a variety of plant species. In Table 1 these various

Broad categorization of factors influencing microspore embryogenesis	Specific effects
Of the anther donor plant	Genotype, variety or species
	Physiological status - influenced by the growth condition, including seasonal effects such as day length, light, temperature etc.
Of the microspore	Developmental stage, pollen genotype
Of the pre-culture environment	Pre-treatments
Of the in vitro culture conditions	Culture media
	Incubation conditions

Table 1. Factors affecting microspore embryogenesis

biotic and abiotic factors, that have a strong bearing on the ability of a microspore to undergo embryogenesis, are grouped into factors of the anther donor plant, the microspore, pre-culture environment and of the in vitro culture condition.

3.1 Genotype

Often, the in vitro androgenic response is genotype-dependent, and culture conditions may have to be optimized for each genotype. Even of the same species, different varieties and genotypes respond vastly differently to induction attempts. For example, in rice, japonica varieties are known to be more responsive than indica types (Yan et al. 1996), while within each ecotype considerable variation exists between varieties in their anther culture responsiveness (Silva and Ratnayake 2009). Plant regeneration occurs from cultured microspores usually in two stages. Initially the microspore divides and proliferates into an undifferentiated cell mass, followed by pattern formation and morphogenesis that lead to shoot regeneration which occurs through embryogenesis (or sometimes organogenesis). Inheritance studies on anther culture ability have shown that in many species these two processes are genetically determined independent events, and whether any of the two processes or both would occur during anther culture is determined to a large extent by the genetic makeup of the cells. The stimulation of these two events often requires different conditions, and is likely controlled by different genes. Experimental evidence suggests that the mode of inheritance is quantitative in nature (Silva 2010 and references therein). No sustained efforts have been made to actually breed the trait, anther culture ability into anther culture-recalcitrant genotypes, perhaps due to the complexities of the genetic control of in vitro response. Nevertheless the possibility of transferring the trait through sexual hybridization remains a viable prospect. On the other hand, the existence of a large non-genetic component of variation for anther culture ability suggests that there is sufficient scope for improving anther culture efficiency through the manipulation of these non-genetic factors that include the culture media components and pre- and post-culture conditions.

3.2 Physiological status of the anther donor plant

The number and the vigor of microspores found within the anther, the nutritional status of the tissues of the anther, the endogenous levels of growth regulators, may to some extent be influenced by the physiological age and growth condition of the donor plant, which in turn will have a bearing on microspore embryogenic competence. Seasonal variations in anther response have been observed in wheat and barley (Datta 2005). Differences in competence of the cultured microspores to assume embryogenic development when anthers are collected from field grown plants and when harvested from pot plants bear further testimony to the influence of the parent plant growth condition on their microspore embryogenic competence (Datta 2005, Silva 2010). It is common practice in anther culture to use anthers from the first or early flush of flowers rather than anthers from buds in later branches or tillers of the plant. The physiology of the donor plant can be altered, and the androgenic response improved, when the donor plant is carefully nurtured and grown under favorable environmental conditions, although pest control measures may have a detrimental effect on in vitro microspore response. Critical environmental factors include light intensity, photoperiod, temperature, nutrition and carbon dioxide concentration. Seasonal variations in the anther response have also been observed, which is probably also related to the overall growth of the donor plant.

3.3 Microspore development stage

The microspore development stage is a complex factor that has a strong influence on the success of anther culture. The exact stage of the microspore that responds to inductive treatment varies with the species, and is restricted to a relatively narrow developmental window. Only microspores that are at a stage sufficiently immature can be induced to change course from a gametophytic pre-programme to embryogenic re-programming leading to sporophytic development. Even though all microspores within an anther would be roughly of a similar age, the incremental differences in the stages of development of individual microspores can be considered significant, in setting each apart with regard to their embryogenic competence. Therefore, even within the same anther only a percentage of pollen cells would undergo divisions leading to their embryogenic development. For many species, the most amenable stage is either the uni-nucleate stage of the microspore or, at or just after the first pollen mitosis, which is the early bi-nucleate stage. At this time the transcriptional status of the microspore may still be proliferative and not yet fully differentiated (Malik et al. 2007). More mature microspores are considered irreversibly committed to gametogenesis and are at points of no return in their programme of maturation into male gametophytes.

In the practical implementation of the anther culture technique, it is necessary to identify easily, the flower buds that need to be harvested in order to collect microspores at the correct stage of maturity that is appropriate for in vitro culture. In many species, a maturity gradient is observed in flower buds within the inflorescence that displays either an acropetal or basipetal developmental succession. Therefore, a series of buds taken from different positions on the inflorescence (or in the case of individual flowers, buds of different sizes) need to be pre-examined in order to identify the stage of microspore development within their anthers. This is done by squashing anthers to release pollen, and staining with acetocarmine to observe the nuclei under the light microscope. DAPI (4', 6-diamidino-2-phenylindole) may be also used to stain the nuclear DNA, which can then be visualized using a fluorescent microscope. Such pre-examination will be very useful to establish a correlation between an easily observable morphological trait, (e.g. the petal to sepal length ratio in tobacco) with the pollen development stage, to be used as a quick guide to identify flowers that would have anthers carrying microspores at the required stage of maturity.

3.4 Pre-treatment stresses

For almost all the plant species in which anther culture has been attempted, it is common to include a physical or chemical pre-culture treatment that is applied to excised flower buds, whole inflorescences or separated anthers. The pre-treatment is required to trigger the induction of the sporophytic pathway, thereby preventing the development of fertile pollen through the gametophytic pathway. A variety of microspore pre-treatment stresses have been tested and found to enhance pollen embryogenesis, although the type, duration and the time of application of these pre-treatments may vary with the species or even the variety. The more commonly used anther or microspore pre-treatment conditions are temperature (cold or heat shock), sucrose and nitrogen starvation, centrifugation, as well as the use of microtubili disruptive agents such as colchicine. Lesser known stress-treatments include irradiation, use of high humidity, anaerobic treatment, electro-stimulation, high medium pH, ethanol and heavy metal treatment. These pre-treatments may be classified into three categories based on their application as widely used, neglected and novel (Shariatpanahi et al. 2006). The more notable stress treatments that promote androgenesis are discussed below.

Temperature pre-treatment: A commonly used pre-treatment to induce androgenesis in microspores is the low temperature shock. Cold pre-treatment is usually carried out at 4 - 10 ^{0}C for a few days up to several weeks. For example in rice, for many varieties, a 10 ^{0}C pre-treatment of harvested panicles for 10 – 30 days is sufficient to trigger androgenesis. Several other cereal species including barley, wheat, oat and triticale require low temperature treatment of the excised spikes or flower buds in order to induce microspore embryogenesis. Different views exist on how cold pre-treatment affects the development of the microspore, some or all of which may be relevant to a given species. It is believed that the cold treatment delays the degeneration of anther wall tissues thereby protecting the microspores within, from toxic compounds released by the degenerating maternal tissues. It has also been suggested that at cold temperatures there is greater survival of embryogenic pollen grains leading to enhanced levels of embryogenesis than would occur without pre-treatment. A further explanation is that microspores in the cold pre-treated anthers disconnect from the tapetum resulting in starvation, causing them to switch from gametophytic pathway to embryogenic development. Also, it has been noted that in cold pre-treated anthers the total content of free amino acids is increased, which is suggestive of metabolic re-programming that a microspore needs to undergo, in preparation for embryogenesis induction. Following cold treatment, small heat shock protein genes have been shown to be expressed in tomato and it has been argued that this is possibly to protect cells against chilling injury. The different explanations (Shariatpanahi et al. 2006 and references therein) are based on observations and experiences with different plant species, and the exact mode of action (or actions) is still to be unraveled convincingly.

High temperature has been also used to trigger microspore embryogenesis in certain species. The method is applied routinely with the isolated microspore culture of rapeseed. Embryogenesis is induced efficiently and synchronously in rapeseed microspores subjected to 32 ^{0}C heat shock for about 8 hours, whereas normal gametogenesis occurs at 18 ^{0}C under otherwise similar conditions. High temperature induced microspore embryogenesis does also occur in wheat (Touraev et al. 1996b) and tobacco (Touraev et al. 1996a). As with cold shock, different mechanisms have been proposed to explain the basis of microspore induction through high temperature treatment. It is known that several heat shock proteins (HSPs) are synthesized in various plant tissues in response to elevated temperatures, which have a role in protecting cells from thermo-damage (Schoffl et al. 1996). This is considered an adaptive mechanism since plants are unable to escape extreme environmental conditions due to their immobility. Amongst the HSPs, members of HSP 70 are the most abundant and evolutionarily conserved while HSP 90 is also abundant and is constitutively expressed in eukaryotes with a specific role in heat stress (Chug and Eudes 2008). Synthesis of HSPs has been observed in heat-stressed microspores also, and HSP 70 was suggested to inhibit apoptosis (Jaattela et al. 1998). Even though much significance was attached earlier to the HSPs, for their considered role in enabling microspore embryogenesis (Cordewener et al. 1995, Zarsky et al. 1995, Touraev et al. 1996), subsequent studies have proved that embryogenesis can be induced in rapeseed microspores even when HSPs are not synthesized. For example, embryogenesis can be induced in rapeseed microspores that are stressed with microtubule de-polymerizing agent, colchicine also (Zhao et al. 2003) and this proves that HSPs are not essentially required. Since plants reportedly produce HSPs in response to a variety of other environmental stresses such as cold, drought, heavy metal stress, and starvation (Schoffl et al. 1998, Zarsky et al. 1995), it is more likely that HSPs have an overall protective role that allows the cells to survive during the periods of adverse environment conditions, and embryogenesis may simply be the consequence of blocked pollen development (Zhao et al. 2003). In rapeseed microspores, the heat shock is believed to also cause microtubule and cytoskeleton rearrangements that lead to altered cell cycle events during microspore culture (Simmonds and Keller 1999).

Nitrogen and sugar starvation: Nutrient starvation, particularly of nitrogen and sugar, has been effective in enhancing the in vitro anther response in some species. Nitrogen starvation may be applied to the anther donor plant or the excised anthers and microspores. The mother plant can be nitrogen-stressed by restricting the application of nitrogen fertilizer to the plant. Excised anthers can be starved of nitrogen by withdrawing or limiting the inorganic and organic nitrogen sources in the initial culture media. In several species, isolated microspores have shown a better embryogenic response with nitrogen starvation. For example, with tobacco pollen, a high rate of embryogenic induction was achieved from starving the microspores of glutamine, by an initial period of growth on glutamine free medium (Kyo and Harada 1986). Also in tobacco, sucrose and nitrogen starvation of pollen of the mid bi-cellular stage resulted in embryogenic induction and yielded a high number of embryos when transferred to a simple medium containing sucrose and nitrogen (Touraev et al.1997). Similarly sucrose starvation at 25 ^{0}C resulted in the efficient formation of embryogenic microspores in tobacco when transferred to sugar containing medium (Touraev et al. 1996a). In wheat, the culture of excised anthers under starvation and heat shock conditions induced the formation of embryogenic microspores at high frequency (Touraev et al. 1996b). During starvation, cytoplasmic and nuclear changes have been observed in the microspores including de-differentiation of plastids, changes in chromatin and nuclear structure, changes in the level of RNA synthesis and protein kinase activity, and the activation of small heat shock protein genes (Shariatpanahi et al. 2006 and refereces therein).

Centrifugation: Subjecting anthers to centrifugation pre-treatment has been useful to improve the efficiency of microspore embryogenesis. In tobacco anthers, a four-fold increase in the regeneration of haploid plants was reported following centrifugation of the anthers prior to culture, although the maturity stage of the microspores determined the degree of success (Tanaka 1973). Experiments with chick pea (*Cicer arietinum*) also appear to suggest that centrifugation may have a positive impact on embryo initiation from cultured anthers, even though the response is variety specific (Grewal et al. 2009). Because the appropriate centrifugal force and the duration of treatment vary with the microspore development stage and variety, it is difficult to standardize protocols for centrifugation pre-treatment.

Colchicine: The use of colchicine for microspore pre-treatment has been reported to be effective in enhancing microspore embryogenesis in rapeseed (Zaki and Dickinson 1991, Zhao et al. 1996), wheat (Barnabas 2003a), coffee (Herrera et al. 2002) and maize (Obert and Barnabas 2004). Colchicine acts by binding to α- and β-tubulin heterodimers thereby inhibiting further dimer addition to microtubules, which causes de-polymerization of the microtubule (Sternlicht et al. 1983). During normal gametogenesis of pollen, the peripheral position of the nucleus is maintained by microtubules and actin filaments (Hause et al. 1992). The de-polymerization of the microtubules causes a shift in the nuclear position from the cell periphery to cell center leading to altered cell polarity. This may eventually result in a symmetric division in place of the asymmetric division that is the standard for first pollen mitosis (Zhao et al. 1996), and maybe the trigger that directs microspores onto the sporophytic pathway. It is also suggested that when colchicine prevents dimer addition to α- and β-tubulin heterodimers, the excess of free tubulins that remain in the cell will inhibit the synthesis of new molecules, some of which may be specifically required for pollen development (Carpenter et al. 1992). Thus, normal gametogenesis is suppressed and embryogenesis is initiated.

From the above account it becomes clear that the microspores can be induced to become embryogenic under a variety of different stresses. The kind of stress to be applied will depend on the plant species. However, more than one stress treatment will work with microspores of some species, particularly in relation to the stage of microspore development at the time of induction. For example, in rapeseed, severe heat stress (41 ^{0}C) is required to induce efficient embryogenic development in late bi-nucleate microspores (Binarova et al. 1997), mild heat stress (32 ^{0}C) to induce late uni-nucleate to early bi-nucleate staged microspores (Ferrie and Keller 1995), whereas uni-nucleate microspores at an earlier stage of development corresponding to vacuolated stage are induced more effectively with colchicine (Zhao et al. 1996). Similarly, in tobacco bi-cellular pollen is easily induced under sucrose and nitrogen starvation pre-treatment conditions (Garrido et al. 1995, Kyo and Harada 1986) whereas microspores at the uni-nucleate stage are induced under heat shock (Touraev et al. 1996a), and centrifugal pre-treatment (Tanaka 1973), although the latter two pre-treatments have little influence in inducing embryogenesis from bi-cellular pollen. Accordingly, through careful experimentation and judicious choice of pre-treatments, it would be possible to induce microspores of a wider developmental range than anticipated before, towards successful embryogenic re-programming.

3.5 Culture medium

As with all other in vitro culture systems, successful induction of embryogenesis from cultured microspores or anthers depends to a large extent on the composition of the culture medium. The requirements of a given species are identified mainly through an empirical process of trial and error. The source of carbon, macronutrients (particularly the form in which nitrogen is supplied in the medium) and micronutrients may determine if embryogenesis will be initiated or not. The type and concentration of the growth regulators, particularly the auxins and cytokinins, as well as their interactive presence can be the deciding media factor that would influence pollen embryogenesis. Standard media have been developed for different species although specific genotypes may have their individual requirements. The more widely used basal media for anther culture are the N6 (Chu 1978), Nitsch and Nitsch (1969), MS (Murashige and Skoog 1962) and B 5 (Gamborg et al. 1968). These media are used often in their original form, but sometimes modified by supplementing or subtracting one or more components to better suite a given plant species or genotype. With more recalcitrant species, new media will have to be formulated that are tailor-made to address their specific requirements.

Carbon source: The carbon source is an important component in tissue culture media that provides the energy required for the growth of the cultured explants. The most frequently used carbon source in tissue culture media is sucrose. Sugars also play a significant role in regulating the osmotic pressure in the culture media, although this would be secondary to its main role as the source of energy. Plants belonging to the families Poaceae and Brassicaceae require fairly high sucrose levels (6 – 17%) for the induction of microspore embryogenesis (Dunwell and Thurling 1985), whereas more regular lower concentrations (2 – 5%) are used with Solanaceous species (Dunwell 2010). Since of late, maltose has come to replace sucrose as the major carbon source in cereal anther culture, usually at 6% in the induction medium and at half this strength in the regeneration medium (Wedzony et al. 2009). Maltose in the anther culture medium is degraded more slowly than sucrose and yields only glucose upon hydrolysis whereas sucrose is metabolized very rapidly into glucose and fructose. Fructose is known to have a detrimental effect on embryoid production in wheat anther culture (Last and Brettell 1990, Navarro-Alvarez et al. 1994). The reduced efficiency of androgenesis observed in the presence of sucrose as the carbon source maybe due to the sensitivity of microspores to fructose that is generated from the hydrolysis of sucrose in the culture medium. Further, the superiority of maltose over other sugars such as glucose, fructose and mannitol, has also been proven with rice anther culture (Bishnoi et al. 2000, Lentini et al. 1995). Exogenously supplied carbohydrates in the culture medium may also in part fulfill the osmotic requirements of the cells growing in vitro. Nevertheless, results indicate that the type of sugar to be used is more important as an energy source rather than in osmotic regulation of the medium. On the other hand, certain sugar alcohols such as mannitol and sorbitol have been used in microspore culture media purely for their role in osmo-regulation. Both mannitol (Raina and Irfan 1998) and sorbitol (Kishore and Reddy 1986) have had beneficial effects on rice anther culture.

Nitrogen source: Nitrogen can be supplied to the culture medium in the inorganic or organic form. The inorganic nitrogen is usually introduced in the form of nitrate or ammonium ions while nitrogen in the organic form can be supplied as vitamins and amino acid supplements. Often, anther cultures may require more than one form of the nitrogen and the correct balance of the different sources of nitrogen may be very important for successful androgenesis. For example, Chu (1978) demonstrated that the level of nitrogen in the form of ammonium ions was critical for androgenesis in rice, on which basis he developed the N6 medium with appropriate concentrations of $(NH_4)_2SO_4$ and KNO_3. While the N6 medium has become the most widely used for rice anther culture, particularly of the japonica types, the nitrogen requirement of the indica rice varieties was proved to be somewhat different. For indica rices, lowering the level of $(NH_4)_2SO_4$ and increasing the concentration of KNO_3 was shown to produce better anther response and green plant regeneration from anther-derived callus (Raina and Zapata 1997).

Micronutrients: Micronutrients play an important and sometimes a crucial role in normal plant growth and development. Deficiency symptoms arise in plants that are grown under sub-optimal levels of micronutrients. As such, the tissue culture media are also formulated with the inclusion of essential micronutrients. However, in depth studies of their influence on in vitro cell culture, particularly microspore embryogenesis, are limited. This neglect is in spite of their absolute requirement for many physiological and biochemical cellular processes, including the catalysis of enzymatic reactions. Two of the micronutrients that have been investigated for their influence on microspore embryogenesis are copper and zinc. The addition of copper sulphate in the anther pre-treatment medium allowed green plant regeneration from an otherwise exclusively albino plant producing recalcitrant barley cultivar (Jacquard et al. 2009, Wojnarowiez et al. 2002).

3.6 Incubation conditions

The temperature at which the cultures are incubated, light / dark conditions, the density of anthers or microspores in a culture vessel and other such post culture environmental conditions may have a subtle effect on the success of microspore embryogenesis, depending on the plant species or genotype. The temperature at which anther cultures are incubated has been shown to be an important factor in rice (Okamoto et al. 2001). However, very few investigations have been carried out to manipulate culture temperatures for enhanced anther culture efficiency. Light is another environmental factor that has a bearing on anther culture success because it is a stimulus that influences in vitro pollen morphogeneis (Reynolds and Crawford 1997). Generally, anther or microspore cultures are maintained in the dark during the initial phase of culture, particularly when regeneration occurs through callus, and transferred to illuminated conditions at a later stage. For some plant species diurnal alternation of incubation for several hours in the light and then darkness has been beneficial (Germana et al. 2005, Sunderland 1971). However, optimal conditions need to be determined for each system.

4. Cellular changes that occur at the initiation of microspore embryogenesis

Basic studies on microspore embryogenesis have been dealt with in detail in a few plant species only, that are considered model systems for generating haploid plants from cultured microspores. The following descriptions pertaining to the changes that occur in the microspores under induction conditions are therefore based largely on the work carried out on these model plant systems, namely; *Brassica napus* (rapeseed), *Hordeum vulgare* (barley), *Nicotiana tobaccum* (tobacco) and *Triticum aestivum* (wheat).

A microspore in which embryogenic development is initiated will display characteristic features that will distinguish it from a normal microspore that is developing into a male gametophyte. These distinguishing features have been observed to be common among many of the plant species that have been subjected to detailed study, and therefore may be adopted as cellular markers of pollen embryogenesis, allowing the development of a unified model for the microspore embryogenic induction process. The changes associated with the de-differentiation process of the microspore include changes in cellular morphology and biochemical processes as well as changes in gene expression profiles.

4.1 Changes in the cellular landscape of microspores during embryogenic induction

Following stress treatment, microspores that have been induced to become embryogenic become enlarged. The enlargement of stress-induced microspores in comparison to the relatively smaller size of the normal microspores is generally taken to be associated with their acquisition of embryogenic competence (Maraschin et al. 2005a), although all the cells that enlarge may not continue in the embryogenic pathway of development.

During early development, a normal microspore that is released from the tetrad has its nucleus located in the center of a cytoplasm–rich cell. Subsequently a vacuole develops and expands to occupy a large part of the cell, which pushes the cytoplasm with the embedded nucleus to the periphery of the cell. In the young microspores that are induced to assume embryogenic development, there is re-positioning of the nucleus from its peripheral location to the center of the cell and the large vacuole gets broken up by radial strands of cytoplasm creating several smaller vacuoles giving the cell a star-like appearance. This stellar morphology of the cell is considered to be one of the early markers that portend re-programming of the microspore for switching from gametogenesis to embryogenic development and has been observed during the embryogenic induction in the four androgenic model systems; barley, wheat, rapeseed and tobacco irrespective of whether the embryogenic induction occurs in the uni- or bi-nuclear stage microspores (Zaki and Dickinson 1991, Touraev et al. 1996, Indrianto et al. 2001, Maraschin et al. 2005).

4.2 Cytoskeletal rearrangements and changes to division symmetry

During normal development of the uni-nucleate microspore, the peripheral position of the pollen nucleus is maintained by cytoskeletal components; the microtubules and actin filaments (Hause et al. 1992). In stress-induced microspores, the re-positioning of the nucleus from the periphery of the cell to its center is considered to be associated with cytoskeletal re-arrangements. The fact that the destruction of the cytoskeleton by colchicine can induce pollen embryogenesis, of which process an initial manifestation is the migration of the nucleus, supports this proposition (Zaki and Dickinson 1991, Zhao et al. 1996, Gervais et al. 2000, Obert and Barnabas 2004). The central nuclear positioning is believed to confer radial polarity in the cell, initiating a symmetric division that results in two equal-sized cells. The plane of division is equatorial and is similar to what is observed in mitotically activated somatic cells (Segui-Simarro and Nuez 2008). Further subsequent divisions eventually lead to the formation of a multi-cellular pro-embryo. This is in contrast to what occurs during the normal process of gametogenesis. In normal pollen development the peripheral nucleus of the uni-nucleate microspore undergoes the first pollen mitosis which is an asymmetric division yielding two cells of distinctly different sizes. The vegetative cell is large and contains bulk of the pollen cytoplasm allowing its nucleus to assume a more central position in the pollen grain. The smaller generative cell, with its arch-shaped cytoplasm, remains close to the intine wall surrounded by the vegetative cell cytoplasm. The nucleus of the smaller generative cell undergoes a second mitotic division to yield the two fertilizing nuclei that carry out the double-fertilization event characteristic of the angiosperms.

However, it must be noted that the division symmetry is not a mandatory requirement for embryogenesis initiation in the microspore. For example, treating maize pollen with colchicine, although allows embryogenic induction, does not result in division symmetry (Barnabas et al. 1999). This appears to suggest that the role of cytoskeletal inhibitors on embryogenic induction in microspores is not confined to the triggering of symmetric divisions alone. Also in some plant species, stress-induced pollen embryogenesis occurs at the bi-nucleate stage, at which time the pollen cell has already undergone an initial asymmetric division, resulting in two very unequal-sized cells. This does not preclude embryogenic development in bi-nucleate B. napus microspores when subjected to heat shock at 32 ^{0}C (Custers et al. 1994) and in late bi-nucleate pollen by an extra heat shock treatment at 41 ^{0}C (Binarova et al, 1997).

As with colchicine pre-treatment, heat and cold shock have been identified to produce cytoskeletal rearrangements in microspores leading to the migration of the nucleus to the cell center, with resultant initiation of embryogenic development (Binarova et al. 1997, Wallin and Stromberg 1995). Several other stress treatments that have the ability to induce embryogenesis also affect cytoskeletal rearrangements in the microspore and the migration of the nucleus to the center of the cell.

4.3 Biochemical changes and cytoplasmic re-modeling

In normal pollen development following the first pollen mitosis, the large vacuole gets re-absorbed. The larger vegetative cell accumulates food reserves such as carbohydrates (starch), lipids and proteins, as well as RNA (Touraev et al. 1997) that powers the pollen tube's growth during its passage through the style to reach the female gametophyte. The generative cell contains much less stored products and fewer organelles. With embryogenic induction, changes occur in the cytoplasm. The cytoplasm becomes more alkaline in contrast to the slightly acidic pH that is observed in the normal microspores (Pauls et al. 2006). There is destruction of cellular organelles such as the plastids, and a decline in the synthesis of ribosomes along with a decrease in the accumulation of starch grains and lipid bodies. Thus there is an overall clearing of the cytoplasm which is suggestive of a state of de-differentiation of the microspore in preparation for the re-programming events that would follow with the initiation of embryogenic development.

Based on these observations it has been proposed that stress leads to the dedifferentiation of microspores by the repression of gametophytic development. It was believed that this re-programming was only possible in microspores that were still at a very early stage of development, and in many experimental systems this stage corresponded to the late uni-nucleate to the early bi-nucleate stage of the microspore. It was also generally accepted that androgenesis could no longer be triggered in bi-nucleate pollen cells in which starch and lipid accumulation has already occurred. The starch-filled cells were deemed irreversibly committed to gametogenesis. However, more recent evidence suggests that this window of opportunity could be made much wider than initially anticipated by precisely timing the application of the stress treatment and the type of stress (Touraev et al. 1997). For example, in *B. napus* an extra heat treatment allows even the late binucleate pollen to undergo embryogenic development.

There are two main pathways by which cytoplasmic remodeling occurs in normal eukaryotic cells. Autophagy is the process by which the destruction of cellular organelles occurs via lysosomes. The destruction of cellular molecules, both long- and short-lived ones, occurs through the ubiquitin 26S proteosomal pathway. Both these pathways are developmentally regulated, but maybe activated by subjecting cells to stress such as heat shock, cold shock and starvation. Autophagial activity mediated by lysosomes in which cellular organelles are destroyed, has been observed to occur in early embryogenesis of *N. tobaccum* microspores (Sunderland and Dunwell 1974). *H. vulgare* microspores in which embryogenic development has been induced by stress have shown expression of genes coding for enzymes of the ubiquitin 26S proteosomal pathway (Maraschin et al. 2005).

4.4 Nuclear rearrangements

In the normal microspores, at the two-celled stage, the vegetative and generative nuclei perform different functions. The vegetative cell concentrates on accumulating metabolites.

In keeping with this function, its nucleus shows high transcriptional activity and contains diffused chromatin. In contrast, the generative cell which is in preparation for a second mitotic division displays greater chromatin condensation in readiness for mitosis and is transcriptionally less active. Changes in the nuclear organization and content are observed in embryogenically induced microspores when compared with microspores that are developing into mature pollen grains. These include changes to patterns of chromatin condensation, nucleolar activity and transcriptional activity of the nucleus. Following the first symmetric division, the two cells of the embryogenic microspores typically display nuclear organization and function that are characteristic of mitotically active cells, such as condensed chromatin patches and compacted nucleoli (Germana 2011). With regard to nuclear DNA content, the doubling of the normal number of chromosomes may be observed. This appears to be a consequence of the in vitro culture conditions, and depending on the species or the genotype, may not compulsorily occur. In the event that chromosomal doubling does occur, the mechanism is believed to be nuclear fusion in the main, and this would normally occur at the initial stages of embryogenic divisions (Segui-Simarro and Nuez 2008).

5. Changes in gene expression

The microspore embryogenesis has been established to work precisely under defined conditions in four plant species, namely *Brassica napus* (rapeseed), *Hordeum vulgare* (barley), *Nicotiana tobaccum* (tobacco) and *Triticum aestivum* (wheat). The reliability with which these species lend themselves to the induction process and the unerring regularity and repeatability of success have allowed these plants to emerge as model systems on which basic studies of the microspore embryogenesis process has been pivoted. In the last decade or so, findings from the burgeoning expansion of genomic, proteomic and metabolomic research and their tools have been applied also to understand the fundamental processes leading to the switching of pathways in pollen development from gametogenesis to embryogenesis, and in the search for genes that are responsible for this turn-around.

Molecular and gene expression studies on microspore embryogenesis have been largely dealt with in the four model plants; wheat, barley, rapeseed and tobacco. This is mainly because of the repeatability and efficiency of the well-established protocols that are available in these plant species for embryogenesis induction from cultured microspores or anthers.

Early studies on molecular aspects have concentrated on identifying genes that are differentially expressed in embryogenic and non-embryogenic microspores. These studies have helped to understand the cellular processes that take place during the transition of the microspore to embryogenic development, from its original gametophytic developmental program. More recently, genomics and bioinformatics approaches have been implemented for gene discovery and functional characterization of candidate genes, whose expression appear to be associated with the embryogenic re-programming in the stress-induced microspores. Once elucidated, the genes that are identified to be expressed in common among different plant species during the initiation of embryogenesis and its progression towards embryo development, may be used as markers for early detection of the embryogenic induction process.

5.1 Genes expressed during barley microspore embryogenesis

In barley microspores, embryogenic induction can be achieved in up to 50% of the microspores by subjecting the cells to a combined mannitol stress and starvation treatment, and thus, the barley microspore embryogenic system provides a very good platform to study differential gene expression (Maraschin et al. 2005b).

In an early study by Vrinten et al. (1999), differential screening of a cDNA library prepared from mannitol-induced 3 day old barley microspores identified three genes that were expressed during embryogenesis initiation. The genes were functionally characterized as encoding a non-specific lipid transfer protein (ECLTP), glutathione-S-transferene (ECGST) and an arabinogalactan-like protein.

A more detailed study involving the screening of mRNA populations of uni-nucleate microspores about to undergo the first pollen mitosis, bi-cellular pollen, and stress-induced embryogenic microspores against microarrays of Expressed Sequence Tags (ESTs) derived from early stages of barley zygotic embryogenesis was carried out by Maraschin et al. (2006) in order to elucidate the gene expression profiles associated with each development programme. Following Principle Component Analysis (PCA) of the normalized gene expression data, it was revealed that in uni-nucleate pollen, mRNAs related to mitotic division and lipid biosynthesis were detected mainly. In bi-cellular pollen induction of genes involved in carbohydrate and energy metabolism were observed. In contrast, the embryogenic pollen displayed the expression of genes involved in protein degradation, starch and sugar hydrolysis, stress responses, inhibition of programmed cell death, metabolism and cell signaling. The gene expression profiles of stress-induced embryogenic microspores point to metabolic changes including proteolysis that appears to relate to the de-differentiation process of the induced microspores. Proteolysis is considered to play an important role in a regulatory mechanism in all cell differentiation and cell cycle progression in plant cells (Hellman and Estelle 2004) and is involved in many aspects of plant development including somatic and zygotic embryogenesis, germination, tissue re-modeling and programmed cell death (Beers et al. 2004). In microspore embryogenesis whether pollen specific proteins are degraded by the expression of genes coding for proteolytic enzymes, allowing the cells to deviate from pre-programmed gametogenesis to assume embryogenic development, remains to be proved.

5.2 Molecular basis of wheat microspore embryogenesis

Differential gene expression between stress-induced microspores and freshly isolated pollen has been studied in wheat using the Suppression Subtractive Hybridization of cDNA clones. These screens have yielded a number of differentially expressed genes (Hosp et al. 2007). Nearly one third of the genes that were differentially expressed could be assigned to functional categories based on similarity with database sequences, while the others were of unknown function or without significant matches to database sequences. A majority of the annotated sequences were found to have a metabolic function which again points to significant biochemical and physiological changes that occur during the switch from gametogenesis to embryogenesis.

A gene that has been identified to be differentially expressed between late-stage microspore derived embryos and mature pollen is one that encodes a cysteine-labeled metallothionine

(EcMt) (Reynolds and Crawford 1996). EcMt was thus indicated as a marker for microspore embryogenesis although it is also induced following Abscisic Acid (ABA) treatment in diverse tissue.

5.3 Tobacco microspore embryogenesis

In tobacco, genes encoding phosphoproteins (NtEP) have been isolated from embryogenic microspores which were shown to have selective expression in de-differentitating pollen (Kyo et al. 2002). In a subsequent study Kyo et al. (2003) isolated 16 cDNAs that were differentially expressed in embryogenic pollen of tobacco. Of these, 13 transcripts were expressed in de-differentiating pollen while the other three were observed in de-differentiating microspores as well as in cell populations undergoing active division. The genes whose expression was confined to de-differentiating pollen included the earlier observed gene coding for NtEP, stress induced genes including ABA-responsive genes, genes coding for Myb transcription factor, glucanase & chitinase, and some unknown genes. Genes that were expressed in both type of cells, the de-differentiating and dividing cells, coded for histones and a mini-chromosome maintaining protein.

Subtraction hybridization of cDNA derived from stress-induced microspores and normal pollen indicated that several genes, involved in metabolism, chromosome re-modeling, (transcription and translation) were up regulated in the induced microspores.

5.4 Rapeseed microspore embryogenesis

The discovery of genes involved in B. napus (rapeseed) microspore embryogenesis has relied to a great extent on the genomic data of its close taxonomic relative, Arabidopsis thaliana. A. thaliana is considered a model organism for genome and transcriptome studies since its genome is fully sequenced and a large number of its genes identified and functionally annotated. Transcriptome analysis of A. thaliana microspores has been carried out in detail and the genes involved in regulating the microspore development towards a male gametophyte have been characterized. However, in the absence of a protocol to initiate embryogenic development in A. thaliana microspores, there has not been the opportunity to leverage the wealth of genome and transcriptome data on normal gametogenesis, in establishing the genes that are differentially expressed during microspore embryogenesis. On the other hand, the availability of such a well-characterized genome resource, with a good deal of genome-wide similarity and identity among orthologous genes between the two species, Arabidopsis and Brassica, has meant that A. thaliana genome resource could be used as a reference database in the discovery of genes that are expressed during embryogenesis re-programming in rapeseed microspores, and differential screens of gene expression profiles between Arabidopsis and Brassica have produced fruitful interpretations.

An early study on differential gene expression between 4 day heat stressed embryogenic and non-embryogenic microspores of rapeseed identified five different cDNAs that were up regulated in the induced microspores (Boutilier et al. 2002). Two of these encoded BURP domain proteins with unknown function (Hattori et al. 1998), the third encoded an AP2/EREBP domain transcription factor which has since been named BABY BOOM (BBM), the fourth cDNA encoded the orthologue of the Arabidopsis AKT1 K^+ channel protein (Sentenac et al. 1992), while the last sequence was one to which no open reading frame had been assigned. Of these, the genes coding for BURP domain proteins were found to be expressed through out microspore and zygotic embryogenesis in rapeseed, with highest intensity during the initial period of storage product accumulation. BBM gene was found to be expressed in the early stages of microspore and zygotic embryogenesis. Over-expressing the BBM gene with a constitutive promoter resulted in the ectopic induction of somatic embryos from seedling tissue in rapeseed and *Arabidopsis*, suggesting a vital role for this gene in embryogenic induction. Further studies have successfully isolated a number of genes that are expressed during later stages of microspore embryo development, one of which is a gene encoding a CLAVATA3/ESR (CLE) family member, CLE 19. In zygotic and microspore embryogenesis the expression of this gene is restricted to globular and heart-shaped embryo development stages. This gene encodes a small secreted protein that is considered to promote cell differentiation and inhibit meristem formation in a range of plant organs (Fiers et al. 2004). Transcriptome analysis of different stages of microspore-derived embryo development has confirmed that there is gradual transition from pollen-dominated expression profiles to embryo-eclipsed profiles in mature cultures (Custers et al. 2001).

Genes that were differentially expressed during microspore embryogenesis in rapeseed have been isolated by subtractive hybridization, and a subtracted cDNA library constructed (Tsuwamoto et al. 2007). After sequencing over 2000 clones that showed differential expression in embryogenically-induced and non-embryogenic microspores, the non redundant sequences selected were searched against the well annotated Arabidopsis Munich Information center for Protein Sequences (MIPS) database to identify functional categories of the differentially expressed genes. When characterized, many sequences related to embryo-specific genes, the most frequent among them being the gene coding for Lipid Transfer Protein (LTP). Although the detailed function of LPTs is mostly unknown, several genes encoding LTPs have been found to be expressed in different embryogenic systems such as zygotic embryos in maize (Sossountzov et al. 1991), somatic embryos in carrot (Sterk et al. 1991), and microspore-derived embryos in barley (Vrinten et al. 1999), suggesting that it has a definite role in the embryogenesis process. In addition, ESTs of genes that code for napin and cruciferine (seed storage proteins in *Brassica*), oleosin (a major component of oil bodies) and phytosulfokine have been identified as being up-regulated during embryogenic induction. A detailed analysis of the expression patterns of 15 selected genes, determined by quantitative real-time reverse transcription-polymerase chain reaction (RT-PCR), found all of them to be highly expressed in the early stages of microspore embryogenesis, but poorly expressed in microspores when freshly isolated (before induction) or cultured under non-embryogenic induction conditions. The Principle Component Analysis based on the expression profiles of the 15 genes placed them in two groups; those having high expression during androgenic initiation and those expressed in the early to middle stage (globular to torpedo stage) of embryogenesis.

Malik et al. (2007) conducted a detailed examination of transcript profiles of embryogenic microspores and pollen-derived embryos in *B. napus* that represented a developmental series, starting from fresh pollen (0 d) to pollen after 3 d (induced), 5 d (dividing microspores), and 7 d stress treatments. Based on the *A. thaliana* information resource, they described ESTs that were most abundant at each stage. According to this study, microspores after a 3 d stress treatment have lost gene transcripts associated with protein synthesis (40 S and 60 S ribosomal proteins, initiation and elongation factors). These transcriptional changes may well relate to the cytoplasmic clearing of the de-differentiation process. In the cDNA libraries derived from microspore cultures following 3 d to 5 d stress treatments, a large number of pollen specific transcripts (pectinesterase, exopolygalactouronase, Bnm1, BP4) were isolated. However, it was not clear if these expression profiles were observed because the cultures may still have contained microspores that were not induced but were developing into mature pollen, or because during the period of transition, from pollen development to early stages of embryo induction, there was parallel expression of both pollen- and embryo-specific genes. A transcript that was abundant in all stages of development after induction (3 d, 5 d, and 7 d) was Bnm1 (invertase / pectin methylesterase inhibitor). This has been earlier shown to be expressed in the early stages of microspore embryogenesis in rapeseed up to globular stages (Treacy et al. 1997). In the 5 d and 7 d cultures, AGP gene transcripts were abundant. AGPs are a large multi-gene family coding for glycosylated Hyp-rich proteoglycans. While no specific function has been determined for AGPs, at least some AGPs have been implicated in the regulation of somatic embryogenesis (Chapman et al. 2000) and microspore embryogenesis (Letarte et al. 2006). In the 7 d microspore cultures, LTP were frequent. The LTP are able to transport phospholipids across cell membranes and are required for functioning of several biological processes such as embryogenesis, defense reaction, adaptations to stress and cutin formation (Kader 1996, Wang et al. 2005). LTPs are expressed in both somatic and zygotic embryos (Sterk et al. 1991). The cDNA libraries constructed from 5-day dividing and 7-day embryogenic microspores contained a number of ESTs for BnCYP78A whose functions are still not known.

From the various analyses of gene expression profiles carried out with stress induced microspores of the four model species, copious amount of data has been generated and is continued to be produced on the different types of genes that are found to be turned on or differentially expressed between the two developmental pathways of the microspore. Although a lot of concentration has been applied in trying to determine the genes that are involved in regulating the stress-induced embryogenic response in microspores and to understand the gene expression patterns that follow, the information gathered from the model species is far from complete at present, and thus elude a unified scheme whereby one could predict the expression of a common cascade of genes that would cause the switching of pathways from gametogenesis to embryogenesis that result in embryo development. Given the complexity of development re-programming that would have had to occur, and considering that there are different pathways by which embryogenesis is induced in microspores of different species, it is not surprising that a clear overall picture has not emerged yet.

However, the genes that have so far been identified as being specifically up-regulated during pollen embryogenesis have been broadly categorized into three main groups; stress responsive genes, gametogenesis-repressive genes and embryogenesis-related genes with functions identified as below.

5.5 Stress responsive genes

During microspore embryogenesis the up regulation of the Head Shock Protein (HSP) genes is observed, not only in response to heat stress but also other stresses applied on microspores such as starvation and colchicine. However, a distinct role for HSP on microspore embryogenesis has yet to be clearly established.

5.6 Gametogenesis repressive genes

Suppression of the gametophytic development must occur in the microspore before the genes for embryogenesis can be switched on (Hosp et al. 2007). Accumulation of starch in the microspore signifies that the cell has embarked on a gametophytic pathway of development where as the inhibition of starch accumulation portends the initiation of embryogenesis. The genes involved in starch synthesis and accumulation are seen to be down regulated during embryogenesis (Maraschin et al. 2006).

5.7 Embryogenesis related genes

Microspore embryogenesis occurs in a pattern of morphogenesis akin to zygotic embryo development. Based on this similarity, embryo pattern regulators such as members of the gene family 14-3-3 of barley have been observed to be up regulated, both in time and space (Maraschin et al. 2003).

6. Development of a multi-cellular pollen embryo

An induced microspore will undergo several rounds of mitotic divisions that result in a number of cells, contained together bounded by the exine or the outer wall of the microspore. There can be different pathways by which muli-cellular structures are produced from an initial star-like microspore. The four basic routes, based on the symmetry of the initial division of the microspore, as described by Razdan (2003) are as follows.

Pathway I: The initial division of the uni-nucleate microspore is symmetrical and results in two cells of equal size. Each of these cells undergoes further divisions producing a ball of cells within the exine.

Pathway II: An asymmetric initial division produces two cells of unequal size. The ball of cells may originate from the continued division of the larger vegetative cell while the smaller generative cell may divide a few times before degenerating.

Pathway III: Following an initial asymmetric division, the generative cell divides successively yielding the multi-cellular state. The vegetative cell remains arrested in division.

Pathway IV: Following an initial asymmetric division, both vegetative and generative cells will be equally active and divide to give rise to the multi-cellular structure.

The first pathway is considered to be the more common, and is observed in many species that are capable of embryogenic initiation from uni-nucleate microspores, prior to the first pollen mitosis (Smykal 2000). Apart from the stage of microspore development (uni- or bi-nucleate) at which it transits from gametogenesis to embryogenesis, it is also possible that the genotype or the plant species, as well as the stress conditions applied to induce embryogenesis in the microspore may also be factors that determine the pathway of cell division that leads to the development of multi-cellular structures from star-like microspores.

Irrespective of the above early pattern of microspore divisions, the embryogenic multi-cellular structures ultimately burst out through the restraining pollen wall, gradually assuming the form of a globular embryo akin to the globular stage of the zygotic embryo. A distinct outer layer of cells that surround a core of meristematic cells is the usual cellular anatomy that is observed in the microspore embryo at early stages of its development. The two types of cells differ in cellular morphology and metabolic processes that distinguish their developmental competences. The cells in the peripheral layer are vacuolated and filled with starch grains, which is suggestive of their differentiated nature, whereas cells to the interior are cytoplasm-filled and devoid of starch deposits which indicates their undifferentiated state and which are in fact capable of further meristematic activity (Barany et al. 2010a). Differences in the organization of cell wall components, particularly the distribution of polysaccharides, have also been noted and linked with the changes that occur in the cellular developmental processes. In microspore-derived embryos of *Capsicum annum* (pepper), cells in the peripheral layer contain a higher level of de-esterified pectins as wall components, a feature that is common with the cell wall of the mature pollen grain which is committed to normal gametogenesis. On the other hand esterified pectins are found in the cell walls of proliferating inner cells of the microspore embryo as with the walls of very young microspores prior to their gametophytic commitment (Barany et al. 2010b). Thus, the esterification of pectins in the cell walls may be considered as a cellular marker of their state of de-differentiation or meristematic activity, and therefore being capable of subsequent re-programming towards embryogenic development.

During further development, the microspore-derived globular embryos will follow the normal stages of post-globular embryogeny as seen in zygotic embryo development. As such, passage through heart-shaped and torpedo-shaped stages will culminate in the development of cotyledonary-staged mature somatic embryos that are ready to germinate into full-fledged sporophytic plants.

Alternatively, the multi-cellular mass that is liberated from the rupture of the microspore wall may proliferate to form a callus, from which organogenesis occurs by the regeneration of shoots and roots following transfer to culture media with appropriate growth regulators. It may sometimes be possible to obtain androgenic haploids either via embryo formation or organogenesis from callus in the same species by the manipulation of the chemical components of the culture medium.

7. References

Barany I, Fadon B, Risueno MC and Testillano PS. 2010a. Microspore reprogramming to embryogenesis induces changes in cell wall and starch accumulation dynamics associated with proliferation and differentiation events. Plant Signaling and Behavior 5(4):341 – 345.

Barany I, Fadon B, Risueno MC and Testillano PS. 2010b. Cell wall components and pectin esterification levels as markers of proliferation and differentiation events during pollen development and pollen embryogenesis in Capsicum annum L. Journal of Experimental Botany 61(4):1159 – 1175.)

Barnabas B. 2003a. Protocol for producing doubled haploid plants from anther culture of wheat (*Triticum aestivum* L.). In: Maluszynaki M, Kasha KJ, Forster BP, Szarejko I.

(eds.). Doubled haploid production in crop plants, a manual. Kluwer Academic Publishers, Dordrecht, The Netherlands, pp 65 – 70.

Barnabas B, Obert B, Kovacs G. 1999. Colchicine, an efficient genome doubling agent for maize (Zea mays L.) microspores cultured in anthero. Plant Cell Reports 18:858 – 862.

Beers EP, Jones AM, Dickerman AW. 2004. The S8 serine, C1A cysteine and A1 aspartic protease families in Arabidopsis. Phytochemistry 65:43 – 58.

Binarova P, Hause G, Cenklova V, Cordewener JHG, van Lookeren Campagne MM. 1997. A short severe heat shock is required to induce embryogenesis in late binuclear pollen of Brassica napus L. Sexual Plant Reproduction 10:200 – 208.

Bonet FJ, Azhaid L & Olmedilla A (1998). Pollen embryogenesis: atavism or totipotency? Protoplasma 202:115 – 121.

Boutilier K, Offringa R, Sharma VK, Kieft H, Ouellet T, Zhang L, Hattori J, Liu CM, van Lammeren AA, Miki BL, Custers JB, van Lookeren Campagne MM. 2002. Ectopic expression of BABY BOOM triggers a conversion from vegetative to embryonic growth. Plant Cell 14:1737 – 1749.

Carpenter IL, Ploense SE, Snustad DP, Silflow CD. 1992. Preferential expression of an α-tubulin gene of Arabidopsis in pollen. Plant Cell 4:557 – 571.

Chu C. 1978. The N6 medium and its application to anther culture of cereal crops. In: Proc Symp Plant Tissue Culture. Science Press, Peking, pp 43 – 50.

Chugh A and Eudes F. 2008. Isolated microspore embryogenesis in cereals: aspects and prospects. In: Recent advances in plant biotechnology and its applications. Ed. KH Neumann, A Kumar and SK Sopory. IK International Publishing House, Pvt. Ltd., New Delhi. p 205 – 226.

Cistue L, Riomas A, Castillo AM and Romagosa L. 1994. Production of a large number of doubled haploid plants from barley anthers protected with high concentration of mannitol. Plant Cell Report 13:709 – 712.

Cordewener JHG, Hause G, Gorgen E, Busink R, Hause B, Dons HJM, Van Lammeren AAM, Van Lookeren Campagne MM and Pechan P. 1995. Changes in synthesis and localization of members of the 70-kDa class of heat-shock proteins accompany the induction of embryogenesis in Brassica napus L. microspores. Planta 196:747 – 755.

Custers JBM, Cordewener JHG, Fiers MA, Maasen BTH, van Lookeren Campagne MM, Liu CM. 2001. Androgenesis in Brassica: a model system to study the induction of plant embryogenesis. In: Bhojwani SS, Soh WY (Eds). Current trends in the embryology of angiosperms. Kluwer Academic Publishers, Dordrecht. Pp. 451 – 470.

Custers JBM, Cordewener JHG, Nollen Y, Dons JJM, van Lookeren Campagne MM. 1994. Temperature controls both gametophytic and sporophytic development in microspore cultures of Brassica napus. Plant Cell Reports 16:267 – 271.

Datta SK. 2005. Androgenic haploids: factors controlling development and its application in crop improvement. Current Science 89(11):1870 – 1878.

Dunwell JM. 2010. Haploids in flowering plants: origins and exploitation. Plant Biotechnol. J. 8:377 – 424.

Dunwell JM and Thurling N. 1985. Role of sucrose in microspore embryo production in Brassica napus ssp. Oleifera. J. Exp. Bot. 36:1478 – 1491.

Fiers M, Hause G, Boutilier K, Casamitjana-Martinez E, Weijers D, Offringa R, van der Geest L, van Lookeren Campagne M, Liu CM. 2004. Mis-expression of the CLV3/ESR-like gene CLE 19 in *Arabidopsis* leads to a consumption of root meristem. Gene 18:37 – 49.

Herrera JC, Moreno LG, Acuna JR, Pena MD, Osorio D. 2002. Colchicine-induced microspore embryogenesis in coffee. Plant Cell Tiss Org Cult 71:89 – 92.

Gamborg OL, Miller RA, Ojima K. 1968. Nutrient requirement suspension cultures of soybean root cells. Exp Cell Res 50:151 – 158.

Garrido D, Vicente O, Heberle-Bors E, Isabel Rodriguez-Garcia M. 1995. Cellular changes during the acquisition of embryogenic potential in Nicotiana tobaccum. Protoplasma 186:220 – 230.

Germana MA. 2011. Anther culture for haploid and doubled haploid production. Plant Cell Tissue and Organ Culture 104:283 – 300.

Germana MA, Chiancone B, Iaconia C, Muleo R. 2005. The effect of light quality on anther culture of Citrus clementina Hort. Ex Tan. Acta Physiol Plant 27(4B):717 – 721.

Gervais G, Newcomb W, Simmonds DH. 2000. Re-arrangement of actin filament and microtubule cytoskeleton during induction of microspore embryogenesis in Brassica napus L. cv. Topas. Protoplasma 213:194 – 202.

Grewal RK, Lulsdorf M, Croser J, Ochatt S, Vandenberg A and Warkentin TD. 2009. Doubled haploid production in chick pea (Cicer arietinum L.): role of stress treatments. Plant Cell Rep 28:1289 – 1299.

Guha S. and Maheshwari SC. 1964. In vitro production of embryos from anthers of *Datura*. Nature 204:497.

Hause G, Hause B, van Lammeren AAM. 1992. Microtubular and actin filament configurations during microspore and pollen development in Brassica napus cv. Topas. Canadian Journal of Botany 70:1369 – 1376.

Hattori j, Boutilier K, van Lookeren Campagne MM, Miki BL. 1998. A conserved BURP domain defines a novel group of plant proteins with unusual primary structures. Mol Gen Genet 259:424 – 428.

Hellmann H and Estelle M. 2004. Plant development: regulation by protein degradation. Science 297(5582):793 – 797.

Hoekstra S, van Zijderveld MH, Louwerse JD, Heidekamp F, van der Mark F. 1992. Anther and microspore culture of *Hordeum vulgare* L. cv. Igri. Plant Science 86:89 – 96.

Hosp J, Maraschin SF, Touraev A and Boutilier K. 2007. Functional genomics of microspore embryogenesis. Euphytica 158:275 – 285.

Indrianto A, Barinova I, Touraev A, Heberle-Bors E. 2001. Tracking individual wheat microspores in vitro: identification of embryogenic microspores and body axis formation in the embryo. Planta 212:163 – 174.

Jaattela M, Wissing D, Kokholm K, Kallunki T, and Egeblad M. 1998. HSP70 exerts its anti-apoptic function downstream of capsase-3-like proteases. EMBO J. 17:6124 – 6134.

Kishor PBK and Reddy GM. 1986. Retention and revival of regenerating ability by osmotic adjustment in long-term cultures of four varieties of rice. J. Plant Physiol. 126:49 – 54.

Kyo M and Harada H. 1986. Control of the developmental pathway of tobacco pollen in vitro. Planta 168:427 – 432.

Kyo M, Yamaji N, Yuasa Y, Maeda T, and Fukui H. 2002. Isolation of cDNA coding for NtEPb1-b3 marker proteins for pollen de-differentiation in a tobacco pollen culture system. Plant Sci 163:1055 – 1062.

Kyo M, Hattori S, Yamaji N, Pechan P and Fukui H. 2003. Cloning and characterization of cDNAs associated with the embryogenic de-differentiation of tobacco immature pollen grains. Plant Sci 164(6):1057 – 1066.

Malik MR, Wang F, Dirpaul JM, Zhou N. Polowick PL, Ferrie AMR, Krochko JE. 2007. Transcript profiling and identification of molecular markers for early microspore embryogenesis in *Brassica napus*. Plant Physiology 144:134 – 154.

Maluszynski M, Kasha KJ, Szarejko I 2003. Published double haploid protocols in plant species. In: Maluszynski M, Kasha K, Forster BP, Szarejko I (eds.). Doubled haploid production in crop plants. A manual. Kluwer, Dordrecht, pp 309 – 335.

Maraschin SF, Priester W de, Spaink HP and Wang M. 2005a. Androgenic switch: an example of plant embryogenesis from the male gametophyte perspective. Journal of Experimental Botany 56:1711 – 1726.

Maraschin SF, Vennik M, Lamers GEM, Spaink HP, Wang M. 2005b. Time-lapse tracking of barley androgenesis reveals position-determined cell death within pro-embryos. Planta 220:531 – 540.

Maraschin SF, Caspers M, Potokina E, Wulfert F, Graner A, Spaink HP and Wang M. 2006. cDNA array analysis of stress-induced gene expression in barley androgenesis. Physiologia Plantarum 127:535 – 550.

Murashige T and Skoog F. 1962. A revised medium for rapid growth and bioassays with tobacco tissue cultures. Physiol. Plantarum 15:473 – 497.

Nitsch JP and Nitsch C. 1969. Haploid plants from pollen grains. Science 163:85 – 87.

Obert B, Barnabas B. 2004. Colchicine induced embryogenesis in maize. Plant Cell Tissue and Organ Culture 77:283 – 285.

Okamoto Y, Kinoshita A, Satake T. 2001. Enhancement of the frequency of callus formation and plant regeneration from rice anther culture by alternating temperature. Breed Res 3:87 – 94 (in Japanese).

Palmer CED and Keller WA. 2005. Overview of haploidy. In: Haploids in Crop Improvement II. Ed. CE Palmer, WA Keller and KI Kasha. Biotechnology in Agriculture and Forestry Vol. 56, Springer Verlag, Berlin, Heidelberg pp: 3 – 7.

Pauls KP, Chan J, Woronuk G, Schulze D, Brazolot J. 2006. When microspores decide to become embryos – cellular and molecular changes. Can J Bot Rev Can Bot 84:668 – 678.

Raina SK and Irfan ST. 1998. High frequency embryogenesis and plantlet regeneration from isolated microspores of indica rice. Plant Cell Rep. 17:957 – 962.

Raina SK and Zapata FJ. 1997. Enhanced anther culture efficiency in indica rice (*Oryza sativa* L.) through modification of the culture media. Plant Breed 116(4):305 – 315.

Razdan MK. 2003. Introduction to plant tissue culture. 2nd Ed. Science Publishers Inc. Enfield, NH, USA.

Reynolds TL and Crawford RL. 1997. Effects of light on the accumulation of abscisic acid and expression of an early cysteine-labeled metallothionein gene in microspores of *Triticum aestivum* during induced embryogenic development. Plant Cell Rep 16(7):458 – 463.

Schoffl F, Prandl R and Reindl A. 1998. Regulation of heat shock response. Plant Physiology 117:1135 – 1141.

Segui-Simarro JM and Nuez F. 2008. How microspores transform into haploid embryos: changes associated with embryogenesis induction and microspore-derived embryogenesis. Physiologia Plantarum 134:1 - 12.

Sentenac H, Bonneaud N, Minet M, Lacroute F, Salmon JM, Gaymard F, Grignon C. 1992. Cloning and expression of a plant potassium ion transport system. Science 256:663 - 665.

Shariatpanahi ME, Bal U, Heberle-Bors E, and Touraev A. 2006. Stresses applied for the re-programming of plant microspores towards in vitro embryogenesis. Physiologia Plantarum 127:519 - 534.

Silva TD. 2010. Indica rice anther culture: can the impasse be surpassed? Plant Cell Tissue and Organ Culture 100:1 -11.

Silva TD and Ratnayake WJ. 2009. Anther culture potential of indica rice varieties, Kurulu thuda and Bg 250. Tropical Agricultural Research and Extension 12(2):53 - 56.

Simmonds DH and Keller WA. 1999. Significance of preprophase bands of microtubules in the induction of microspore embryogenesis in *Brassica napus*. Planta 208:383 - 391.

Smykal P. 2000. Pollen embryogenesis - the stress mediated switch from gametophytic to sporophytic development. Current status and future prospects. Biol. Plant. 43:481 - 489.

Sternlicht H, Ringel I, Szasz J. 1983. Theory for modeling the copolymerization of tubulin and tubulin-colchicine complex. Biophys J 42:255 - 267.

Sunderland N. 1971. Anther culture: a progress report. Sci Prog (Oxford) 59:527 - 549.

Tanaka M. 1973. The effect of centrifugal treatment on the emergence of plantlets from cultured anther of tobacco. Japan J Breed 23:171 - 174.

Toureav A, Illham A, Vincente O, Heberle-Bors E (1996a). Stress induced microspore embryogenesis from tobacco microspores: an optimized system for molecular studies. Plant Cell Reports 15:561 - 565.

Toureav A, Indrianto A, Wratschko I, Vincente O, Heberle-Bors E (1996b). Efficient microspore embryogenesis in wheat (*Triticum aestivum* L.) induced by starvation at high temperatures. Sex Plant Reprod 9:209 - 215.

Touraev A, Pfosser M, Vicente O, Heberle-Bors E. 1996. Stress as the major signal controlling the developmental fate of tobacco microspores: towards a unified model of induction of microspore/pollen embryogenesis. Planta 200:144 - 152.

Touraev A, Vicente O and Heberle-Bors E. 1997. Initiation of micropsore embryogenesis by stress. Trends in Plant Science 2(8):297 - 302.

Vrinten PL, Nakamura T, Kasha KJ. 1999. Characterization of cDNAs expressed in the early stages of microspore embryogenesis in barley (*Hordeum vulgare* L.). Plant Mol. Biol. 41(4):455 - 463.

Wallin M, Stromberg E. 1995. Cold-stable and cold-adapted microtubules. International Review of Cytology 157:1 - 31.

Wedzony M, Forster BP, Zur I, Golemiec E, Szechynska-Hebda M, Dubas, Gotebiowska G. 2009. Progress in doubled haploid technology in higher plants. In: Touraev A, Forster BP, Jain SM, (eds.). Advances in haploid production in higher plants. Springer, Berlin, pp 1 - 34.

Yan J, Xue Q and Zhu J. 1996. Genetic studies of anther culture ability in rice (*Oryza sativa*). Plant Cell Tissue and Organ Culture 45:253 - 258.

Zaki MAM, Dickinson HG. 1991. Microspore-derived embryos in *Brassica*: the significance of division symmetry in pollen mitosis I to embryogenic development. Sexual Plant Reproduction 4: 48 – 55.

Zarsky V, Garrido D, Eller N, Tupy J, Vicente O, Schoffl F and Heberle-Bors E. 1995. The expression of a small heat shock gene is activated during induction of tobacco pollen embryogenesis by starvation. Plant Cell Environ. 18:139 – 147.

Zhao JP, Simmonds DH, Newcomb W. 1996. Induction of embryogenesis with colchicine instead of heat in microspores of *Brassica napus* L. cv. Topas. Planta 189:433 – 439.

10

Presumed Paternal Genome Loss During Embryogenesis of Predatory Phytoseiid Mites

Shingo Toyoshima[1] and Hiroshi Amano[2]
[1]*NARO Institute of Vegetable and Tea Science,*
[2]*Graduate School of Agriculture, Kyoto University,*
Japan

1. Introduction

Predatory phytoseiid mites are classified into the family Phytoseiidae (Acari: Mesostigmata), the most diverse group of mesostigmatic mites (Kranz & Walter, 2009). More than 2000 species of phytoseiid mites have been described (Chant & McMurtry, 2007), almost all of which are small (0.3 mm–0.4 mm; Fig. 1a) and eat other mites, insects, pollen, and fungi. Since they also prey upon pest insects and mites in agricultural fields, they are considered to be a key agent in an integrated pest management system (Gerson et al., 2003). To understand their role in agriculture, their morphology (external and internal), life history characteristics, and behavioral traits have been studied for more than 50 years (e.g., Helle & Sabelis, 1985). Several species of phytoseiid mites, mentioned below, are useful agents and are the most studied species (the recent name is given in parentheses). Because of their small size, they are not considered as an experimental animal in anatomical analyses examining their life cycle characteristics.

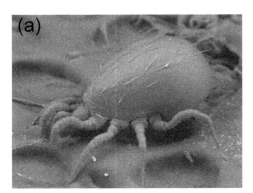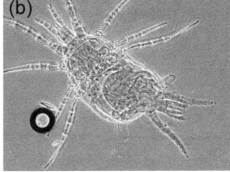

Fig. 1. External and internal appearance of phytoseiid mites. (a) *Neoseiulus womersleyi*. Body length is ca. 400 μm (scanning electron microscope image). (b) An example of delayed oviposition in *N. womersleyi*. Embryonic development proceeds regardless of the nutritional condition of the mother, and the larvae hatch in the mother's body (phase contrast microscope image).

The reproductive system of predatory phytoseiid mites has interested scientists for many years. While adult females start oviposition only after copulation, males and females were found to be haploids and diploids, respectively, by the karyotyping of eggs (Wysoki, 1985). Heterochromatinization in the part of the chromosomes was observed in eggs just after deposition (Nelson-Rees et al., 1980). Although cytological evidence is insufficient, this has been presented as pseudo-arrhenotoky: male-destined oocytes need to be fertilized to begin embryogenesis, and the heterochromatinization of the paternal genomes occurs in male-destined eggs, inducing the elimination of paternal genomes during embryogenesis, resulting in functionally haploid males (Schulten, 1985). This system was also described as "parahaploidy" in previous studies (e.g., Hoy, 1979) and was recently investigated as a form of "paternal genome loss" presented in several insects (Burt & Trivers, 2006). An explanation of several terms in relation to the reproductive system of phytoseiid mites is presented below.

"Pseudo-arrhenotoky" is described as a mode of reproduction, differing from true "arrhenotoky" in which haploid males in the arrhenotokous insects and mites are emerged parthenogenetically from unfertilized eggs. "Deuterotoky" and "thelytoky" are also known as parthenogenetic reproductive modes whereby mothers produce offspring of both sexes from unfertilized eggs (deuterotoky) and only female offspring from unfertilized eggs (thelytoky) (Bell, 1982; Norton et al., 1993). "Thelytoky" is also known in a few species of phytoseiid mites. "Parahaploidy" is described as a ploidy of the chromosomes, differing from true "haploidy" in males, which the haploid males are derived from haploid eggs (Hartl & Brown, 1970). "Paternal genome loss (PGL)" is a phenomenon of chromosome behavior and is divided into two classes, embryonic PGL and germ-line PGL (Ross et al., 2010b). In embryonic PGL, the paternal genome is eliminated during early embryonic development of males and is found in some armored scale insects. In germ-line PGL, the paternal genome is deactivated during male embryogenesis and is eliminated from the germline during or just before spermatogenesis. Germ-line PGL is found in most scale insects, sciarid flies, and in the coffee borer beetle. Since the prefixes "pseudo-" and "para-" are ambiguous, one may better describe the reproductive system of phytoseiid mites in terms of "PGL in diploid arrhenotoky," as in the case of scale insects (Ross et al., 2000a). Although "paternal genome elimination (PGE)" is a more accurate description of the same phenomenon (Herrick & Seger, 1999; Ross et al., 2010a, b), PGL is used in this article. Other terms should be referred to Bell (1982).

PGL in mites is known in three families: Anoetidae (phoretic with insects), Dermanyssidae (blood-feeding ectoparasites), and Phytoseiidae (Oliver, 1983). PGL in Anoetidae and Dermanyssidae has not been well studied and additional information is not available. In phytoseiid mites, although cytological evidence for PGL is insufficient to understand it thoroughly, elaborate hypotheses were developed to investigate their evolutionary significance (e.g., Nagerkerke & Sabelis, 1998). PGL in mites is believed to be an intermediate step from diplodiploid bisexual reproduction to haplodiploid arrhenotoky (Bull, 1983), as noted previously for PGL in scale insects (Schrader & Hugher-Schrader, 1931). Also, the evolutionary constraint of PGL in phytoseiid mites was examined from the viewpoint of sex ratio control (Sabelis et al., 2002). In recent studies, the evolution of PGL in scale insects was considered a consequence of genetic conflict between males and females, as well as between parents and offspring (Ross, 2010a). More genetic and cytological evidence of PGL in mites is needed to incorporate it into the general framework of genetic conflict in scale insects to better understand the mechanisms of sex determination and evolutionary significance of PGL in mites.

In this article, previous researches and additional evidence in relation to PGL in phytoseiid mites are briefly reviewed with a plenty of literatures, in order to interest scientists working on the different research fields. Authors hope that these scientists will contribute new ideas toward the completion of the reproductive biology of phytoseiid mites. Further understanding of the PGL in phytoseiid mites will make a contribution, hopefully, to the comprehensive researches on the reproductive biology and embryogenesis of various creatures.

2. Experimental evidence for the notion, pseudo-arrhenotoky

Recent knowledge of PGL in phytoseiid mites, referred to by many scientists working on reproductive biology, is based upon experimental evidence published in the 1970s and reviewed in the 1980s. Almost all data are well documented and reliable. For instance, behavioral observation established some basic information as follows: (1) Phytoseiid females need to copulate with males to produce female and male offspring. (2) After copulation, they drastically change their prey-search activity and prey consumption. (3) Around 24–36 h after copulation, they start oviposition when fed abundant prey under comfortable conditions (Fig. 2). Karyotyping revealed that males are haploid and females are diploid. However, these data and experimental evidence are insufficient to support the mechanism of pseudo-arrhenotoky. In this section, the experimental evidence to support pseudo-arrhenotoky is summarized, and problems are pointed out.

2.1 Indirect evidence for insemination in male eggs

Based upon the idea that sperm only serve to activate the ovary to start egg production, a belief was held that all eggs are deposited with a male-biased sex ratio even when a few sperms are accepted by females during copulation. However, Amano and Chant (1978) and Schulten et al. (1978) showed that the amount of egg production is correlated with the amount of sperm accepted by females during copulation in *Phytoseiulus persimilis, Amblyseius andersoni,* and *Amblyseius* (= *Neoseiulus*) *bibens.* For instance, in a study of *P. persimilis,* in which copulation was artificially interrupted (Amano & Chant, 1978), the average number of eggs deposited gradually increased from 2.7 to 66.3 when the duration of copulation increased from 30 min to 130 min. It was experimentally confirmed in advance that the duration of copulation was nearly correlated with the amount of sperms accepted by females.

These experimental results suggest not only that insemination activates the ovary for egg production, but also that each egg requires fertilization for its embryonic development. However, the presumption is made that the role of sperm is similar to what is known as "pseudogamy" or "gynogenesis." In pseudogamy or gynogenesis, eggs only develop after penetration by sperm, but the sperm nucleus degenerates without fusing with the egg nucleus so that it makes no genetic contribution to the developing embryo (White, 1973). It may be partially true that such "borrowed sperms" help to produce male offspring in phytoseiid mites, because male offspring were produced by crossbreeding between closely related species, and these individuals (sons) had only maternal characters (Congdon & McMurtry, 1988; Ho et al., 1995).

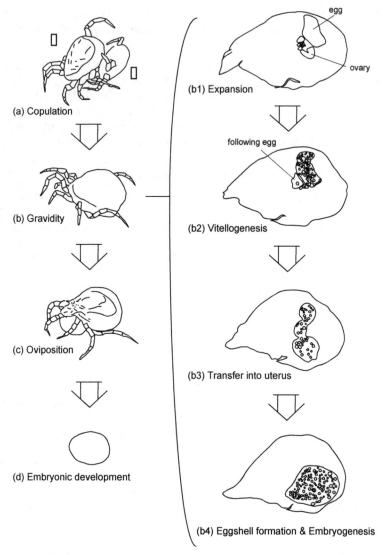

Fig. 2. Schematic illustration of reproductive events and the egg forming process during the gravid period. **Schematic flow on the left (a–d):** A virgin female copulates for 2–3 h with a male. Mated females consume more prey than virgin females and become fat with a roundish body. Gravid females lay an egg 24–36 h after copulation and continue deposition of eggs, one by one, at intervals of ca. 8 h during which they consume large amounts of prey under ideal conditions. The egg is laid in the middle of embryogenesis. **Schematic flow on the right (b1–b4):** After copulation, an oocyte expands toward the dorsal region from an ovary located in the center of the body; it is fertilized there and begins vitellogenesis. After expansion of other eggs, the first egg moves into the uterus in the ventral region and forms an eggshell there.

It was also shown at the almost same time that X-ray irradiation on adult males (fathers) induced sterility in male offspring (sons) in *P. persimilis*, *A. bibens* (Helle et al., 1978), and *Metaseiulus* (= *Galendromus*) *occidentalis* (Hoy, 1979). No daughters emerged, and total developing offspring was low in these experiments. A significant reduction in the expected number of sons was also seen. Many died in the early embryonic stages, while the survivors were sterile, even though mortality is not expected in the case of arrhenotoky. Thus, the irradiated paternal genomes are presumed to affect insemination and/or embryogenesis of daughters and sons, also suggesting that male offspring (sons) possess paternal genomes in their germ plasma, since the effect of radiation on the paternal genomes was seen in the male offspring as sterility.

2.2 Chromosome observations

According to studies on the karyotyping of many phytoseiid species (Hansell et al., 1964; Wysoki & Swirski, 1968; Wysoki, 1973; Blommers-Schlosser & Blommers, 1975; Wysoki & McMurtry, 1977; Wysoki & Bolland, 1983), most have a haploid number of 4 and a diploid number of 8 chromosomes, and 3 thelytokous species have 8 chromosomes, except for 3 (haploid) and 6 (diploid) chromosomes in 5 species (Wysoki, 1985). The chromosomes are generally acrocentric except for some metacentric one in a few species, and differ from each other only in size (1–4 μm). Since the ratio of haploid to diploid eggs was equal to that of males to females (Hansell et al., 1964), a haplodiploid genetic system was presumed in phytoseiid mites except for the thelytokous species. Although male diploidy was not confirmed in the karyotype investigations, heterochromatinized (heteropicnotic) chromosomes were observed in eggs immediately after deposition in *M. occidentalis* (Nelson-Rees et al., 1980).

Based on chromosome observation, pseudo-arrhenotoky was proposed (Schulten, 1985). According to this proposal, all eggs start their development from syngamy (2n = 6). The process of heterochromatinization starts within 24 h after egg deposition by arrangement of the 6 chromosomes into 2 groups: 3 heterochromatic chromosomes and 3 euchromatic chromosomes. The following stage strongly resembles the formation of bivalents, which is normally found at the diplotene stage in meiotic division. The heterochromatic chromosomes (n = 3) are eliminated from some or all cells, but almost certainly from the germ line. Thereafter, the male germ line and most somatic cells are haploid. Spermatogenesis in deutonymphs (immature stage) starts with a single equatorial mitotic division, and 2 sperm are produced.

Unfortunately, several events in the proposal have not been elucidated cytologically. In addition, the observation of heterochromatic chromosomes was not conducted in male eggs (the sexes of eggs were not specified), and the number of eggs observed was not sufficient to confirm diploidy in the early embryonic stages of deposited eggs. Furthermore, *M. occidentalis* is not a common species with regard to its chromosome number. Wysoki (1985) reported that 47 of 55 species examined have a basic number of n = 4 and that *M. occidentalis* has n = 3, which is exceptional and known in only 5 species. Therefore, further cytological evidence in a species with the common number of chromosomes (n = 4) is required to confirm the existence of pseudo-arrhenotoky or PGL in phytoseiid mites. In addition, no observations were made of the internal process of egg formation (Fig. 2b1–4) and embryogenesis in the proposal.

3. Presumed Paternal Genome Loss (PGL) in phytoseiid mites

In this section, the unrecognized and unpublished evidence for PGL in phytoseiid mites is briefly reviewed. The fusion of pronuclei and the elimination of the paternal genome during the early stages of embryogenesis in eggs destined to be male are shown as histological evidence for PGL. Inheritance of genetic markers from father to son is also explained as evidence for PGL. To clarify the evidence, the internal morphology, process of egg formation, and embryogenesis in phytoseiid mites are summarized. Hereafter, the eggs destined to be males and females are referred to as "male eggs" and "female eggs," respectively.

3.1 Internal morphology, process of egg formation and embryogenesis

The female genital system consists of a pair of spermathecae, a lyrate organ, an ovary, an oviduct, a uterus, a vagina, and a genital opening (see details in Di Palma & Alberti, 2001), which was determined by morphological comparisons between female and male specimens, as well as previous findings (Michael, 1892; Alberti & Hänel, 1986; Alberti, 1988). The spermathecae are a temporal storage of sperms just after copulation. The transfer of sperm from spermathecae into the ovary and the shape of sperm in the ovary have not yet been clearly elucidated (Di Palma & Alberti, 2001). It is hard to believe that the spermatozoa can find their opening and pass through the duct, since a lumen is hard to detect (G. Alberti, personal communication).

The morphology of the large lyrate organ consists of paired, distinct flattened arms separated indistinctly into several segments. The function of the lyrate organ is as a trophic (nutrimentary) tissue to support the rapid growth of oocytes and vitellogenesis. A nucleus and many mitochondria are distinguished in each segment. The lyrate organ is distinct in dermanyssid mites (Alberti & Hänel, 1986) and also in *P. persimilis* (Alberti, 1988). The ovary is located at the center of the two arms of the lyrate organ. The oocytes in the ovary are connected via nutritive cords (fusomes) to the nutritive tissue of the lyrate organ (Alberti & Hänel, 1986), although such connections are difficult to detect.

One of among several oocytes in the ovary expands toward the dorsal region of the body just after copulation (Fig. 2b) and starts vitellogenesis in the dorsal region (also see details in Toyoshima et al., 2000). Although the exact timing has not been confirmed, the penetration of sperm into the oocyte seems to occur when the oocyte occupies the dorsal region (Toyoshima et al., 2009). Following the completion of vitellogenesis (Fig. 2b2), the egg (an inseminated oocyte) passes via the short oviduct into the uterus (Fig. 2b3) and remains there, forming an eggshell and starting embryonic development (Fig. 2b4). Subsequently, the next oocyte expands and enters vitellogenesis in the dorsal region.

Superficial cleavage occurs in the centrolecithal egg cell in the early stages of ontogenesis. However, 2-, 4-, 8-, 16-, 32-cell stages are not well discriminated in the following steps because the egg in the uterus forms chorion of poor permeability, preventing the penetration of fixative solution for histological observation. After several nuclear divisions, spindle-shaped blastodermal cells are spherically distributed to form a continuous layer. The process of germinal band formation begins on the ventral surface of the periblastula. Entodermal cells have a rather loose distribution. Details of subsequent stages of embryogenesis are given in Yastrebtsov (1992).

The eggs, in the blastula stage, are deposited one by one (Fig. 3). Young females deposit an average of one to five eggs daily at 25°C (Sabelis, 1985a). The degree of embryonic development in deposited eggs depends upon internal and external factors (Sanderson & McMurtry, 1984). As an extreme case, a hatched larva was seen in a female body (Fig. 1b).

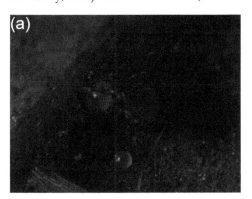 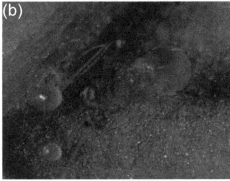

Fig. 3. Ovipositional behavior of *Phytoseiulus persimilis*. The first (a) and the second (b) eggs are laid, one by one, at 6-h intervals. As with all small phytoseiid mites, the egg is huge relative to the size of the gravid female. The rapid growth of large eggs may be supported by the lyrate organ, which is also large relative to the body.

3.2 Fusion of two pronuclei

As a first step in observing the insemination of male eggs, male eggs should be selected from the continuum of egg production during rearing experiments. The first egg should be the focus as a male egg candidate because over 90% of the first eggs develop into males (Toyoshima & Amano, 1998). According to the process of egg formation described above, the first egg in females just after copulation is determined easily and precisely by internal observation of the female. In contrast, unambiguously discriminating sperm in the ovary and recognizing their penetration into the oocyte (the first egg) are difficult (Di Palma & Alberti, 2001), because the sperms in the ovary have not yet been detected with full evidence.

Therefore, eggs with two pronuclei were sought thoroughly (Toyoshima et al., 2000). Then, two pronuclei were detected in the first egg before it had completed its expansion (at ca. 8 h after the end of copulatory behavior). The pronuclei were different in size, which was confirmed in a series of sliced specimens, and joined at the center of the egg. While yolk granules were developing and accumulating gradually around the joining pronuclei, the pronuclei were discriminated precisely by the double-membrane structure. When yolk granules filled the egg (at ca. 10.7 h after the end of copulatory behavior), the joining pronuclei began to change shape and finally fused. This observation was indirect evidence for insemination of male eggs, but histologically elucidated that the paternal genome fuses with the maternal genome in male eggs before embryogenesis.

After a period of time, the first egg moves into the uterus in the ventral region of the body (Fig. 2). A nucleus appears at the center of the egg and it later divides at the same position. The process of nuclear division following embryogenesis was not observed histologically because the eggshell prevented the fixation of the interior of the egg. Therefore, eggs were extracted from the female body for observation of chromosomal behavior, as described below.

3.3 Elimination of paternal genome during early embryogenesis

While male diploidy was confirmed in the first egg during vitellogenesis in *P. persimilis*, male diploidy has not been detected in eggs after deposition. By karyotyping just deposited eggs, the first and the following (presumed) male eggs were haploid in *P. persimilis* and in *Amblyseius* (=*Neoseiulus*) *womersleyi* determined from a previous study (Toyoshima & Amano, 1999). The time needed for the elimination of chromosomes differed from that of *M. occidentalis*, as reported by Nelson-Rees et al. (1980). Therefore, male diploidy in the early embryonic stages must be confirmed by extracting the first egg (male egg) from the female body cavity. In the first egg extracted from females of *P. persimilis*, diploid cells were observed in an early stage of embryogenesis (the exact stage was not confirmed), and the coexistence of haploid and diploid cells in the same egg was also seen at a later stage of embryogenesis (Toyoshima & Amano, 1999). Finally, only haploid cells were observed in eggs just before deposition. Although heterochromatinized chromosomes were not identified in this experiment, paternal genomes in male eggs were confirmed to be eliminated during the early stages of embryogenesis just before deposition. The stage of embryogenesis was not determined for the extracted eggs.

The difference in timing for chromosome elimination between *M. occidentalis* and *P. persimilis* seems to be due to the difference in the number of chromosomes: *M. occidentalis* has three whereas *P. persimilis* has four. In other words, one *speculates* that one of the three chromosomes in *M. occidentalis* is a result of the combination of two of the four chromosomes. If the total DNA content in the haploid genome appears to be equal among all species, one of the three chromosomes in *M. occidentalis* must be larger than those in other species with the basic number. Since the DNA C-value affects cell cycles (Cavalier-Smith, 1978), the size of chromosomes may also affect cell cycles. As a result, the chromosome elimination in male eggs of *M. occidentalis* seems to occur at a later stage of embryogenesis, which is observed at the time the egg is deposited. This speculation should be confirmed experimentally in the future.

3.4 Inheritance of genetic markers from father to sons

Inheritance of paternal genetic elements from father to son was partially elucidated in *Typhlodromus pyri* by the random amplified polymorphic DNA (RAPD) markers (Perrot-Minnot & Navajas, 1995) and in *Neoseiulus californicus* by the direct amplification of length polymorphism (DALP) markers (Perrot-Minnot et al., 2000). Of two RAPD markers, one marker (330 base pairs: bp) was paternally transmitted to male and female offspring, and the other (990 bp) was paternally transmitted to all females and to some male offspring. Although RAPD markers sometimes showed ambiguous inheritance, the conclusion was made that obligate fertilization could account for the inheritance of nuclear genetic material from father to son, and the *speculation* was made that the paternal genomes were partially retained in some tissues. On the other hand, the inheritance of codominant genetic markers, which is detected by DALP, provided evidence for selective elimination of the paternal genome among male tissues (Perrot-Minnot et al., 2000), also suggesting that sperm contained exclusively maternal genes whereas some male somatic tissues retained most paternal chromosomes.

The inheritance of a genetic marker from father to son was also confirmed in *N. womersleyi* (Toyoshima & Hinomoto, unpublished data). The sequence characterized amplified region (SCAR)-PCR marker was designed from a fragment amplified by using arbitrary primers (ca. 20 bases long) and used to confirm the inheritance of this marker. When a genetic marker was amplified from the extracted DNA of two populations of *N. womersleyi* with a primer set (Awa18A1&Awa18B3) by PCR and was digested by a restriction enzyme (*EcoRI*), the inheritance of paternal genetic material by sons was confirmed (Fig. 4). Since the paternal band was relatively weak in sons, only a small amount of template DNA was available for the primers in the extracted DNA solution. Whether the paternal genetic material is retained in a specific tissue or if the heterochromatinized fragments of paternal genomes are randomly dispersed in the entire body has not yet been clarified. According to the X-ray experiments mentioned above, as well as the weak presence of genetic markers, entire paternal genomes are probably retained only in the germ plasma of male offspring. Further investigation is necessary to confirm the maintenance of the effective paternal genomes in spermatocytes of male offspring.

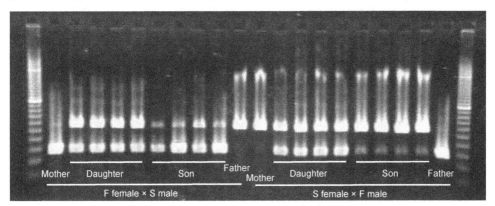

Fig. 4. An electropherogram of a SCAR-PCR marker in an agarose gel (Toyoshima & Hinomoto, unpublished data). A female of a strain of *Neoseiulus womersleyi* with a marker that ran fast on the gel (F) was crossed with a male with a slow marker (S), and vice versa. Daughters and sons have markers from both parents, but the paternal marker in the sons of both crosses is thin.

4. Presumed mechanisms of sex determination in phytoseiid mites

The sex of an individual is *determined* by an initial set of factors (genotypic or environmental). Then, sex *develops* (or *differentiates*) during an integrated series of genetic and physiological steps (Bull, 1983). The sex becomes *fixed* at a certain step during development. The initial set of factors (or "primary sex-determining signal") should be the focus in understanding the relationship between sex determination and PGL in phytoseiid mites.

The sex determination of phytoseiid mites was initially investigated by inbreeding experiments. While Poe and Enns (1970) indicated strong depression by inbreeding in two phytoseiid species, Hoy (1977) indicated weak or scant depression in *M. occidentalis*, and suggested that sex determination in *M. occidentalis* was not based on the multiple allele mechanism (complementary sex determination: CSD) that has been demonstrated for parasitoid wasps and the sawfly (e.g., Cook & Crozier, 1995). Although the sex of an individual is determined and fixed at fertilization in hymenopteran insects, sex may not be determined at fertilization in phytoseiid mites. Therefore, the elimination of the paternal genome in male eggs is not caused by the combination of alleles from the parents. Sex may be controlled by the mother and determined by the accumulation of certain substances in male eggs. In this section, a resultant phenomenon of maternal control is shown with a presumed mechanism of sex control.

4.1 Maternal control of offspring sex

Maternal control of offspring sex was demonstrated by sex ratio control under certain conditions. While the mother deposits eggs with female-biased sex ratios when prey is abundant, she deposits eggs with an even sex ratio ($♀:♂=1:1$) when prey abundance is insufficient (Friese & Gilstrap, 1982; Momen, 1996). Even in a prey patch with abundant prey, mothers also adjust the sex ratio of offspring in a manner that fits the sex ratio to the prediction of sex allocation theories (Hamilton, 1967). While the mother deposits eggs with a female-biased sex ratio when in isolation in a prey patch, she deposits eggs with an even sex ratio when in a crowd with abundant prey (Dinh et al., 1988; Nagerkerke & Sabelis, 1991). To estimate the mechanism to control the sex of the offspring by mothers, the nutritional condition of mothers should be considered as a proximate mechanism rather than the mating structure as a relatively ultimate (evolutionary) mechanism.

Female and male eggs deposited by gravid females of phytoseiid mites can be lined up in a sequence because the females deposit eggs one by one (Fig. 5). Several female eggs were shown to exist between male eggs in a sequence when gravid females deposit eggs with a female-biased sex ratio under abundant prey conditions. The number of female eggs between male eggs in a sequence decreased when gravid females adjusted the sex ratio of offspring in relation to the number of prey items available. Finally, only one female egg was present between male eggs in a sequence when gravid females deposited eggs with an even sex ratio under insufficient prey conditions (Toyoshima & Amano, 1998). It should now be presumed how to determine the sex of eggs.

The female eggs produced under abundant prey conditions may be changed into phenotypic males in response to the nutritional condition when produced. Paternal genomes would remain in the phenotypic males if PGL were genetically controlled but flexible. However, phenotypic sex is consistent with genetic sex (karyotype) in eggs produced with an even sex ratio under minimal prey conditions (Fig. 5; see also Toyoshima & Amano, 1999). Actually, the number of male eggs is similar among different prey conditions, although the total number of eggs decreased gradually in relation to the number of prey items available to gravid females (Toyoshima & Amano, 1998). The number of female eggs may have decreased rather than having changed into phenotypic males.

Abundant food

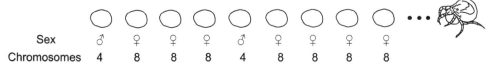

Sex	♂	♀	♀	♀	♂	♀	♀	♀	♀
Chromosomes	4	8	8	8	4	8	8	8	8

Minimum food

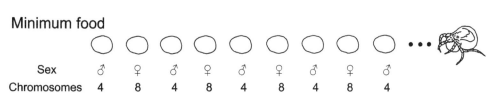

Sex	♂	♀	♂	♀	♂	♀	♂	♀	♂
Chromosomes	4	8	4	8	4	8	4	8	4

Fig. 5. Relationship between karyotypes of eggs and sex of offspring produced by mothers under abundant and minimum food condition.

Mothers may be able to choose sperm with a gene for maleness at a certain locus to control sex ratios under poor nutritional conditions. In this case, the paternal genomes are eliminated after the maleness of the egg is determined genetically. On the basis of genomic conflict between mother and father (e.g., Ross et al., 2010a), however, the utilization of sperm for maleness under poor prey conditions is a great disadvantage to the father because the genetic material in the sperm is eliminated during embryogenesis, and as a result does not contribute to the son's characteristics. Therefore, fathers probably do not produce sexually dimorphic sperm. Unfortunately, it is completely difficult to elucidate, anatomically, the sexual dimorphism of sperms in the ovary or to discriminate, genetically, the sperm with a gene for maleness. Detecting sperm for maleness modified by the mother in her ovary also presents a problem.

The possibility exists that the control of oocytes or eggs by the mother (without paternal contribution) is a simple process for maternal control of offspring sex. Mothers may be able to absorb female eggs under poor prey conditions, known as oosorption (resorption of oocytes), which is an effective mechanism to save resources on eggs in various insect species (Bell & Bohm, 1975). This is based on the idea that the sex of oocytes is already determined in the ovary upon emergence. According to this idea, mothers do not choose a male-destined oocyte for fertilization to develop into a male egg but absorbs female eggs when encountering a poor prey patch.

Gravid mothers under starvation were observed at 12-h intervals to confirm the absorption of female oocytes and/or eggs in the mother's body (Toyoshima et al., 2009). When gravid mothers just after depositing the first (male) egg were restricted to no prey items, the mothers did not hold an egg in the uterus but held 1–2 eggs in the dorsal region (refer to Fig. 2). Since mothers laid an average of 1.8 eggs during starvation, the 1–2 eggs in the dorsal region were transferred into the uterus, one by one, to form an eggshell and were deposited regardless of the nutritional condition of the mothers. An oocyte expanded during the deposition of the eggs and was maintained for at least 72 h in the dorsal region after vitellogenesis (Fig. 6). Two pronuclei were conjugated at first but later fused when the egg was in the dorsal region. The egg was deposited as a female egg when abundant prey items were provided for the starving mothers. According to this starvation experiment, female eggs may not be absorbed to control the sex ratio during food limitation. The confirmation was also made that oocyte expansion and vitellogenesis, as well as embryogenesis in the eggshell, advances until the nutrients in the starving mother are depleted. In this case, fertilization occurred after the sex of oocytes was determined, and the sex of the oocytes (and eggs) in the body was not influenced by the nutritional condition of the mothers. Paternal genomes in the egg may be eliminated during early embryonic development if the sex of the egg has been determined as male.

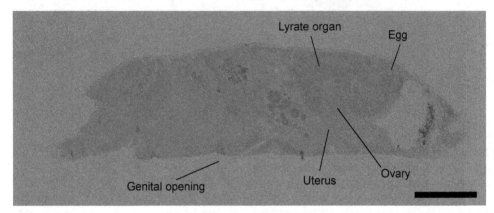

Fig. 6. Sagittal section of an adult female *Neoseiulus californicus* starved for 36 h. The starved female holds an egg in the dorsal region even after depositing an average of 1.8 eggs. The female is facing left (light microscope image). Bar = 50 μm.

4.2 Female-biased nutritional allocation in eggs

Mothers change their investment of nutrients into eggs in relation to the sex of eggs, as well as the prey consumption rate and their own age (Toyoshima & Amano, 1998). They produce larger eggs when consuming abundant prey, and gradually decrease egg size when aged and when prey consumption is restricted. In addition, they produce larger female eggs and smaller male eggs, and maintain the difference in size between sexes even when prey consumption is restricted. Although the evolutionary significance of sexual size dimorphism was discussed in a previous study, the sex determining mechanism of eggs was not yet determined from the size difference between sexes because of insufficient evidence.

The difference in egg size between sexes may be determined before fusion of the paternal and maternal pronuclei in the egg. According to an internal observation of starving females (Toyoshima et al., 2009), the vitelline membrane forms around the egg in the body of a starving female (Fig. 7c). The envelope was still not complete but interrupted. The adjacent, but not fused, pronuclei were visible in the egg enveloped with the vitelline membrane (Fig. 7a & 7b). The membrane is usually formed in oogenesis at the end of vitellogenesis. Since the size of the egg is determined when the vitelline membrane is formed, the combination of maternal and paternal genetic material did not influence size determination. In turn, the size of eggs was determined only by maternal control. More precisely, the size of eggs was influenced by sex, which was determined in response to the nutritional condition of the mothers.

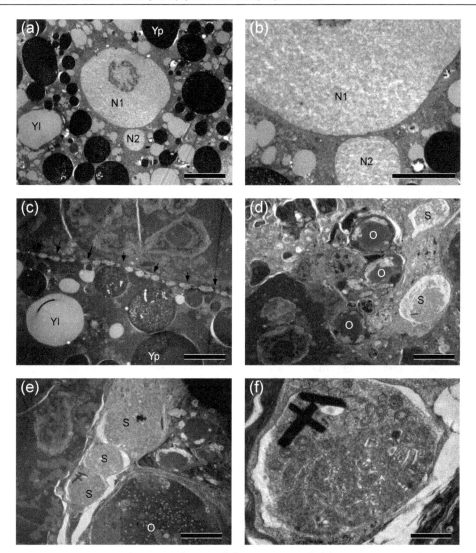

Fig. 7. Fine structures in the egg and around the ovary revealed by transmission electron microscopy. (a) Adjacent pronuclei in an egg filled with yolk granules. Bar = 10 µm. (b) Adjacent part of the pronuclei in (a). Bar = 5 µm. (c) The vitelline membrane (indicated by seven arrows). The envelope is still not completed, but interrupted. The vitelline membrane separates the egg with yolk granules (below) from the lyrate organ (above). Bar = 5 µm. (d) Ovary with several oocytes. Presumed sperms are visible around the ovary. Bar = 5 µm. (e) Presumed sperms between an oocyte and lyrate organ. Bar = 5 µm. (f) Detailed structure of a presumed sperm of (e). The detailed structure in the sperm cell is different from that shown in Di Palma & Alberti (2001). Black bars are not a staining artifact but an unknown structure. Bar = 1 µm. Abbreviations: N1, N2, nuclei; O, oocyte; S, presumed sperm; Yl, lipid-yolk granule; Yp, protein-yolk granule (unpublished micrographs by Toyoshima & Alberti).

How nutrition is invested into the eggs of each sex is still unclear. When female eggs are significantly different in size from male eggs, the size of eggs should be influenced only by the sex of the offspring. Thus, the sex of the eggs should be converted from male to female at a certain level of nutritional accumulation in the eggs. However, the size of the eggs is influenced not only by the sex but also the age and the nutritional condition of the mothers (Toyoshima & Amano, 1998). Female eggs deposited by mothers under poor prey conditions are smaller than male eggs deposited under abundant prey conditions, although the difference of egg size between the sexes (female eggs are larger than male eggs) is maintained at each prey condition. The switch from one sex to the other occurs at a relative criterion rather than an absolute criterion. The accumulation of a certain nutritional substance in addition to the minimum requirement of nutrition may lead to conversion of eggs from male to female. However, how the ratio of female to male is controlled in the ovary also remains unclear.

Sexual size dimorphism of eggs is also observed in an arrhenotokous phytophagous mite, *Tetranychus urticae* (Mache et al., 2010; Toyoshima, 2010), in which virgin mothers of this species lay only haploid male eggs, and fertilized mothers lay female (diploid) and male (haploid) eggs with female-biased sex ratios under good conditions on host plants. Male eggs produced by fertilized mothers are smaller not only than female eggs but also than male eggs produced by virgin females. If the sex of eggs produced by fertilized mothers is ignored, the eggs of fertilized females are not different from that of virgin females. From this comparison between eggs of fertilized and virgin females, virgin females are suspected to also produce concealable cytologically female eggs, which would be fundamentally destined as females but developed to males when not fertilized. This idea is not yet supported by data, but, if this is true, the evolutionary position of PGL in phytoseiid mites and arrhenotokous gynogenesis in certain animals may be understood in the course of the evolutional succession of sexual reproduction.

5. Future perspective

PGL in phytoseiid mites is still wrapped in mystery, although cytological and genetic evidence has accumulated since the 1980s. To understand the *evolutionary process* and *significance* of PGL in phytoseiid mites, we should clarify several events during oogenesis and embryogenesis (Fig. 8), as well as similar reproductive systems in closely related groups. The sex determining mechanism in eggs during oogenesis, sperm behavior in the ovary, and sperm penetration into oocytes are important events in the early reproductive process, which may lead to insights when exploring the signals for PGL. According to Di Palma & Alberti (2001), putative sperm cells extend a thick projection for insertion into an oocyte, and projections are visible around oocytes in the ovary (Fig. 7d–f). However, more investigation is necessary to determine when the sperm inserts into the oocyte and how the sperm nucleus behaves in the oocyte. Sperm behavior in the ovary and oocytes may also be influenced by the qualitative and/or quantitative differences between female-destined and male-destined oocytes. The difference in oocytes at the starting point in the ovary should be investigated histologically and genetically.

Thelytokous species are also known in phytoseiid mites. The thelytokous reproductive system is presumed to be derived from PGL because the number of thelytokous species is small compared to those with PGL. Comparative morphology of oogenesis and oocyte expansion in the ovary between thelytokous and PGL species will shed new light on the *evolutionary flexibility* and *constraints* of PGL in phytoseiid mites.

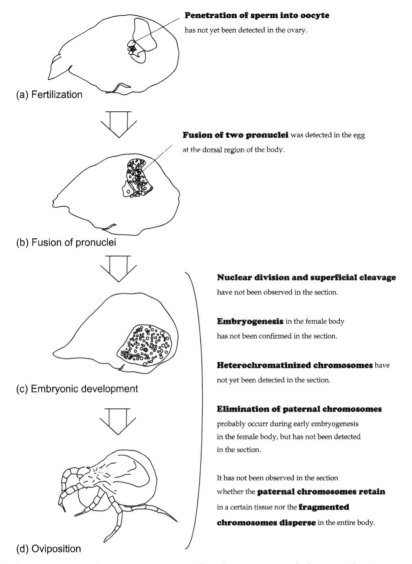

Penetration of sperm into oocyte
has not yet been detected in the ovary.

(a) Fertilization

Fusion of two pronuclei was detected in the egg
at the dorsal region of the body.

(b) Fusion of pronuclei

Nuclear division and superficial cleavage
have not been observed in the section.

Embryogenesis in the female body
has not been confirmed in the section.

Heterochromatinized chromosomes have
not yet been detected in the section.

(c) Embryonic development

Elimination of paternal chromosomes
probably occurr during early embryogenesis
in the female body, but has not been detected
in the section.

It has not been observed in the section
whether the **paternal chromosomes retain**
in a certain tissue nor the **fragmented
chromosomes disperse** in the entire body.

(d) Oviposition

Fig. 8. Unknown events during oogenesis and embryogenesis of phytoseiid mites.

The process of the elimination of paternal genomes during embryonic development should be visualized to better understand the events of PGL. It is difficult, but not impossible, to penetrate the fixatives into the egg when wrapped in an eggshell in the female body. It is most important that the chromosome behavior in each cell during the blastula stage be observed to follow the heterochromatinized chromosomes, as a central point of the study of PGL. It may be able to reveal when, where, and how genomic material is eliminated during the embryonic development of males. Finally, we can start to understand why the paternal genome is eliminated and why this reproductive system is maintained in phytoseiid mites.

6. Acknowledgment

We are really grateful to Prof. em. Dr. Dr. h.c. G. Alberti (Ernst-Moritz-Arndt-Universität Greifswald) for his corrections and thoughtful suggestions to our manuscript. We also thank Drs. J. Abe (Kanagawa University), N. Hinomoto (NARO Agricultural Research Center), H. Kishimoto (NARO Institute of Fruit Tree Science), and Y. Tagami (Shizuoka University) for their valuable comments to the manuscript. Finally, we are grateful to the Publishing Process Manager for giving us an opportunity to review our previous researches.

7. References

Alberti, G. (1988) Genital system of gamasida and its bearing on phylogeny. In: *Progress in Acarology*, Volume I, Basavanna, G. P. C. & Viraktmath, C. A. (Eds.), 479-490, Ellis Horwood, Chichester

Alberti, G. & Hänel, H. (1986) Fine structure of the genital system inthe bee parasite, Varroajacobsoni (Gamasida: Dermanyssina) with remarks on spermiogenesis, spermatozoa and capacitation. *Experimental & Applied Acarology*, Vol.2, pp. 63-104

Amano, H. & Chant, D. A. (1978) Mating behaviour and reproductive mechanisms of two species of predacious mites, *Phytoseiulus persimilis* Athias-Henriot and *Amblyseius andersoni* (Chant) (Acarina: Phytoseiidae). *Acarologia*, Vol.20 pp. 196-213

Bell, G. (1982) *Masterpiece of Nature: The Evolution and Genetics of Sexuality*, 635p., Croom Helm, London

Bell, W. J. & Bohm, M. K. (1975) Oosorption in insects. *Biological Reviews*, Vol.50, pp. 373-396

Blommers-Schlosser, R. & Blommers, L. (1975) Karyotypes of eight species of Phytoseiid mites (Acarina: Mesostigmata) from Madagascar. *Genetica*, Vol.45, pp. 145-148

Bull, J. J. (1983) *The Evolution of Sex-Determining Mechanisms*, 316p., Menlo Park, Benjamin Cummings

Burt, A. & Trivers, R. (2006) *Genes in Conflict: The Biology of Selfish Genetic Elements*, 602p., Belknap Press of Harvard Univ. Press, Cambridge.

Cavalier-Smith, T. (1978) Nuclear volume control by nucleoskeletal DNA, selection for cell volume and cell growth rate, and the solution of the DNA C-value paradox. *Journal of Cell Science*, Vol.34, 247-278

Chant, D. A. & McMurtry, J. A. (2007) *Illustrated Keys and Diagnoses for the Genera and Subgenera of the Phytoseiidae of the World*, 220p., Indira Publishing House, Michigan

Congdon, B. D. & McMurtry, J. A. (1988) Morphological evidence establishing the loss of paternal chromosomes in males of predatory phytoseiid mites, genus *Euseius*. *Entomologia Experimentalis et Applicata*, Vol.48, pp. 95-96

Cook, J. M. & Crozier, R. H. (1995) Sex determination and population biology in the Hymenoptera. *Trends in Ecology & Evolution*, Vol.10, pp. 281-286

Di Palma, A. & Alberti, G. (2001) Fine structure of the female genital system in phytoseiid mites with remarks on egg nutrimentary development, sperm access system, sperm transfer, and capacitation (Acari, Gamasida, Phytoseiidae). *Experimental & Applied Acarology*, Vol.25, pp. 525-591

Dinh, N. V., Janssen, A. & Sabelis, M. W. (1988) Reproductive success of *Amblyseius idaeus* and *A. anonymus* on a diet of two-spotted spider mites. *Experimental & Applied Acarology*, Vol.4, pp. 41-51

Gerson, U., Smiley, R. L. & Ochoa, R. (2003) *Mites (Acari) for Pest Control*. 539p., Blackwell Science Ltd., Oxford

Hamilton, W. D. (1967) Extraordinary sex ratios. *Science*, Vol.156, pp.477-488

Hansell, R. I. C., Mollison, M. M. & Putman, W. L. (1964) A cytological demonstration of arrhenotoky in three mites of the family Phytoseiidae. *Chromosoma (Berl.)*, Vol.15, pp. 562-567

Hartl, D. L. & Brown, S. W. (1970) The origin of male haploid genetic systems and their expected sex ratio. *Theoretical & Population Biology*, Vol.1, pp. 165-190

Helle, W., Bolland, H. R., van Arendonk, R., De Boer, R. & Schulten, G. G. M. (1978) Genetic evidence for biparental males in haplo-diploid predator mites (Acarina: Phytoseiidae). *Genetica*, Vol.49, pp. 165-171

Helle, W. & Sabelis, M. W. (1985) *Spider Mites Their Biology, Natural Enemies and Control, Volume 1B*. 458p., Elsevier, Amsterdam

Herrick, G. & Seger, J. (1999) Imprinting and paternal genome elimination in insects. In: *Genomic Imprinting*, R. Ohlsson, (Ed.), 41-71, Springer Verlag, Amsterdam

Ho, C.-C., Lo, K.-C. & Chen, W.-H. (1995) Comparative biology, reproductive compatibility, and geographical distribution of *Amblyseius longispinosus* and *A. womersleyi* (Acari: Phytoseiidae). *Environmental Entomology*, Vol.24, pp. 601-607

Hoy, M. A. (1979) Parahaploidy of the "arrhenotokous" predator, *Metaseiulus occidentalis* (Acarina: Phyotoseiidae) demonstrated by X-irradiation of males. *Entomologia Experimentalis et Applicata*, Vol.26, pp. 97-104

Krantz, G. W. & Walter, D. E. (2009) *A Manual of Acarology (3rd. ed.)*. 807p., Texas Tech University Press, Texas

Macke, E., Magalhaes, S., Khan, H. D.-T., Luciano, A., Frantz, A., Facon, B. & Olivieri, L. (2010) Sex allocation in haplodiploid is mediated by egg size: evidence in the spider mite *Tetranychus urticae* Koch. *Proceedings of the Royal Society B*, Vol.278, No.1708, pp. 1054-1063

Michael, A. D. (1892). On the variations in the internal anatomy of the gamasinae, especially in that of the genital organs, and on their mode of coition. *Transactions of the Linnean Society of London. 2nd Series: Zoology*, Vol.5, No.9, pp. 281-317

Nagerkerke, C. J. & Sabelis, M. W. (1991) Precise sex-ratio control in the pseudo-arrhenotokous phytoseiid mite *Typhlodromus occidentalis* Nesbitt. In: *The Acari, Reproduction, Development and Life-History Strategies*. Schuster, R. & Murphy, P. W. (Eds.), 193-207, Chapman and Hall, London

Nagerkerke, C. J. & Sabelis, M. W. (1998) Precise control of sex allocation in pseuso-arrhenotokous phytoseiid mites. *Journal of Evolutionary Biology*, Vol.11, pp. 649-684

Nelson-Rees, W. A., Hoy, M. A. & Roush, R. T. (1980) Heterochromatinization, chromatin elimination and haploidization in the parahaploid mite *Metaseiulus occidentalis* (Nesbitt) (Acarina: Phytoseiidae). *Chromosoma (Berl.)*, Vol.77, pp. 263-276

Norton, R. A., Kethley, J. B., Johnston, D. E. & OConnor, B. M. (1993) Phylogenetic perspectives on genetic systems and reproductive modes of mites. In: *Evolution and Diversity of Sex Ratio in Insects and Mites*, Wrensch. D. L. & Ebbert, M. A. (Eds.), 8-99, Chapman & Hall, New York

Oliver, J. H. (1983) Chromosomes, genetic variance and reproductive strategies among mites and ticks. Bulletin of the ESA, Vol.29, No.2, pp.8-17

Perrot-Minoot, M.-J., Lagnel, J., Migeon, J. & Navajas, M. (2000) Tracking paternal genes with DALP markers in a pseudoarrhenotokoys reproductive system: biparental transmission but haplodiploid-like inheritance in the mite *Neoseiulus californicus*. *Heredity*, Vol.84, pp. 702-709

Perro-Minnot, M.-J. & Najavas, M. (1995) Biparental inheritance of RAPD markers in males of the pseudoarrhenotokous mite *Typhlodromus pyri*. *Genome*, Vol.38, pp. 838-844

Ross, L., Pen, I. & Shuker, D. M. (2010a) Genomic conflicts in scale insects: the causes and consequences of bizarre genetic systems. *Biological Reviews*, Vol.85, pp. 807-828

Ross, L., Shuker, D. M. & Pen, I. (2010b) The evolution and suppression of male suicide under paternal genome elimination. *Evolution*, Vol.65, pp. 554-563

Sabelis, M. W. (1985) Development. In : *Spider Mites, Their Biology, Natural Enemies and Control, Volume 1B*, Helle, W. & Sabelis, M. W. (Eds.), 43-53, Elsevier, Amsterdam

Sabelis, M. W., Nagerkerke, C. J. & Breeuwer, J. A. J. (2002) Sex ratio control in arrhenotokous and pseudo-arrhenotokous mites. In: *Sex Ratios, Concepts and Research Methods*, Hardy, I. C. W. (Ed.), 235-253, Cambridge University Press, Cambridge.

Sanderson, J. P. & McMurtry, J. A. (1984) Life history studies of the predaceous mite *Phytoseius hawaiiesis*. *Entomologia Experimentalis et Applicata*, Vol.35, pp. 27-234

Schrader, F. & Hughes-Schrader, S. (1931) Haploidy of metazoan. *Quarterly Review of Biology*, Vol.6, pp. 411-438

Schulten, G. G. M. (1985) Pseudo-arrhenotoky. In: *Spider Mites, Their Biology, Natural Enemies and Control, Volume 1B*, Helle, W. & Sabelis, M. W. (Eds.), 67-71, Elsevier, Amsterdam

Schulten, G. G. M., van Arendonk, R. C. M., Russell, V. M., & Roorda, F. A. (1978) Copulation, egg production and sex-ratio in *Phytoseiulus persimilis* and *Amblyseius bibens* (Acari: Phytoseiidae). *Entomologia Experimentalis et Applicata*, Vol.24, pp. 145-153

Schultz, R. J. (1969) Hybridization, unisexuality, and polyploidy in the teleost *Poeciliopsis* (Poeciliidae) and other vertebrates. *The American Naturalist*, Vol.103, pp. 605-619

Toyoshima, S. (2010) Effects of fertilization status and age of gravid females on the egg size in *Tetranychus urticae* Koch. *Journal of the Acarological Society of Japan*, Vol.19, pp. 107-112

Toyoshima, S. & Amano, H. (1998) Effect of prey density on sex ratio of two predacious mites, *Phytoseiulus persimilis* and *Amblyseius womersleyi* (Acari: Phytoseiidae). *Experimental & Applied Acarology*, Vol.22, pp. 709-723

Toyoshima, S. & Amano, H. (1999) Cytological evidence of pseudo-arrhenotoky in two phytoseiid mites, *Phytoseiulus persimilis* Athias-Henriot and *Amblyseius womersleyi* Schicha. *Journal of the Acarological Society of Japan*, Vol.8, pp. 135-142

Toyoshima, S., Michalik, P., Talarico, G., Klann, A. E. & Alberti, G. (2009) Effects of starvation on reproduction of the predacious mite *Neoseiulus californicus* (Acari: Phytoseiidae). *Experimental & Applied Acarology*, 47: 235-247.

Toyoshima, S., Nakamura, M., Nagahama, Y. & Amano, H. (2000) Process of egg formation in the female body cavity and fertilization in male eggs of *Phytoseiulus persimilis* (Acari: Phytoseiidae). *Experimental & Applied Acarology*, Vol.24, pp. 441-451

White, M. J. D. (1973) *Animal Cytology and Evolution (3rd ed.)*. 961p., University Press, Cambridge

Wysoki, M. (1973) Further studies on karyotypes an sex determination of phytoseiid mites (Acarina: Mesostigmata). Genetica, 44: 139-145.

Wysoki, M. (1985) Karyotyping. In: *Spider Mites, Their Biology, Natural Enemies and Control, Volume 1B*, Helle, W. & Sabelis, M. W. (Eds.), 191-196, Elsevier, Amsterdam

Wysoki, M. & Bolland, R. (1983) Chromosome studies of phytoseiid mites (Acari: Gamasida). *International Journal of Acarology*, Vol.9, pp. 91-99

Wysoki, M. & McMurtry, J. A. (1977) Karyotypes of eight species of phytoseiid mites of the genus *Amblyseius* Berlese (Acarina: Mesostigmata). *Genetica*, Vol.47, pp. 237-239

Wysoki, M. & Swirski, E. (1968) Karyotypes and sex determination of ten species of phytoseiid mites (Acarina: Mesostigmata). *Genetica*, Vol.39, pp. 220-228

Yastrebtsov, A. (1992) Embryonic development of gamadis mites (Parasitiformes: Gamasida). *International Journal of Acarology*, Vol.18, pp. 121-141

11

Somatic Embryogenesis in Recalcitrant Plants

Laura Yesenia Solís-Ramos[1,2,*], Antonio Andrade-Torres[3,4],
Luis Alfonso Sáenz Carbonell[3], Carlos M. Oropeza Salín[3] and
Enrique Castaño de la Serna[1]

[1]*Unidad de Bioquímica y Biología Molecular de Plantas,
Centro de Investigación Científica de Yucatán, (CICY), Mérida, Yucatán,
Studies on habanero chilli biotechnology*
[2]*Escuela de Biología, Universidad de Costa Rica, San Pedro,
Genetic Plant Transformation and Biotechnology*
[3]*Unidad de Biotecnología, Centro de Investigación Científica de Yucatán, (CICY),
Mérida, Yucatán, Biotechnological studies on coconut*
[4]*INBIOTECA- Instituto de Biotecnología y Ecología Aplicada, Universidad Veracruzana,
Xalapa, Veracruz, México, Genetic Plant Transformation and Biotechnology*
[1,3,4]*México*
[2]*Costa Rica*

1. Introduction

There are two types of embryogenesis in plants: zygotic and somatic (Figure 1). Zygotic embryogenesis is one of the most important steps in the life cycle of plants. The process begins with double fertilization, followed by determination of the three axes of embryos (longitudinal, lateral, and radial) and morphologic changes of the embryos (globular, heart-shaped, and torpedo-shaped; Figure 1). Subsequently, seed storage proteins accumulate in the embryos, and finally, the embryos become desiccated and dormant. These processes are regulated by numerous factors, including phytohormones, enzymes, and other substances related to embryogenesis.

Somatic embryogenesis (SE) is the process by which somatic cells, under induction conditions, generate embryogenic cells, which go through a series of morphological and biochemical changes (Quiróz-Figueroa *et al.*, 2006), that result in the production of bipolar structure without vascular connection with the original tissue. The development of somatic embryos closely resembles the development of zygotic embryos both morphologically and physiologically (Figure 1). The process is feasible because plants possess cellular totipotency where by individual somatic cells can regenerate into a whole plant. Since the first reports on carrot in 1958 (Reinert 1958; Steward *et al.*, 1958), somatic embryogenesis has been reported in various plant species.

In addition to natural *in vivo* forms embryogenesis (apomixis), there exist at least three ways to induce embryo development from *in vitro* cultured plant cells: *in vitro* fertilization, from microspores and *in vitro* somatic embryogenesis (Féher *et al.*, 2003). *In vitro* SE can develop either from callus (indirect SE) or directly from the explant without any intermediate callus stage (direct SE). Somatic embryogenesis is also induced directly, or through callus, in the culture of somatic embryos, and this process is called secondary SE in contrast to primary SE induced from explant cells (Gaj, 2004).

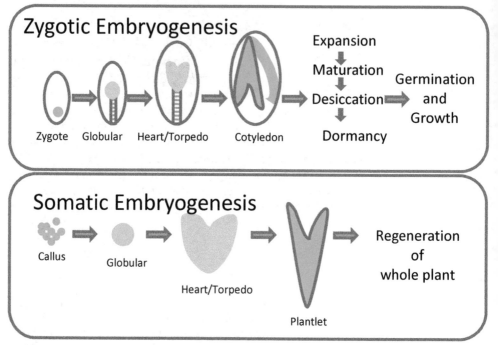

Fig. 1. Zygotic vs somatic embryogenesis. Modified from Zimmerman, 1993.

Somatic embryos originate by two pathways, unicellular or multicellular. When embryos have a unicellular origin, coordinated cell divisions are observed and the embryo sometimes connected to the maternal tissue by a suspensor-like structure. In contrast, multicellular-origin embryos are initially observed as a protuberance, with no coordinated cell divisions observable, and those embryos in contact with the basal area are typically fused to the maternal tissue (Quiroz-Figueroa *et al.*, 2006).

Somatic embryogenesis with a low frequency of chimeras, a high number of regenerates and a limited level of somaclonal variation (Ahloowalia, 1991; Henry *et al.*, 1998) is more attractive than organogenesis as a plant regeneration system (e.g., in genetic transformation, *in vitro* mutagenesis and selection). However there are several factors that influence the initiation of somatic embryogenesis in plants.

In the context of this paper, *in vitro* plant recalcitrance is defined as the inability of plant tissue cultures to respond to *in vitro* manipulations. In its broadest terms, tissue culture recalcitrance also concerns the time-related decline and/or loss of morphogenetic competence and totipotent capacity (Benson, 2000).

2. Some factors influencing somatic embryogenesis induction

The process of acquisition of embryogenic competence by somatic cells must involve reprogramming of gene expression patterns as well as changes in the morphology, physiology, and metabolism of plant cells. Studies on factors controlling *in vitro* plant morphogenesis are highly important not only for the development of improved regeneration systems, but also for the analysis of molecular mechanisms underlying plant embryogenesis.

In vitro development of cells and tissues depends on different factors such as: genotype, type of plant, age and developmental stage of an explant, physiological state of an explant-donor plant, and the external environment which includes composition of media and physical culture conditions (light, temperature) (Gaj, 2004).

The embryogenic potential is largely defined by the developmental program of the plant as well as by environmental cues (Féher, 2005). The key role of endogenous hormone metabolism affected by genetic, physiological and environmental cues is well accepted in the induction phase of somatic embryogenesis (Jiménez, 2005).

The cells which represent an intermediate state between somatic and embryogenic cells are called competent. Cellular competence is associated with the dedifferentiation of somatic cells that allows them to respond to new developmental signals. It is well accepted that embryogenic competent cells can be morphologically recognized as small, rounded cells with rich cytoplasm and small vacuoles. In this respect they are very similar to meristematic cells or zygotes and this similarity is further emphasized by their asymmetric division (Féher, 2005).

Wounding, high salt concentration, heavy metal ions or osmotic stress positively influenced somatic embryo induction in diverse plant species (reviewed by Dudits *et al.*, 1995). These procedures were accompanied by increased expression of diverse stress-related genes, evoking the hypothesis that somatic embryogenesis is an adaptation process of *in vitro* cultured plant cells (Dudits *et al.*, 1995).

Endogenous hormone levels however can be considered as major factors in determining the specificity of cellular responses to these rather general stress stimuli (Féher *et al.*, 2003). The temporal and spatial changes in endogenous auxin levels are important factors controlling the embryogenic cell fate (Féher *et al.*, 2003).

Among different external stimuli that induce an embryogenic pathway of development plant growth regulators (PGRs) such as auxins and cytokinins used for *in vitro* media have been the most frequently considered, as they regulate the cell cycle and trigger cell divisions (Francis and Sorrell, 2001). The high efficiency of 2,4-dichlorophenoxy acetic acid (2,4-D) for induction of embryogenic response found in different *in vitro* systems and plant species indicates a specific and unique character of this PGR. This synthetic growth regulator and an auxinic herbicide appear to act not only as an exogenous auxin analogue but also as an effective stressor (Gaj, 2004).

The polar transport of auxin in early globular embryos is essential for the establishment of bilateral symmetry during plant embryogenesis. Interference with this transport causes a failure in the transition from axial to bilateral symmetry and results in the formation of embryos with fused cotyledons (Liu *et al.*, 1993).

The chromatin remodelling plays two major roles during the early stages of somatic embryogenesis. Differentiation requires unfolding of the supercoiled chromatin structure, in order to allow the expression of genes inactivated by heterochromatinization during differentiation, and subsequent chromatin remodelling can result in the specific activation of a set of genes required for embryogenic development (Féher et al., 2003).

Also a wide and complex variety of molecules can now be enlisted, including polysaccharides, amino acids, growth regulators, vitamins, low molecular weight compounds, polypeptides, etc. (Chung et al., 1992). Some such compounds are derived from the cell wall, whereas others originate inside the cells (Quiroz-Figueroa et al., 2006).

3. Some genes related to somatic embryogenesis

3.1 WUSCHEL (WUS)

WUS is a homeobox gene which encodes a transcription factor that regulates the pool of stem cells in the shoot meristem and is regulated by a feedback loop involving the CLAVATA (CLV) genes (Weigel and Jurgens, 2002; Bhalla and Singh, 2006). WUS expression can be first localized to the shoot meristem in the heart stage embryo, and the shoot meristem of the plant by regulating the stem cell pool can continue to produce organs throughout the life of the plant. The stem cells are specified by a WUS-dependent signal produced in the organizing center cells, which lie below the stem cell niche of the central zone and CLV3 is in turn produced by the stem cells of the central zone (Baurle and Laux, 2005; Reddy and Meyerowitz, 2005). Increases in the number of stem cells lead to an increasing amount of the secreted CLV3 protein, which acts via the CLV1/CLV2 receptor complex to reduce WUS expression and the number of stem cells, thus maintaining a constant pool of stem cells (Weigel and Jurgens, 2002).

3.2 Baby Boom (BBM)

The Baby Boom (BBM) gene, which was isolated from microspore embryo cultures of Brassica napus (Boutilier et al., 2002), encodes a transcriptional factor belonging to the AP2/ERF family. BBM expression was observed during zygotic and pollen-derived somatic embryogenesis. The ectopic expression of BBM and Arabidopsis BBM (AtBBM) in transgenic plants induced the formation of somatic embryo-like structures on the edges of cotyledons and leaves, as well as additional pleiotropic phenotypes, including neoplastic growth, phytohormone-free plant regeneration from explants, and abnormal leaf and flower morphology. Therefore, BBM is likely to promote cell proliferation and morphogenesis during embryogenesis (Boutilier et al., 2002).

3.3 SERK (Somatic Embryogenesis Receptor Kinase)

Among the genes involved in somatic embryogenesis, Somatic Embryogenesis Receptor Kinases (SERKs) genes has been detected in the early stages of the process, which form a subgroup in the Leucine-Rich Repeat-Receptor-Like Kinases (LRR-RLKs) comprising the largest subfamily of RLKs in plants and are also related to key processes in plant growth (Sharma et al., 2008).

The first SERK gene was identified in competent cells of carrot (*Daucus carota*) *in vitro* cultured (Schmidt *et al.*, 1997), this gene encodes a transmembrane receptor kinase type with leucine-rich repeat (LRR). DcSERK has been considered as a marker of cells competent to form embryos in culture (Schmidt *et al.*, 1997). DcSERK has been found be expressed in somatic and zygotic embryos but in no other plant tissues at very early stages of somatic embryo development, i.e., from the single-cell stage to the globular stage (Schmidt *et al.*, 1997). Genes homologous to DcSERK were isolated from *Arabidopsis* (AtSERK1), maize (ZmSERK1, ZmSERK2), *Medicago truncatula* (MtSERK1) (Nolan *et al.*, 2003), *Hieracium* (HpSERK), *Helianthus annuus* (Thomas *et al.*, 2004), *Oryza sativa* (Hu *et al.*, 2005), *Theobroma cacao* (Santos *et al.*, 2005), Citrus unshui (Shimada *et al.*, 2005), y *Solanum tuberosum* (Sharma *et al.*, 2008) suggesting the ubiquity of a small family of SERK in all species of plants, in addition to the functional conservation of a specific role in embryogenesis. Their expressions were detected during somatic embryogenesis (Somleva *et al.*, 2000; Baudino *et al.*, 2001; Hecht *et al.*, 2001; Shah *et al.*, 2002; Nolan *et al.*, 2003; Tucker *et al.*, 2003; Thomas *et al.*, 2004), as well as in developing ovules and early-stage embryos of *Arabidopsis*, *Hieracium* and maize.

4. Genetic transformation to abate recalcitrance

Genetic transformation has proven to be an alternative to abate recalcitrance to *in vitro* morphogenesis and to increase resistance to pathogenic microorganisms (Cai *et al.*, 2003; Shin *et al.*, 2002; Zuo *et al.*, 2002; Herrera-Estrella *et al.*, 2004). This has been achieved by insertion and over-expression of genes related to the control of morphogenesis, such as the heterologous gene WUSCHEL in *Arabidopsis thaliana* and *Coffea canephora* cultures that promoted the transition from vegetative to embryogenic state, and eventually led to somatic embryo formation (Zuo *et al.*, 2002; Arroyo-Herrera *et al.*, 2008). In *Capsicum chinense*, the induced expression of WUSCHEL in segments of transformed stems began to form globular structures, suggesting that heterologous WUSCHEL was active and involved in the process of morphogenesis (Solís-Ramos *et al.*, 2009). It has been demonstrated in *Arabidopsis*, that over-expression of a SOMATIC EMBRYOGENESIS RECEPTOR-like KINASE (SERK) gene (AtSERK1) increases the embryogenic competence of callus derived from transformed seedlings 3 to 4-fold when compared with the wild-type callus (Hecht *et al.*, 2001).

Most of the important crops and grasses are recalcitrant for *in vitro* culturing, which hampers the development of reliable regeneration techniques. This document is focused in the somatic embryogenesis of recalcitrant plants, showing the particular cases of two plant species: habanero chili (*Capsicum chinense* Jacq.) and coconut palm (*Cocos nucifera* L.).

5. Studies in habanero chili (*Capsicum chinense* Jacq.)

5.1 Introduction

All chili peppers belong to the genus *Capsicum* of the Solanaceae family and are important horticultural crops. Members of the *Capsicum* genus have been shown to be recalcitrant to differentiation and plant regeneration under *in vitro* conditions, which in turn makes it very difficult or inefficient to apply recombinant DNA technologies via genetic transformation aimed at genetic improvement against pests and diseases (Ochoa-Alejo and Ramírez-Malagón, 2001). *Capsicum chinense* Jacq. (habanero chili) (Fig. 2-G), a species of economic importance for Mexico is no exception (Santana-Buzzy *et al.*, 2005; López-Puc *et al.*, 2006), and no efficient, reproducible somatic embryogenesis regeneration system has yet been developed for this species. A dependable system is indispensable for their genetic improvement and regeneration of transformed tissue (Solís-Ramos *et al.*, 2009).

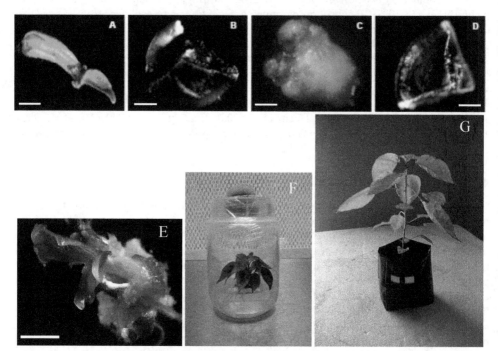

Fig. 2. A to G: Different responses of *C. chinense* explants after induction: A, cotyledon. Bar= 1mm; B, Zygotic embryo segment with radicle; C, Zygotic embryo segment forming callus; D, Zygotic embryo segment without embryogenic response. Bar= 0.5mm. E: somatic embryo germinating. Bar= 1mm. F: seedling obtained from somatic embryo. G: seedling after acclimatization under greenhouse conditions.

5.2 Indirect somatic embryogenesis protocol

Direct organogenesis has been the most frequently used morphogenic route for *in vitro* regeneration of *Capsicum* plants; however, the major problem faced to achieve this goal has been the failure of elongation of the induced shoot buds (Ochoa-Alejo and Ramírez-Malagón, 2001). Shoot buds and rosettes are not well formed during the induction step, perhaps because of a lack of true apical meristems (Binzel *et al.*, 1996; Ochoa-Alejo and Ramírez-Malagón 2001; Steinitz *et al.*, 2003).

In recent years, a number of investigators have developed methods in order to increase the efficiency of the somatic embryogenesis process for chili pepper micropropagation via direct somatic embryogenesis (DSE) (Harini and Sita 1993; Binzel *et al.*, 1996; Khan *et al.*, 2006) and indirect somatic embryogenesis (ISE) (Binzel *et al.*, 1996; Buyakalaca and Mavituna, 1996, Kintzios *et al.*, 2001, Zapata-Castillo *et al.*, 2007, Solís-Ramos *et al.*, 2010b). *Capsicum chinense* Jacq. is a recalcitrant species for *in vitro* morphogenesis, and up to date there is no efficient system for genetic transformation and regeneration of this species via somatic embryogenesis. However an ISE protocol was developed using mature *C. chinense* zygotic embryo segments (ZES) (Solís-Ramos *et al.*, 2010b) (Figure 2 C, E-G). The ZES cultured in semi-solid MS-3R medium (MS medium with 8.9 μM NAA, 11.4 μM IAA and 8.9 μM BAP) developed an embryogenic callus and 8% of these explants developed somatic embryos (Figure 2-E). Torpedo-stage somatic embryos were detached from the callus and subcultured in semi-solid MS medium without growth regulators, producing a 75% conversion rate to plantlets with well-formed root tissue. Histological analysis showed the developed structures to have no vascular connection to the source tissue and to be bipolar, confirming that this protocol induced formation of viable somatic embryos from mature *C. chinense* zygotic embryo segments, and seedlings can be obtained (Figure 2 F-G).

5.3 Endogenous GUS-like activity in *C. chinense* tissues

The gene *uidA* codes for β-glucuronidase which is utilized as a reporter in plant genetic transformation because it is generally believed that higher plants do not show GUS-like endogenous activity (Jefferson *et al.*, 1987; Martin *et al.*, 1991; Sudan *et al.*, 2006). However, several studies have demonstrated that some plant species show endogenous GUS-like activity in vegetative tissues as well as reproductive organs (Cervera, 2005; Sudan *et al.*, 2006). Therefore, in order to avoid undesirable effects in interpreting genetic transformation results, it is recommended to evaluate potential endogenous GUS-like activity in tissues that will be targeted to genetic transformation by using *uidA* as a reporter. The pH of the assay buffer is very critical for detection of the GUS activity in plants. The *E. coli*-derived GUS has optimum activity at pH 7.0 and hence plant tissues are assayed at neutral pH after transformation (Sudan *et al.*, 2006).

Segments of mature zygotic embryos of *C. chinense* were used as explants for transient transformation with *Agrobacterium tumefaciens* LBA4404 (pCAMBIA2301) and C58C1 (pER10W-35S Red) (Solís-Ramos *et al.*, 2010a, Solís-Ramos *et al.*, 2010b). T-DNA in pCAMBIA2301 (Center for the Application of Molecular Biology to International Agriculture, Canberra, Australia) includes a copy of *Escherichia coli uidA* gene under the control of CaMV35S promoter and the NOS terminator. In this binary vector, *uidA* gene coding sequence is interrupted by a Castor Bean catalase intron, which has to be removed for eukaryotic expression and prevents bacterial transcriptions of the gene coding sequence. Transient transformation of *C. chinense* explants and plant regeneration were carried out following the protocol previously described by Solís-Ramos *et al.* (2009). In addition, as a positive control leaves explants of *Nicotiana tabacum* were transient transformed via *A. tumefaciens* LBA4404 (pCAMBIA2301), to verify that the protocol used for GUS activity was done properly. Histochemical staining of *C. chinense* explants was carried out following a protocol reported by Jefferson (1987). Presence of blue spots was recorded and interpreted as transient GUS expression (Figure 3). Also the transient expression of red fluorescent protein was detected using a Leica MZFLIII stereoscopic microscope equipped with appropriate filtres (546/10 nm, 600/40 nm).

Successful transient transformed *C. chinense* zygotic embryo explants were achieved with *A. tumefaciens* LBA4404 (pCAMBIA2301) and the bacteria were eliminated with 1 g/L cefotaxime and 500 mg/L timentin (Solís-Ramos *et al.*, 2009). The calli of *C. chinense* transient transformed with pER10W-35S Red (used as control for transformation efficiency) expressed the red fluorescent protein (DsRFP), but not the non-transformed calli (data not shown) (Solís-Ramos *et al.*, 2010a). A screening for endogenous GUS-like activity in *C. chinense* tissues was performed in phosphate buffer adjusted to pH 6, 7, 7.5 and 8. At pH 6 and 7 the 100% of all samples (vegetative and reproductive tissues) presented endogenous GUS-like activity (Figure 3-C) (Solís-Ramos *et al.*, 2010a). At pH 7.5 no GUS-like activity was observed in all of the petals, root, stem or leaves. However, in septum, stamen and calli some GUS-like activity was observed. A substantial decrease, or even a total absence, of GUS-like activity was observed in phosphate buffer pH 8 in almost all tissue analyzed with an exception for a slight activity in stamens (Figure 3-A) (Solís-Ramos *et al.*, 2010a). Our results of histochemical staining in phosphate buffer pH 8, suggest that *uidA* gene was introduced in regenerants of *C. chinense* and *N. tabacum* and the gene was transcriptional active as it can be inferred from the blue stain observed in tissues of regenerated plantlets. The main problem during initial steps of transformation is just to get an assay conditions which can provide an initial screening. This problem has been solved by adjusting the pH to 8 for *C. chinense* (Solís-Ramos *et al.*, 2010a).

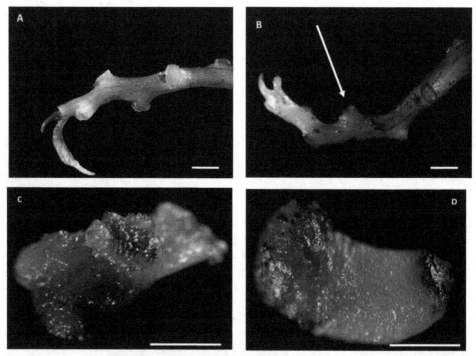

Fig. 3. *C. chinense* explants showing endogenous GUS-like activity at pH 7 (C), and without endogenous GUS-like activity at pH 8 (A). B and D transformed explants showing GUS expression. A, B bar= 5mm. C, D bar= 1mm.

5.4 Protocol for genetic transformation

Habanero chili plants were transformed via *Agrobacterium tumefaciens* co-cultivation with reporter genes: *uiDA, DsRFP,* and *WUSCHEL* (Solís-Ramos *et al.*, 2009, Solís-Ramos *et al.*, 2010a, Solís-Ramos *et al.*, 2010b). *WUSCHEL* (WUS) has been shown to promote the transition from vegetative to embryogenic state when overexpressed in *Arabidopsis thaliana* (Zuo *et al.*, 2002). The hypothesis tested is that the genetic transformation of Habanero chili and overexpression of heterologous gene *WUS* will promotes an embryogenic response in this species (Solís-Ramos *et al.*, 2009). The transformed chimeric plants where used for induction of expression of heterologous gene *WUS*. After 15 days of induction, the segments of transformed stems begun to form globular structures, and the wild type did not show development, suggesting that heterologus *WUS* was active and involved in the process of morphogenesis. The induced transformed explants showed the expression of *WUS* by Northern reverse analysis, and none *WUS* transcripts had detectable in the wild type. The histological analysis of induced transformed stems showed the development of meristematic nodules and the formation of globular somatic embryos, which presented necrosis after 45 days of in vitro culture, which did not continue development into other embryonic stages or in plants. The results showed that overexpression of gene *WUS* in stems of Habanero chili promote the formation of embryogenic structures but these stagnate in their growth suggesting that other signals may be need it for induction of proper development in this species (Solís-Ramos *et al.*, 2009). In addition this suggests that *WUS* encourages the development of undifferentiated tissue in species that may help as an alternative to solve the recalcitrance from this plant species (Solís-Ramos *et al.*, 2009).

6. Studies for coconut (*Cocos nucifera* L.) somatic embryogenesis

6.1 Introduction

Coconut (*Cocos nucifera* L.) is widely distributed throughout the humid tropics where it is cultivated over an estimated twelve million ha. It is a very important perennial crop, since it significantly contributes to food security, improved nutrition, employment and income generation. Coconut is a monospecific palm species consisting of numerous ecotypes and hybrids all possessing desirable agronomic properties. There is a great ethnic diversity in the ways that various coconut resources are produced and used (Foale, 2005). It is often referred to as "the tree of life" because of the many uses that have been developed for all parts of the palm. More recent uses of economic importance include fibre-derived products for the automobile industry; activated charcoal; virgin oil; bottled water; and oil for production of coco-biodiesel. In the Philippines, an industrial plant was launched in 2006 for the production of 75 million liters / year of coco-biodiesel where it is being used as a fuel additive (Lao, 2009). A blend at 2% coconut oil with diesel has been shown to reduce harmful exhaust emissions (opacity, K value) by as much as 63% (Lao, 2008).

However, most coconut groves require replanting because of loss due either to palm senescence or to diseases such as lethal yellowing in America (Harrison and Oropeza, 2008), the lethal diseases in Africa (Eden-Green, 1997) and cadang-cadang in Asia (Hanold and Randles, 1991). Unfortunately, improved disease resistant planting materials are scarce and seed propagation does not yield sufficient material to satisfy the rapidly growing demands.

Therefore, alternative approaches for the propagation of improved planting materials must be considered and *in vitro* propagation or micropropagation *via* somatic embryogenesis seems to provide a convenient alternative for the future due to its potential for massive propagation.

Several explants have been tested with diverse results, being the most responsive immature infloresencenes and plumules in increasing order (Blake and Hornung, 1995; Chan *et al.*, 1998; Pérez-Nuñez *et al.*, 2006). For this reason plumules have been more extensively used to improve on the different developmental changes in the process: callogenesis, embryo formation, germination and conversion.

6.2 *In vitro* culture of coconut palm

6.2.1 Coconut micropropagation using plumule explants

In order to increase the efficiency of somatic embryogenesis in coconut, two different approaches were evaluated, secondary somatic embryogenesis and multiplication of embryogenic callus. Primary somatic embryos obtained from plumule explants were used as explants and formed both embryogenic callus and secondary somatic embryos. The embryogenic calluses obtained after three multiplication cycles were capable of producing somatic embryos. The efficiency of the system was evaluated in a stepwise process beginning with an initial step for inducing primary somatic embryogenesis followed by three steps for inducing secondary somatic embryogenesis followed by three steps for embryogenic callus multiplication, and finally production of somatic embryos from callus (Pérez-Nuñez *et al.*, 2006). The actual process of somatic embryogenesis by embryogenic callus multiplication and secondary somatic embryogenesis is shown in Figure 4. The total calculated yield from one plumule was 98,000 somatic embryos (SEs). Comparing this to the yield obtained from primary somatic embryogenesis results in about a 50,000-fold increase (Pérez-Nuñez *et al.*, 2006).

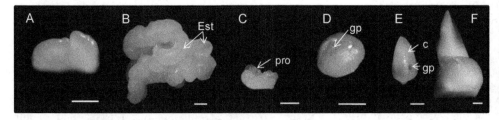

Fig. 4. An embryogenic structure derived from an embryogenic callus used as explant (A), developed an embryogenic callus (B) after 90 days of culture in medium I. After transferring embryogenic callus to medium II, callus with somatic embryos at different stages occurred. Piece of callus with pro-embryos (C), globular embryo (D), coleoptilar embryo (E) and germinating embryo (F). Bar= 1mm. Coleoptile: [c], germinative pore: [gp], pro-embryo [pro]

This protocol represented an important progress towards practical application by showing a way to improve the efficiency of coconut somatic embryo production. However has still some bottlenecks, as the relative low percentage of formation of embryogenic calli (40-

60%) and calli with somatic embryos (12-24%) and the low number of somatic embryos formed (2-10) per callus. In order to increase these figures and optimize this protocol to avoid many steps of multiplication, different plant growth regulators and compounds has been tested.

6.3 Exogenous plant growth regulators

6.3.1 Brassinosteroids

The effect of the brassinosteroid 22(S), 23(S)-homobrassinolide on initial callus, embryogenic callus and somatic embryo formation in coconut plumule explants was tested. The explants were exposed (during a 3 or 7 d pre-culture) to different concentrations (0.01, 0.1, 1, 2 and 4 µM) of the brassinosteroid. The explants responded favorably to the brassinosteroid increasing their capacity to form initial callus, embryogenic callus and somatic embryos. The largest amount of somatic embryos formed, 10.8 somatic embryos / explant, was obtained exposing the explants for 3 d to the brassinosteroid at 0.01 or 0.1 µM, whereas 3.8 somatic embryos / explant were obtained from untreated explants. Efficiency-wise the overall effect of HBr increases the total amount of somatic embryos formed per explant 2.8 times (Azpeitia *et al.* 2003).

6.3.2 Gibberellic acid

In some reports of coconut the GA_3 is added into the culture medium to promote the germination of somatic embryos (Perera *et al.*, 2009). However the effect of addition of this phytohormome had not been tested on the formation of somatic embryos. The results obtained with GA_3 were positive at 0.5 µM using the protocol of embryogenic calli multiplication from plumule explants. This concentration promoted 1.5 fold the number of the embryogenic calli forming somatic embryos. The number of somatic embryos per callus also increased, about 5 fold at day 30 (globular embryos) and 2 fold afterwards (coleptilar embryos). Also when the effect of GA_3 was evaluated on the germination of somatic embryos, the results were positive. The proportion of calli with germinating embryos was 2 fold higher than in the control treatment with no phytohormone. The number of germinating somatic embryos *per* callus was also higher under phytohormone treatment, also a 2 fold increase in relation to the control treatment. Therefore, a combined 4 fold increase in the overall number of germinating embryos (Montero-Cortés *et al.*, 2010). Then altogether, the use of GA_3 was positive both for the formation of somatic embryos and on their germination, so this could be a very useful approach to improve the performance of coconut micropropagation.

6.3.3 Uptake of auxins

6.3.3.1 Uptake of 2, 4-D

As previously reported for inflorescence coconut explants (Oropeza and Taylor, 1994), ^{14}C-2,4-D was taken up by plumular explants. The rate was faster during the first week of culture, and then reduced until reaching a plateau at day 90. The ^{14}C-2,4-D concentration in the explants reached its maximum values within the first 20 d of culture, prior to the appearance of any morphogenic response. It is interesting to note that when radioactivity was steadily taken up, calli were formed and once the calli started to form embryogenic structures, uptake practically stopped. This result suggests that the uptake of 2,4-D may be related to the induction of these morphogenic responses.

6.4 Characterization of genes related to somatic embryogenesis

6.4.1 Shoot apical meristem formation and maintaining (*KNOX* family genes)

The expression the class I *KNOX* (KNOTTED-like homeobox) genes seem to play an important role during somatic embryogenesis. In *Picea abies* overexpression of *HBK3*, a class I *KNOX* homeobox gene improves the development of somatic embryos and lines in which *HBK3* was down-regulated had reduced ability to produce immature somatic embryos and were not able to complete the maturation processes (Belmonte *et al.*, 2007).

The complete sequences of two *KNOX* like genes were obtained *CnKNOX1* and *CnKNOX2*. The deduced aminoacid sequence of both showed the highly conserved domains characteristic of *KNOX* genes. *CnKNOX1* showed high homology with KNOX class I proteins. *CnKNOX1* expression was detected throughout the embryogenesis process except in somatic embryos at the pro-globular stage, becoming highest in somatic embryos at the coleoptilar stage. No detection of *CnKNOX1* expression occurred in calli with aberrant embryos. The addition of gibberellic acid stimulated the expression of *CnKNOX1* earlier and the relative expression at all stages was higher. *CnKNOX2* expression occurred at all stages peaking at globular stage but gibberellic acid treatment decreased expression (Montero-Cortés *et al.*, 2010).

6.4.2 Somatic embryogenesis (Somatic Embryogenesis Related Kinase-*SERK*)

Somatic embryogenesis involves different molecular events including differential gene expression and various signal transduction pathways for activating or repressing numerous genes sets (Chugh and Khurana, 2002). Genes involved in somatic embryogenesis are stage specific and one of the genes identified in early somatic embryogenesis is SOMATIC EMBRYOGENESIS RECEPTOR-LIKE KINASE (DcSERK) that was originally isolated from embryogenic cells in suspension cultures of the dicot *Daucus carota* (Schmidt *et al.*, 1997). It was found to be expressed in embryogenic but not in non-embryogenic cultures, in cells predicted to be embryogenic, in tissue explants induced by placing them under embryogenic culture conditions, and during somatic embryogenesis up the globular stage. During *D. carota* zygotic embryogenesis, SERK expression occurred up to the early globular stage, but no expression was found in any other plant tissues, and cells transformed with a SERK promoter-luciferase reporter gene were able to form somatic embryos (Schmidt *et al.* 1997). Similar findings have been obtained in other dicots. In *A. thaliana*, the *AtSERK1* gene was expressed during the formation of embryogenic cells in culture, early embryogenesis, and in plant in developing ovules, specifically in all cells of the embryo sac up to fertilization, and in all cells after fertilization of the developing embryo until the heart stage (Hecht *et al.*, 2001). *A. thaliana* seedlings overexpressing AtSERK1 exhibited a three- to fourfold increase in efficiency for initiation of somatic embryogenesis; therefore, an increase in the level of the AtSERK1 conferred embryogenic competence in culture (Hecht *et al.*, 2001).

The complete sequence one *SERK* like gene was obtained and referred as *CnSERK*. Predicted sequence analysis showed that CnSERK encodes a SERK protein with the domains reported in the SERK proteins in other species. These domains consist of a signal peptide, a leucine zipper domain, five LRR, the Serine- Proline-Proline domain, which is a distinctive domain of the SERK proteins, a single transmembrane domain, the kinase domain with 11 subdomains and the C terminal region. Analysis of its expression showed that it could be detected in embryogenic tissues before embryo development could be observed. In contrast it was not detected or at lower levels in non-embryogenic tissues, thus suggesting that *CnSERK* expression is associated with induction of somatic embryogenesis and that it could be a potential marker of cells competent to form somatic embryos in coconut tissues cultured *in vitro* (Pérez-Nuñez et al., 2009).

6.5 Protocol for genetic transformation

We have developed a protocol for genetic transformation of this palm species (Andrade-Torres *et al.*, 2011); evaluating reporter genes, transformation methods, and conditions for the use of antibiotics to select transformed plant cells. The gene *uidA* was first used for *A. tumefaciens* mediated transformation of coconut embryogenic calli. However, endogenous GUS-like activity was found in calli not co-cultured with bacteria. Then essays for *Agrobacterium*-mediated transformation were developed using green and red fluorescent genes. Both genes are suitables as reporter genes for coconut transformation. In order to establish a protocol for coconut genetic transformation, an approach was used that combined biobalistics to generate micro-wounds in explants, vacuum infiltration and co-culture with *A. tumefaciens* (C58C1+ pER10W-35SRed containing the embryogenesis related gene *WUSCHEL*). Calli treated with the combined protocol showed red fluorescence with greater intensity and greater area than calli treated with either biobalistics or infiltration, followed by bacteria co-culture. PCR amplification of DNA extracts from transformed embryogenic callus produced a band with the expected size using *WUSCHEL* primers (862 bp). No band was obtained using the *VirE2* primers. This is the first report of transient genetic transformation of *C. nucifera* and it is the first step toward a protocol that will be useful for the study of the role of genes of interest and for practical applications, such as the improvement of coconut micropropagation via somatic embryogenesis (Andrade-Torres *et al.*, 2011).

7. Conclusions and perspectives

The majority of the mechanisms that regulate plant embryogenesis still remain to be clarified. In the higher plants, some genes and factors related to important mechanisms of embryogenesis are plant-specific. The availability of model systems of plant somatic embryogenesis has created effective tools for examining the details of plant embryogenesis. However, studies that used no model plants for somatic embryogenesis systems also revealed the molecular mechanisms in charge of controlling the expression of some genes during somatic embryogenesis, and with practical applications. So the molecular mechanisms of plant embryogenesis might be clarified by experiments using somatic and zygotic embryogenesis either from model or not model plants.

Numerous protocols on successful somatic embryogenesis induction and plant regeneration in different plant species, published last years, suggest that nowadays SE can be achieved for any plant provided that the appropriate explant and culture treatment are employed

(Gaj, 2004). A prerequisite for the successful establishment of a SE system is a proper choice of plant material -the explants being a source of competent cells, and, on the other hand, determination of physical and chemical factors which switch on their embryogenic pathway of development (Gaj, 2004).

The process of acquisition of embryogenic competence by somatic cells must involve reprogramming of gene expression patterns as well as changes in the morphology, physiology, and metabolism (Namasivayam, 2007). These alterations reflect dedifferentiation, activation of cell division and a change in cell fate.

Although few genes have been associated with embryogenesis induction, the search for genes involved in embryogenesis, such as SERK (Hecht *et al.*, 2001), LEC (Lotan *et al.*, 1998; Stone *et al.*, 2001), BABY BOOM (Boutilier *et al.*, 2002), WUSCHEL (Zuo *et al.*, 2002), and PICKLE (Ogas *et al.*, 1999), is a major field of research today (Quiroz-Figueroa *et al.*, 2006).

The characterization and functional analysis of protein markers for somatic embryogenesis offer the possibility of determining the embryogenic potential of plant cells in culture long before any morphological changes have taken place, and of gaining further information on the molecular basis of induction and differentiation of plant cells (Tchorbadjieva *et al.*, 2005).

The genetic transformation is certainly an important goal to facilitate genetic improvement against several diseases caused by phytopathogenic fungi, bacteria, and viruses, as well as for improvement against different pests (Ochoa-Alejo and Ramírez-Malagon, 2001). However, development of a reproducible tissue culture regeneration protocol is the first step in utilizing the power and potential of this new technology. The system established for *Capsicum chinense* is a promising alternative for cell or transformed plant regeneration through indirect somatic embryogenesis, and may contribute to genetic improvement of *C. chinense* Jacq. by incorporating reporter and interest genes (Solís-Ramos *et al.* 2009; Solís-Ramos *et al*, 2010a; Solís-Ramos *et al.*, 2010b).

The protocol for micropropagation of coconut from plumule explants based on embryogenic callus multiplication provides an option not available before for massive propagation of coconuts (Figure 5). However, although it allows the propagation of the progenie of known selected palms, it cannot be used for the cloning of palm individuals with known desirable agronomic traits. On the other hand, the recent developments to obtain embryogenic callus and somatic embryos from immature ovary and anther explants (Perera *et al.*, 2007; 2008; 2009), provide an opportunity to try to use these calli as a source of explants (the embryogenic structures) an integrate them into the callus multiplication scheme used with plumule explants. This has already been attempted in CICY using also floral tissue explants, but in this case rachillae slices from immature inflorescences (Oropeza and Chan, unpublished results). The callus obtained was tested for multiplication and although it responded poorly at the beginning though a series of multiplications the percentage of callus formation from embryogenic structure explants was above 40%. Therefore, although preliminary, this is a very promising result that shows that massive propagation from somatic tissue explants from adult plants is attainable in the near future. Finally we should continue with the studies to understand somatic embryogenesis in coconut. The study of genetic control is central for this purpose; therefore it is very important to learn more about the role of those genes that have been isolated and to extend the study to other genes and components of the genetic control of somatic embryogenesis. The study of these processes, will allow us not only to understand a phenomenon but it might open new avenues of opportunity for further improvement for a more efficient and better quality clonal propagation of coconuts.

Fig. 5. Final stages of micropropagation process of coconut palm. (A) *In vitro* plantlets ready to be transferred to *ex-vitro* conditions. (B) Acclimatization of plantlets in greenhouse covered with transparent perforated bags and (C) plantlets ready to be transferred to field conditions.

8. Acknowledgments

Laura Solís-Ramos thanks the Ph.D. fellowship from Dirección de Intercambio Académico de la Secretaría de Relaciones Exteriores (SRE), Mexico (Academic Exchange Office of the Secretariat of Foreign Affairs, Mexico) and Centro de Investigación Científica de Yucatán

(CICY) for facilities and laboratory support. A. Andrade-Torres thanks to CONACYT for the Ph.D. scholarship (204774), and Universidad Veracruzana for the support through the Dirección General de Desarrollo Académico (DGDA) and PROMEP (UV-491). The authors would like to thank to V. Hocher and J-L. Verdeil IRD/CIRAD Montpellier, France respectively, where the isolation of *CnKNOX* was carried out. Partial funding of the research reported here was from CONACyT, México (Grant no. 43834-Z).

9. Abbreviations

SE: somatic embryogenesis

ZES: zygotic embryo segments.

MS: Murashige & Skoog medium (1962)

BAP, 6-benzylaminopurine

NAA: naphthaleneacetic acid

IAA: indoleacetic acid

ISE: indirect somatic embryogenesis

MS-3R: MS medium with BAP+IAA+NAA

DsRFP: Red fluorescent protein

GUS: β-glucuronidase (gene *uidA*)

pH: hydrogen potential

2, 4-D: 2, 4-dichlorophenoxyacetic acid

GA_3: gibberellic acid

10. References

[1] Andrade-Torres, A., C. Oropeza, L. Sáenz, T. González-Estrada, J. E. Ramírez-Benítez, K. Becerril, J. L. Chan, and L. C. Rodríguez-Zapata. 2011. Transient genetic transformation of embryogenic callus of *Cocos nucifera* L. Biologia. 66 (5):790-800.
[2] Ahloowalia BS. 1991. Somatic embryos in monocots. Their genesis and genetic stability. Rev. Cytol. Biol. Veget-Bot. 14: 223-235.
[3] Arroyo-Herrera A. Ku-González A. Canche-Moo R. Quiróz-Figueroa F.R. Loyola Vargas V. Rodríguez Zapata L.C. Burgeff D'Hondt C. Suárez-Solís V.M. & Castaño E. 2008. Expression of WUSCHEL in *Coffea canephora* causes ectopic morphogenesis and increases somatic embryogenesis. Plant Cell Tiss. Org. Cult. 94: 171-180.
[4] Azpeitia A, Chan J L, Sáenz L, Oropeza C. 2003. Effect of 22(S), 23(S)-homobrassinolide on somatic embryogenesis in plumule explants of *Cocos nucifera* (L.) Cultured *in vitro*. J of Hort Sci and Biotech. 78: 591-596.

[5] Baudino S, Hansen S, Brettschneider R, Hecht VRG, Dresselhaus T, Lorz H, Dumas C, Rogowsky PM. 2001. Molecular characterization of two novel maize LRR receptor-like kinases, which belong to the SERK gene family. Planta 213:1-10.

[6] Baurle, I. and Laux, T. 2005. Regulation of WUSCHEL transcription in the stem cell niche of the *Arabidopsis* shoot meristem. Plant Cell 17: 2271-2280.

[7] Belmonte MF, Tahir M, Schroeder D, Stasolla C. 2007. Overexpression of *HBK3*, a class I *KNOX* homeobox gene, improves the development of Norway spruce (*Picea abies*) somatic embryos. J Exp Bot 58: 2851-2861.

[8] Benson, E. 2000. Special Symposium: *In vitro* Plant recalcitrance do free radicals have a role in plant tissue culture recalcitrance. In vitro Cell. Dev. Biol.-Plant. 36: 163-170.

[9] Bhalla, P. L. and Singh, M. B. 2006. Molecular control of stem cell maintenance in shoot apical meristem. Plant Cell Rep. 25:249–256.

[10] Binzel ML, Sankhla N, Joshi S, Sankhla D. 1996. Induction of direct somatic embryogenesis and plant regeneration in pepper (*Capsicum annuum* L.). Plant Cell Reports 15: 536-540.

[11] Blake J and Hornung R. 1995. Somatic embryogenesis in coconut (*Cocos nucifera* L.) In: Jain S, Gupta P, Newton R (eds). Somatic Embryogenesis in Woody Plants, v. 2. Kluwer Academic Publishers Dordrecht pp 327-340.

[12] Boutilier, K.; Offringa, R.; Sharma, V. K.; Kieft, H.; Ouellet, T.; Zhang, L. M.; Hattori, J.; Liu, C. M.; van Lammeren, A. A. M.; Miki, B. L. A.; Custers, J. B. M.; Campagne, M. M. V. 2002. Ectopic expression of BABY BOOM triggers a conversion from vegetative to embryonic growth. Plant Cell 14:1737-1749.

[13] Buyukalaca S, Mavituna F. 1996. Somatic embryogenesis and plant regeneration of pepper in liquid media. Plant Cell, Tissue and Organ Culture 46: 227-235.

[14] Cai, W-Q, Fang, X., Shang, HS, Wang, X. and Mang, K.Q. 2003. Development of CMV- and TMV resistant transgenic chilli pepper: field performance and biosafety assessment. Molecular Breeding. Vol: 11. pp: 25-35.

[15] Cervera M. 2005. Histochemical and fluorometric assays for *uidA* (GUS) gene detection. Chapter 14. In: PEÑA, Leandro ed. Transgenic plants: Methods and protocols. Humana Press. 286: 203-213.

[16] Chan JL, Sáenz L, Talavera C, Hornung R, l Robert M, Oropeza C. 1998. Regeneration of coconut (*Cocos nucifera* L.) from plumule explants through somatic embryogenesis. Plant Cell Rep 17:515–521

[17] Chugh A. and P. Khurana. 2002. Gene expression during somatic embryogenesis – recent advances. Current Science. 83 (6): 715-730.

[18] Chung W, Pedersen H, Chin C-K. 1992. Enhanced somatic embryo production by conditioned media in cell suspension cultures of Daucus carota. Biotechnol Lett. 14: 837-840

[19] Dudits D, Gyugyey J, Bugre L, Bakó L. 1995. Molecular biology of somatic embryogenesis. In: Thorpe TA (ed) *In vitro* Embryogenesis in Plants. pp: 267-308. Kluwer Academic Publisher, Dordrecht.

[20] Eden-Green SJ. 1997. History and World distribution of lethal yellowing-like diseases of palms. In: Eden-Green SJ, Ofori F (eds). Proceeding of an International

Workshop on Lethal Yellowing Diseases of Coconut, Elmina, Ghana, November, 1995. Natural Resources Institute, Chatham, UK. pp 9-25.

[21] Fehér A, Pasternak TP, Dudits D. 2003. Review of Plant Biotechnology and Applied Genetics. Transition of somatic plant cells to an embryogenic state. Plant Cell, Tissue and Organ Culture 74: 201-228.

[22] Fehér A. 2005. Why Somatic Plant Cells Start to form Embryos? In. Mujib A., Samaj J. (eds). Plant Cell Monographs (2). Somatic Embryogenesis. Springer-Verlag Berlin Heidelberg. 85-101.

[23] Foale M. 2005. An introduction to the coconut palm. In: Batugal P, Ramanatha V, Rao GP, Oliver J (eds.). Coconut Genetic Resources. International Plant Genetic Resources Institute – Regional Office for Asia, the Pacific and Oceania (IPGRI-APO), Serdang, Selangor DE, Malaysia. pp 1-8.

[24] Francis D. and Sorrell D.A. 2001. The interference between the cell cycle and plant growth regulators: a mini review. Plant Growth Regul. 33: 1-12.

[25] Gaj MD. 2004. Factors influencing somatic embryogenesis induction and plant regeneration with particular reference to Arabidopsis thaliana (L.) Heynh. Plant Growth Regulation 43: 27-47.

[26] Hanold D and Randles JW. 1991. Detection of coconut cadang-cadang viroid-like sequences in oil and coconut palm and other monocotyledons in the south-west Pacific. Ann of Appl Biol 118: 139-151.

[27] Harini I, Sita L. 1993. Direct somatic embryogenesis and plant regeneration from inmature embryos of chilli (Capsicum annuum L.). Plant Science 89: 107-112.

[28] Harrison NA and Oropeza C. 2008. Phytoplasmas associated with coconut lethal yellowing. In: Harrison NA, Rao GP and Marcone C (eds). Characterization, Diagnosis and Management of Phytoplasmas. Studium Press LLC, Houston, USA pp: 219-248.

[29] Hecht V, Vielle-Calzada J-P, Hartog MV, Schmidt Ed DL, Boutilier K, Grossniklaus Ueli, de Vries SC. 2001. The Arabidopsis SOMATIC EMBRYOGENESIS RECEPTOR KINASE 1 Gene is expressed in developing ovules and embryos and embryo and enhances embryogenic competence in culture. Plant Physiol 127:803-816.

[30] Hecht V, Vielle-Calzada JP, Hartog MV, Schmidt DL, Boutilier K, Grossniklaus U, Vries SC. 2001. The Arabidopsis SOMATIC EMBRYOGENESIS RECEPTOR KINASE 1 Gene is expressed in developing ovules and embryos and enhances embryogenic competence in culture. Plant Physiology, november. 127: 803-816.

[31] Henry RJ. 1998. Molecular and biochemical characterization of somaclonal variation. In: Jain SM, Brar DS & Ahloowalia BS (eds) Somaclonal Variation and Induced Mutations in Crop Improvement (pp 485-499). Kluwer Academic Publishers, Dordrecht.

[32] Herrera-Estrella, L., Simpson, J., and Martínez-Trujillo, M. 2004. Transgenic Plants. An historical perspective. In: Peña, L. Transgenic Plants. Methods and Protocols. Humana Press. 286. Pp.: 3-31.

[33] Hu H, Xiong L, Yang Y. 2005. Rice SERK1 gene positively regulates somatic embryogenesis of cultured cell and host defense response against fungal infection. Planta 222:107–117.

[34] Jefferson R. 1987. Assaying chimeric genes in plants: the GUS gene fusion system. Plant Molecular Biology Reporter, December 5 (4): 387-405.

[35] Jiménez VM. 2005. Involvement of plant hormones and plant growth regulators on in vitro somatic embryogenesis. Plant Growth Regulation. 47:91–110.

[36] Khan H., Siddique I, Anis, M. 2006. Thidiazuron induced somatic embryogenesis and plant regeneration in *Capsicum annuum*. Brief communication. Biologia Plantarum, 50 (4): 789-792.

[37] Kintzios S., Drossopoulos J, Lymperopoulos CH. 2001. Effect of vitamins and inorganic micronutrients on callus growth and somatic embryogenesis from leaves of chilli pepper. Plant Cell, Tissue and Organ Culture, 67: 55-62.

[38] Lao DA. 2008. Coco-biodiesel more than a diesel replacement. Bioenergy Forum, Bangkok, April 2008.

[39] Lao DA. 2009. Coco-biodiesel in the Philippines. In: Coconut Philippines published by Asia Outsourcing

[40] Liu C, Xu Z, Chua Nh. 1993. Auxin Polar Transport is essential for the establishment of bilateral symmetry during early plant embryogenesis. The Plant Cell. 5: 621-630.

[41] López-Puc G, Canto-Flick A, Barredo-Pool F, Zapata-Castillo P, Peniche-Montalvo M, Barahona-Pérez F, Iglesias-Andreu, Santana-Buzzy N. 2006.   Direct somatic embryogenesis: a highly efficient protocol for in vitro regeneration of habanero pepper (*Capscium chinense* Jacq.). HortScience 41(7):1645–1650.

[42] Lotan T, Ohto M, Yee MK, West MA, Lo R, Kwong RW, Yamagishi K, Fischer RL, Goldberg RB, Harada JJ. 1998. Arabidopsis LEAFY COTYLEDON 1 is sufficient to induce embryo development in vegetative tissue. Cell. 93: 1195-1205.

[43] Martin T, Wohner RV, Hummel S, Willmitzer L and Frommer WB.1991. The GUS reporter system as a tool o study plant gene expression. In: GALLAGER, S. ed. GUS protocols. Academic Press, Inc. California. pp. 23-43.

[44] Montero-Córtes M, Sáenz L, Córdova I, Quiroz A, Verdeil J-L, Oropeza C 2010. GA$_3$ stimulate the formation and germination of somatic embryos and the expression of a KNOTTED-like homeobox gene of *Cocos nucifera* (L.) Plant Cell Rep 29: 1049-1059

[45] Namasivayam P. 2007. Acquisition of embryogenic competence during somatic embryogenesis. Review paper. Plant Cell Tiss Organ Cult. 90: 1-8. DOI 10.1007/s11240-007-9249-9

[46] Nolan KE, Irwanto RR, Rose RJ. 2003. Auxin up-regulates MtSERK1 expression in both Medicago truncatula root-forming and embryogenic cultures. Plant Physiol 133:218–230.

[47] Ochoa-Alejo N and Ramírez-Malagon, R. 2001. In vitro chili pepper biotechnology. Invited review. In vitro Cell. Dev. Biol-Plant. 37: 701-729. DOI: 10.1079/IVP2001216.

[48] Ogas J, Kaufmann S, Henderson J, Somerville C. 1999. PICKLE is a CHD3 chromatin-remodeling factor that regulates the transition from embryogenic to vegetative development in *Arabidopsis*. Proc Natl Acad Sci USA. 96: 13839-13844.

[49] Oropeza C and Taylor HF. 1994. Uptake of 2,4-D in coconut (*Cocos nucifera* L.) explant. In: Lumsden PJ, Nicholas JR, Davies WJ (eds). Physiology, growth and development of plant in culture. Kluwer Academics Publishers, The Netherlands pp 284-288.

[50] Perera PIP, Hocher V, Verdeil JL, Doulbeau S, Yakandawala DMD, Weerakoon LK. 2007. Unfertilized ovary: a novel explant for coconut (*Cocos nucifera* L.) somatic embryogenesis. Plant Cell Rep 26: 21-28.

[51] Perera PIP, Perera L, Hocher V, Verdeil JL, Yakandawala DMD, Weerakoon LK. 2008. Use of SSR markers to determine the anther-derived homozygous lines in coconut. Plant Cell Rep 27: 1697-1703.

[52] Perera PIP, Yakandawala DMD, Hocher V, Verdeil J-L, Weerakoon LK. 2009. Effect of growth regulators on microspore embryogenesis in coconut anthers. Plant Cell Tiss Org Cult 96:171-180.

[53] Pérez-Nuñez MT, Chan JL, Sáenz L, González T, Verdeil JL, Oropeza C. 2006. Improved somatic embryogenesis from *Cocos nucifera* (L.) plumule explants. *In Vitro* Cell Dev Biol Plant 42:37-43.

[54] Pérez-Núñez MT. Souza R. Sáenz L. Chan J.L. González T. Zúñiga J.J. & Oropeza C. 2009. Detection of a *SERK*-like gene in coconut in vitro cultures and analysis of its expression during the formation of embryogenic callus and somatic embryos. Plant Cell Rep. 28: 11-19.

[55] Quiroz-Figueroa F., Rojas-Herrera R, Galaz-Avalos RM, Loyola-Vargas VM. 2006. Embryo production through somatic embryogenesis can be used to study cell differentiation in plants. Review paper. Plant Cell Tiss Organ Cult. 86: 285-301. DOI 10.1007/s11240-006-9139-6

[56] Reddy, G. V. and Meyerowitz, E. M. 2005. Stem-cell homeostasis and growth dynamics can be uncoupled in the *Arabidopsis* shoot apex. Science 310:663-667.

[57] Reinert J. 1958. Untersuchungen über die Morphogenese an Gewebekulturen. Ber Dtsch Bot Ges 71: 15.

[58] Santana-Buzzy N, Canto-Flick A, Barahona-Pérez F, Montalvo-Peniche MC, Zapata-Castillo P, Solís-Ruíz A, Zaldívar-Collí A, Gutiérrez-Alonso O, Miranda-Ham L. 2005. Regeneration of habanero pepper (*Capsicum chinense* Jacq) via organogenesis. HortScience 40(6): 1829-1831.

[59] Santos MD, Romano E, Yotoko KSC, Tinoco MLP, Dias BBA, Aragao FJL. 2005. Characterization of the cacao somatic embryogenesis receptor-like kinase (SERK) gene expressed during somatic embryogenesis. Plant Sci 168:723-729.

[60] Schmidt EDL, Guzzo F, Toonen MAJ, de Vries SC. 1997. A leucine rich repeat containing receptor-like kinase marks somatic plant cells competent to form embryos. Development 124:2049-2062.

[61] Schmidt EDL, Guzzo F, Toonen MAJ, de Vries SC. 1997. A leucine-rich repeat containing receptor-like kinase marks somatic plant cell competent to form embryos. Development 124:2049-2062.

[62] Shah, K.; Russinova, E.; Gadella, T. W. J.; Willemse, J.; de Vries, S. C. 2002. The Arabidopsis kinase-associated protein phosphatase controls internalization of the somatic embryogenesis receptor kinase 1. Genes Dev. 16:1707-1720.

[63] Sharma, S.K., S. Millam, I. Hein and G.J. Bryan. 2008. Cloning and molecular characterization of a potato SERK gene transcriptionally induced during initiation of somatic embryogenesis. Planta. 228: 319-330.

[64] Shimada T, Hirabayashi T, Endo T, Fujii H, Kita M, Omura M. 2005. Isolation and characterization of the somatic embryogenesis receptor-like kinase gene homologue (CitSERK1) from *Citrus unshui* Marc. Sci Hortic 103: 233-238.

[65] Shin R., Han J-H., Lee G.J., and Peak, K.H. 2002. The potencial use of viral coat protein genes as transgene screening marker and multiple virus resistance of pepper plants coexpressing coat proteins of cucumber mosaic virus and tomato mosaic virus. Transgenic Reseach. 11: 215-219.

[66] Solís-Ramos L.Y., González-Estrada T., Nahuath-Dzib S., Rodríguez-Zapata, L., Castaño E. 2009. Overexpression of WUSCHEL in *C. chinense* causes ectopic morphogenesis. Plant Cell, Tissue and Organ Culture. DOI: 10.1007/s11240-008-9485-7. 96:279-287.

[67] Solís-Ramos LY, González-Estrada T, Andrade-Torres A, Godoy-Hernández G, Castaño de la Serna E. 2010a. Endogenous GUS-like activity in *Capsicum chinense* Jacq. Electronic Journal of Biotechnology. July 15, 13 (4). DOI: 10.2225/vol13-issue4-fulltext-3.

[68] Solís-Ramos LY, Nahuath-Dzib S, Andrade-Torres A, Barredo-Pool F, González-Estrada T, Castaño de la Serna E. 2010b. Indirect somatic embryogenesis and morphohistological analysis in *Capsicum chinense*. Section Cellular and Molecular Biology. Biologia. 65(3): 504-511. DOI: 10.2478/s11756-010-0049-z.

[69] Somleva MN, Schmidt EDL, de Vries SC. 2000. Embryogenic cells in *Dactylis glomerata* L. (Poaceae) explants by cell tracking and by SERK expression. Plant Cell Rep 19:718–726.

[70] Steinitz B., Kusek M., Tabib Y., Paran I. & Zelcer A. 2003 Pepper (*Capsicum annuum* L.) regenerates obtained by direct somatic embryogenesis fail to develop a shoot. In vitro Cell. Dev. Biol. Plant 39: 296-303.

[71] Steward FC, Mapes MO, Mears K .1958. Growth and organized development of cultured cells. II. Organization in cultures grown from freely suspended cells. Am J Bot 45: 705–708

[72] Stone SL, Kwong LW, Yee KM, Pelletier J, Lepiniec L, Fischer RL, Goldberg RB, Harada JJ (2001) LEAFY COTYLEDON2 encodes a B3 domain transcription factor that induces embryo development. Proc Natl Acad Sci USA 98: 11806-11811.

[73] Suda Ch, Prakash S, Bhomkar P, Jain S and Bhalla-Sarin, N. 2006. Ubiquitous presence of β-glucuronidase (GUS) in plants and its regulation in some model plants. Planta, September 224 (4): 853-864.

[74] Tchorbadjieva M, Kalmukova R, Pantchev I, Kyurkchiev S. 2005. Monoclonal antibody against a cell wall marker protein for embryogenic potential of *Dactylis glomerata* L. suspension cultures. Planta 222:811–819.

[75] Thomas C, Meyer D, Himber C, Steinmetz A. 2004. Spatial expression of a sunXower SERK gene during induction of somatic embryogenesis and shoot organogenesis. Plant Physiol Biochem 42:35–42.

[76] Tucker, M. R.; Araujo, A. -C. G.; Paech, N. A.; Hecht, V.; Schmidt, E. D. L.; Rossell, J. -B.; de Vries, S. C.; Koltunow, A. M. G. 2003. Sexual and apomictic reproduction in *Hieracium subgenus* Pilosella are closely interrelated developmental pathways. Plant Cell 15:1524–1537.

[77] Weigel, D. and Jurgens, G. 2002. Stem cells that make stems. Nature 415: 751–754.

[78] Zapata-Castillo P, Canto-Flick A, López-Puc G, Solís-Ruíz A, Pérez-Barahona F, Iglesias-Andreu L, Santana-Buzzy N. 2007. Somatic embryogenesis in Habanero pepper (*C. chinense* Jacq.) from cell suspensions. Hort Science.

[79] Zuo J, Niu QW, Frugis G, Chua N.H. 2002. The WUSCHEL gene promotes vegetative-to-embryonic transition in *Arabidopsis*. The plant Journal 30(3): 349-359. doi: 10.1046/j.1365-313X.2002.01289.x.

12

Liquid-Crystal in Embryogenesis and Pathogenesis of Human Diseases

MengMeng Xu[1] and Xuehong Xu[2,3,4,*]

[1]Medical Scientist Training Program, Duke University School of Medicine,
[2]Department of Physiology Center for Biomedical Engineering Technology,
[3]Center for Stem Cell Biology and Regenerative Medicine,
University of Maryland School of Medicine,
[4]Shaanxi Normal University School of Life Science,
[1,2,3]USA
[4]China

1. Introduction

In 1979, a systematic publication summarizing the state of research on liquid-crystals in biological organisms was published [Brown GH *et al* 1979]. After this historic publication on liquid-crystals and biology, the field remained largely dormant for more than two decades. However in 1978 and 1979, Haiping He and Xizai Wu, who had continued pursuing this field despite international disinterest, reported their findings on liquid-crystal involvement during chicken development. For the first time, they revealed that massive quantities of liquid-crystals in the liver, yolk sac, blood, and many other developing tissues and organs of chicken during embryogenesis. Their later studies also reported similar liquid-crystalline structures during fish development. In 1988, another group reported the existence of vaterite $CaCO_3$ within the liquid-crystals found in yolk fluid, identifying the spherical calcified structures first reported in 1979 as one of three iso-forms of calcium carbonate [Feher G 1979, Li M *et al* 1988]. Subsequent studies have identified liquid-crystalline structures to be omnipresent in the liver during avian development [Xu XH *et al* 1995a, 1995b, 1997]. Recent studies have revealed that liquid-crystals play a critical role in the preservation of calcium and other trace elements required for embryo development [Xu MM *et al* 2009, 2010, 2011; Xu XH *et al* 2009, 2011a].

In recent years, more and more human diseases have been related to liquid-crystals. Amongst these diseases are genetic disorders, such as Age-related Macular Degeneration [Haimovici R *et al* 2001], steatohepatitis and atherosclerosis [Goldstein JL *et al* 2008], and Anderson-Fabry Disease [Xu MM *et al* 2009]. For Fabry patients, the accumulation of liquid-crystal or concentric lamellar bodies glycosphingolipids in neurons can cause severe neuroradiological abnormalities, including periventricular white-matter signal intensity abnormality and single/multiple lacunar infarction, large ischaemic cerebral infarct and posterior thalamic involvement [Ginsberg L *et al* 2006; Lidove O *et al* 2006; Moore DF *et al* 2003]. This accumulation also occurs in the cardiac vascular system, resulting in angina and varied complications, ranging from arrhythmia to myocardial infarction and heart failure [Pieroni M *et al* 2006; Linhart A *et al* 2007]. Although a large volume of publications indicate an irreplaceable role for liquid-crystals in both normal physiological development and pathogenesis, the exact function of liquid-crystal is uncertain.

In this chapter, we will summarize current research around liquid-crystal involvement in embryogenesis and how these normal embryonic events, when triggered at inappropriate times, can lead to pathogenic events. Current publications indicate that liquid-crystals are often involved in normal embryogenesis, but the appearance of liquid-crystals in post-natal development often heralds pathogenesis in mature tissue. Normally embryogenic events can be trigger by a variety of factors, such genetic predispositions or bacterial infection. As part of normal embryogenesis, the formation of liquid-crystals indicates erroneous initiation of normal growth, often leading to disease.

2. Methodology for liquid-crystal function in medical biology and embryogenesis

Liquid crystal is refers to a material's physical state. As the term suggests, liquid-crystals display both liquid and solid tendencies. This fluid characteristic perfectly matches the properties of a living organism. However, this same adaptability to life introduces some critical attention to experimental approaches when studying liquid-crystals *in vitro*. In this section, we will focus on research approaches used for studying liquid-crystals in biological embryogenesis and human pathogenesis. The preeminent combination of all procedures could perform excellent inspection on the liquid crystal [Xu XH et al 2011a].

2.1 Animal sample collection according to animal IACUC

In the United States, any research application involving laboratory animals is required to be approved by the Institutional Animal Care and Use Committee (IACUC) prior to experimentation, if the research is funded by a federal agency such as the National Institute of Health. All animal procedures must also follow guidelines approved by the home institution's Animal Care and Use department. For mouse tissue and organ harvest, mice must first be euthanized via CO_2 asphyxiation followed by cervical dislocation. After the necessary tissues have been harvested, the mouse remains should be sealed within plastic bagging and frozen for temporal storage in the animal core facility. Permanent treatment of animal body will be carried out by the core.

In most instances of commercial mice, such as 129SvEv and DBA/2J breeds from the Jackson Laboratory (Bar Harbor, ME), Harland Laboratories (Indianapolis, IN), and other commercial sources were bred under standard conditions and sacrificed by cervical dislocation according to IACUC regulation. All tissues were harvested immediately and snap frozen with liquid nitrogen upon dissection. For timing experiments using mouse embryo, appearance of the vaginal plug was designated as embryonic day (E) 0.5, and embryos were either collected from the stage of interest, or pregnancies were allowed to reach term. For postnatal studies, newborn mice were collected at the day of birth (postnatal day, P).

For avian experiments, fertilized chick eggs such as White Leghorn (*Gallus domesticus L.*) are normally incubated at 37⁰C with relative air humidity of 60%. Experiments are conducted and samples harvested at time-points according to experimental procedure. The age of embryos is documented as day of incubation (D) and the postnatal age of chicks is documented as Postnatal (P). Collections of tissues and organs are the same as described above and can be found within references [Xu MM *et al* 2009, 2010, 2011; Xu XH *et al* 2009, 2010, 2011ab].

2.2 Cryo-section histology analysis

For purely biological studies of embryogenesis, both paraffin section and cry-section are good options for well-preserved tissue and organ samples. However, for liquid-crystal functional studies during embryo development and pathological events, cryo- or frozen-section are preferred over paraffin sections, as the cryo/frozen methods preserve more of the sample's original characteristics than the more chemically intensive paraffin preservation method. If possible, fresh sample smear-slides are also preferred. [Xu MM *et al* 2009, Xu XH *et al* 2009].

Though the previous methods allow for long term storage, fresh smear-slide preparation arguably retains the most fidelity to *in vivo* systems as the tissue is not processed in any way prior to observation. In smear-slide preparation, samples are harvest from the embryos at different stages. Each sample is then immediately smeared on a slide wetted with PBS (PH 7.4) buffer then mounted with a cover slip. This method is best observed by polarization microscopy, which can be conducted immediately following the sample smear preparations.

When in need of a method for long term storage or if retaining the physical structure of the tissue is desired, sryosection preparation is preferred. Embryo samples are submerged in the cryomatrix embedding agent (OCT) and placed in an aluminum foil basket or other suitable container to be dipped. The samples are then frozen by dipping the foil basket or the container into liquid nitrogen. The now frozen tissues embedded in the OCT block can now be placed on the cryostat microtome and sections cut for experimentation. Thicker cuts 10~30 mm are preferred for polarization microscopy and thinner cuts of ~5mm are preferred for H&E staining. The samples collected using these two procedures should be mounted with 20% of glycerol in PBS (PH 7.2) and sealed before proceeding to further analysis.

2.3 Immunohistochemistry and confocal microscopy

After collection and smear-section preparation, samples should be washed with PBS, fixed in 4% formaldehyde in phosphate-buffered saline (PBS), and permeabilized for 10 min in PBS containing 0.25% Triton X-100. Immunocytochemical staining can be perform performed by incubating the samples with primary antibodies in PBS/Tween (PBS containing 0.1% Tween 20 and 3% BSA) for 1–2 hr, followed by incubation with appropriate secondary antibodies diluted according to the manufacturer's recommendation in PBS/Tween solution for 1–2 hr. Stained samples can then be visualized with confocal microscope.

Immunohistochemistry is a powerful approach to unveil distribution of protein of interest, which can be used to locate liquid-crystal related proteins in tissues and organs. However, because this method requires the use of damaging solutions such as formaldehyde, Triton X-100, and Tween containing blocking buffer, results should be compared to smear-samples for confirmation of signal localization corresponding to liquid crystal.

2.4 Polarization microscopy and thermal phase-transition

Polarization or polarized microscopy is an irreplaceable tool for studying materials with refractory activity. In the case of liquid crystal study, two states of material, crystal and liquid crystal, have birefringent activity. In our studies, we have found microscopy to be an invaluable tool in examining the light activity of liquid crystals in animal tissue.

Natural light such as light from an ordinary light source is called non-polarized because it vibrates in random directions. Polarized light on the other hand, travels within a single plane and presents with vertical vibrations that can produce linear, circular, and elliptical polarized light. A polarizing plate or polarizing prism is often used as a polarizing filter to remove all but one wave with the same directional vibration.

Observing liquid-crystals under polarized light requires a basic understanding of light polarization. To create a polarized light, two devices, the primary and secondary polarizing devices, are oriented perpendicular to each other as crossed nicols to filter polarized light from normal lighting. The primary device will filter a polarized light from the light source, while the secondary device cuts the light depending on orientation of the two devices. These primary polarizing device and secondary polarizing device are called Polarizer (P) and Analyzer (A), respectively. In perpendicular nicols, the analyzer is rotated to be perpendicular to the polarizer. Since both nicols act as filters, the analyzer cancels out the polarized light from the polarizing lense, to yield no light to the observer (Figure 1A). In parallel nicols, the analyzer is rotated so that the direction of the transmitting polarized light is parallel with the polarizer. This allows polarized light transmitted via the polarizer to travel through the Analyser, maximizing the amount of light transmitance (Figure 1B).

The light bending ability of liquid crystals can be thought of as an additional nicol. When polarized light launching through a crystal or liquid crystal materials is divided into two linearly polarized light rays, these two rays possess mutually crossing vibration directions, called birefringence (double refraction). A crystal or liquid crystal that refracts in this way is called anisotropy. When an anisotropic crystal or liquid crystal is inserted between a polarizer and an analyzer in a crossed nicols state, the crystal or liquid crystal changes the state of the polarized light and the light to pass through partially (Figure 1C). These changes are different depending on various crystal or liquid crystal, which can be utilized to determine characteristics of an anisotropic material, liquid crystal in this case.

Using this light-bending property of liquid crystals, thermal-probe sample stage in conjunction with polarization microscope, can be used to monitor and record phase transition of liquid crystals. When a sample is observed with crossed-nicols, the anisotropic texture of crystal or liquid crystal will vanish once the temperature of stage reaches the point of phase transition from liquid crystal to isotropic states. This technique is an effective method for observing the liquid crystal properties of biological samples containing liquid crystals [Xu MM *et al* 2010, 2011; Xu XH *et al* 2011a].

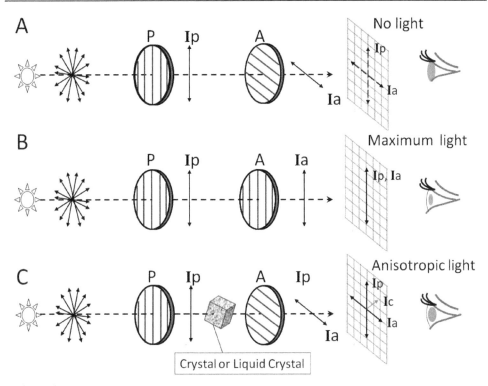

A. Crossed nicols;
B. Parallel nicols;
C. Anisotropy between crossed nicols.
P: polarizer
A: analyzer

Fig. 1. Light path of polarization and anisotropy

2.5 Small angle X-ray scattering (SAXS) and X-ray diffraction (XRD)

When X-ray beam of a particular wavelength diffracts from atoms in a crystalline structure, the wavelength of the x-ray (λ), scattering angle (θ), integer representing the order of the diffraction peak (n), and inter-plane distance (d), usually the distance between atoms, ions, molecules, follow the Bragg's Law (Figure 2A).

$$2d \sin\theta = n\lambda$$

This equation predicts that different layers of atoms in lattice planes will generate various distances corresponding to peaks. Crystal samples, multiple peaks will be present in a wide-spread diffraction angle (2θ, XRD; Figure 2B), while liquid crystal exhibit fewer peaks within an area of small scattering angles (2θ, SAXS; Figure 2C). For crystals or liquid-crystals within biological samples, once liquid-crystals or crystals have been isolated or extracted, temperature and other conditions must be tightly controlled to retain the original characteristics and diffraction pattern of samples.

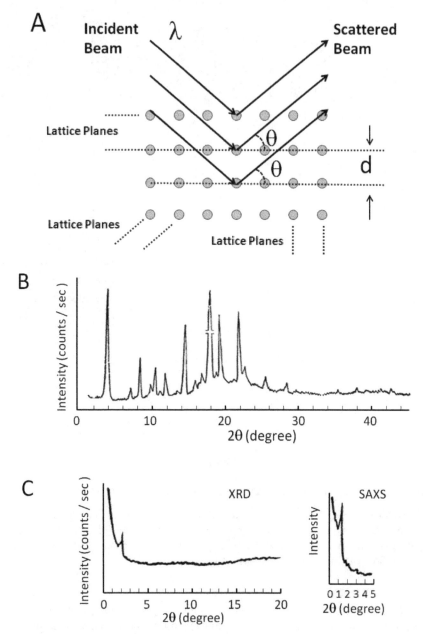

Fig. 2. Bragg's law and the diffraction patterns of crystal and liquid crystal. Relationship between wavelength, scattering angle, and distance of lattice planes (A). XRD pattern of crystal cholesteryl oleate within 50 degree of diffraction angle (2θ) (B), and SAXS pattern of liquid crystal in embryonic liver within 50 degree (2θ) (left in C) and within 5 degree of scattering angle degree (2θ) (right in C).

2.6 Generation of mouse model using daily diet procedure

Mice are obtained from Jackson Laboratories or another commercial facility. Male or female pups are split into two groups. Group one is fed on the standard chow diet (Harlan Teklad #2018 rodent chow) with 48% carbohydrate, 16% protein and 4% fat. Group two is on a low carbohydrate high protein diet with 15% carbohydrate, 58% protein and 26% fat, which we was designed to mimic the typical Western diet.

3. Comparative summary of liquid-crystals in embryonic tissue development and post-natal pathology

As discussed earlier in the chapter, embryonic tissues and organs of several animal models have exhibited traces of liquid-crystals during embryonic development stage that did not persist postnatally. In liver and yolk sac, massive liquid-crystals are present from embryogenesis to early post-natal development. In this section, we will summarize the characteristics of liquid-crystals in different tissues in comparison to the liquid-crystals found in human diseases (Table 1).

3.1 General characteristics of embryonic liquid-crystal

During embryogenesis, liquid crystals are widely distributed in the tissues of vertebrates and invertebrates, including Apis cerana chrysalis, fish, reptile, avian and mammal early embryo in vitro [X XH *et al* 1993, 2009, 2011a, Xu MM et al 2009 2011]. In chicken development, more than twenty different organs and tissues exhibit liquid crystal droplets including liver, meso- and metanephros, lungs, blood in heart, and brain. The presence of liquid crystal normally appears at different developmental stages depending on the tissue type, and lasts until early postnatal stages. The earliest liquid crystal droplets appear on the inner embryonic disc during the second day of development [He H et al 1978]. Regardless of their distribution, however, the liquid crystal droplets eventually vanish within three to four weeks into the postnatal period, also depending on tissue type maturation [X XH *et al* 2009, 2011a].

During chicken development, two particular organs, the liver and yolk sac, exhibit massive birefringent liquid crystal at higher levels than all over tissues in the developing embryo. The hepatic birefringent particles are mainly composed of cholesteryl oleate, cholesterol, lecithin and an unidentified component [Xu XH *et al* 1992, 1995a, 2011a]. These liquid crystal droplets are situated in hepatocytes of the hepatic cord region. In the kidney development, LC droplets can exist in the cytoplasm of epithelial cells and the lumen of proximal tubules in the mesonephros and metanephros. The existence of LC in two very different organs indicates that the liquid crystal likely plays many different roles during the development or a similar role in many tissues.

3.2 Decrease-rate dependent thermal phase transition

Under polarization microscope, liquid crystal exhibit Maltese-crosses optical textures, while crystals produced more angular (needle-like, rhombus, or dot-shape) diffraction patterns The two states also reacted differently to pressure experiments, with the liquid-crystals dividing into smaller Maltese-cross droplets, while crystals fractured uner duress [Xu XH et al 2009, 2011a].

Thermal phase transitions have also been revealed in liquid crystal obtained from various tissues (Table 1). Not surprisingly, with thermal stage temperature increase, the birefringent liquid crystal droplets transit to non-refracting isotropic droplets. With temperature decrease, the liquid crystal droplets transit into crystal. However, when the isotropic droplets cool, two different results, controllable by rate of temperature decrease, were possible. If the rate of temperature drop is fast (the slide is placed on a 4ºC plate), then the isotropic droplets will transition into liquid-crystals. However, if the rate of temperature decrease is slow (temperature is allowed to drop in step with the slowly cooling copper thermo-controller) then the isotropic droplets will transit to crystal (Figure 3). This finding

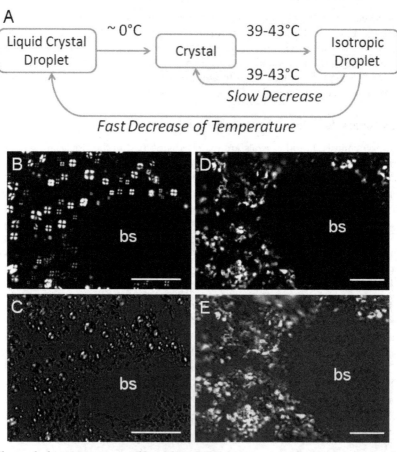

Fig. 3. Thermal phase transitions of liquid crystal droplet, crystal and isotropic droplet. The phase transition from isotropic droplet to crystal or liquid crystal depends on the rate of temperature decrease (A). B and C exhibit the hepatic liquid crystal droplets in crossed nicols with 90 degree of angle (B) and 45 degree (C). D and E show the hepatic crystal in crossed nicols at 90 degree of angle (D) and 45 degree (E), which transited from liquid crystals. Anisotropic liquid crystals locate in hepatocytes in the cords and are absent in the blood sinus (bs). Bars, 60 μm.

was initially established in embryonic hepatic liquid crystal then [Xu XH *et al* 1995] proven to be a general phenomenon in embryonic liquid crystal of other tissues and organs [Xu XH *et al* 2011 and Table 1].

Embryonic Tissue/Organ*		Temperature of Phase Transition (°C)			Liquid Crystal in Diseases	Ref.
		LC to Isotopic	Crystal to Isotopic	Isotopic to Crystal **		
Liver	E14	37.6 ~41.5	39.4~42.0	35.6~36.3	Steatohepatitis Gaucher disease	12, 13, 40, 43, 44, 51
Kidney Mesone-phros	E8	36.6~40.5	38.1~40.8	38.6~40.8	Fabry-Anderson	1, 6, 33, 36
Metane-phros	E14	36.2~40.2	37.9~41.1	38.6~40.7	disease	
Lung	E17	37.1~41.3	38.6~41.6	35.7~35.6	Gram-positive bacteria sputum	37
Aorta	E	37.8 ~41.1	NE	NE	Atherosclerotic lesions	10, 11, 19, 20
Vein	E	37.2 ~41.4	NE	NE	Foam cell abnormality	4, 5, 12, 13, 48
Heart	E	36.8 ~40.0	NE	NE	-	15
	P 10	Non-detectable	Non-detectable	Non-detectable		
Eye (retina)	E19	37.2 ~41.1	39.3~41.1	35.4~36.5	ARMD	14, 31, 32,34
	P6	Non-detectable	Non-detectable	Non-detectable		
York sac	E2~3	38.5~42.5	44.3~46.3	46.2~46.6	-	38, 47
	E17~9	38.9~42.3	44.1~46.3	46.3~47.2		
Blood	E9	NE	NE	NE	Gaucher disease	13, 26, 48

*The data from domestic fowl Taihe
**These temperatures are documented with slow decrease of thermal stage.
NE: Not examined

Table 1. Liquid-crystal characteristics in embryonic tissues/organs and post-natal diseases

3.3 SAXS and XRD measurements of liquid-crystal and crystal

SAXS pattern can be documented with massive hepatic liquid crystals. Within 50 degree of 2θ, only one scattering peak can be detected corresponding to 35 Å in small angle area, indicating the distance of molecular layers of liquid crystal droplets. Using this approach, the period distance of liquid crystal from the fat body of Apis cerana chrysalis [Xu XH *et al* 1994a] and hepatic liquid crystals of various avians have been documented [Xu XH et al

1995, 2009, 2011]. Crystals transited from liquid-crystals can generate many more diffraction peaks corresponding to more crystalgraphic planes. Oddly, XRD pattern reveals more orders on the hepatic crystals.

SAXS and XRD measurements elucidated significant differences between hepatic liquid crystals obtained from different species of avian. In Taihe fowl, the SAXS scattering of the hepatic liquid crystal expresses a strong peak at 2θ. But the peak is weak or absent in its crystal XRD diffraction. In pigeon, the SAXS hepatic liquid crystal peak does have a significant XRD diffraction pattern. This difference indicates that although liquid crystal can be found in the same tissues of the two avians, they likely contain different chemical components (Table 2).

3.4 Liquid crystals in human disease

Unlike during embryogenesis, no liquid crystals are found in postnatal development or in normal physiological systems. However, liquid crystalline structures have been reported in different tissues during pathological processes, including atherosclerosis, abnormal lipid depositions, Age-Related Macular Degeneration, and active monocytes.

In 95 patient samples of atherosclerotic lesions, liquid crystals composed of cholesterol, cholesterol ester, and phospholipid were observed [Lang PD et al 1970]. This data has been confirmed by another group [Goldstein JL et al 1977] and mimicked within *in vitro* systems [Goldstein JL et al 1979]. Maltese-crosses, indicating liquid-crystals, have also been found in lipid depositions accumulating in smooth muscle and foam cells [Kruth H 2001]. These liquid crystal depositions in the vascular wall were found to be low density lipoprotein-cholesteryl esters mediated by cell surface receptors [Goldstein JL et al 1977, 1979, 1997, 2008; Brown MS et al 1974, 1974].

Age-Related Macular Degeneration (ADM) is the leading cause of severe vision loss in adults over 50. The Center for Disease Control and Prevention estimates that 1.8 million people are suffering AMD and over 7 million are at substantial risk for vision loss from AMD in United State. Liquid crystal Maltese's-crosses and crystals structure were observed in the drusen of retina in ARMD patients [Small DM 1970, 1986, 1988; Haimovici R et al 2001]. In these patients, drusen are much bigger than normal and are filled with accumulated anisotropic structures.

Recently, cytoplasmic accumulation of liquid-crystal like droplets have also been found in monocytes, macrophages, and squamous epithelial cells of sputum from a patient affected with Gram-Positive Bacteria [MM Xu et al 2011]. In sputum collected during the recovery phase of respiratory infection, massive Maltese-crosses were fully loaded in host cells. Though the mechanism of formation for these liquid-crystal like droplets has discovered, further study could lead to new perspectives on post-infection removal of infectious agents.

In addition to Fabry-Anderson Disease [MM Xu et al 2009], birefringent particle accumulation are also observed in Gaucher disease [Goodman ZD et al 2009, Hillman RS et al 2005]. This disease is a lysosomal storage disorder, in which deficiency of glucocerebrosidase causes a buildup of fatty substance glucocerebroside in the monocyte and macrophages of certain organs. As the observations were made on biopsy samples, the birefringent fragments observed are likely crystals generated from native liquid crystals as an artifact of freezing. This birefringent accumulation was observed in liver and blood of the patients as well (Table 1).

Newborn Chicken				Newborn Pigeon			
Crystal from LC		Liquid Crystal		Crystal from LC		Liquid Crystal	
XRD		SAXS		XRD		SAXS	
I / I_0	d (Å)	I / I_0	d (Å)	I / I_0	d (Å)	I / I_0	d (Å)
-	-	100	37.0	46	38.10	100	38.02
100	19.55	-	-	32	19.48	-	-
5	17.63	-	-	11	12.10	-	-
1	10.59	-	-	10	10.71	-	-
6	9.44	-	-	30	7.21	-	-
1	7.70	-	-	17	5.89	-	-
2	5.90	-	-	12	5.38	-	-
1	5.22	-	-	100	5.13	-	-
1	5.12	-	-	33	5.01	-	-
18	4.91	-	-	56	4.82	-	-
2	4.60	-	-	22	4.48	-	-
3	4.08	-	-	43	4.39	-	-

Notes: No corresponding parameters between XRD and SAXS show with "-".

Table 2. SAXS and XRD comparison of liquid-crystal and crystal during avian development

4. Comparative study on mouse models of steatohepatitis and embryonic hepatic liquid crystal

4.1 High protein and fat diet mouse models

Using a diet high in protein and fat, a Steatohepatitis mouse model was generated. After three months, the liver of these animals turned light yellow and were larger in comparison to the liver of control animals. After 9 months, the livers were two to three times the size of mice being fed the normal diet. These livers also exhibited plaque and had significantly enlarged spleens. Using X-ray diffraction, small angle X-ray scattering, and phase transition, previous reports have characterized the distribution ultrastructure, and chemical composition of chicken hepatic LCLDs. Using this information and the well-established Steatohepatitis animal model, three conclusions were made: (1) The liquid-crystals in Steatohepatitis liver were produced in a pathological process similar to hepatic liquid-crystal formation in avian embryogenesis; (2) Small angle X-ray scattering revealed that liquid-crystal are distributed every 38Å in the hepatic tissue of Steatohepatitis animals. This distribution matched that reported in avian embryonic livers; (3) In Steatohepatitis animals, the liquid-crystals are distributed on the hepatic cords, which match the localization of liquid-crystals in embryonic liver.

4.2 Gene manipulation of related protein expression

Gene manipulation has become powerful tool for exploring gene function on molecular mechanisms of human diseases. In one instance, Shimano and colleges generated a transgenic mouse overexpressing truncated SREBP-1a (sterol-regulated proteolysis), a SREBP a membrane-bound transcription factor released by sterol-regulated proteolysis [Wang X et al 1994]. This gene abnormality resulted in lipid deposition in atherosclerotic lesions originating from liver accumulation of HLCDs [Shimano H et al 1996].

Stimulating a high level expression of the promoter in liver, the transgene was generated encoding the nuclear fragment of SREBP-1a, the most potent of the three SREBP isoforms. The truncated SREBP-1a is synthesized as a cytosolic protein instead of trafficking to cell surface, and it enters the nucleus without proteolysis, resulting in massive lipid overproduction. The livers of these transgenic mice dramatically enlarged and filled with fat, consisting of a cholesteryl esters and triglycerides mixture. The amounts of fat can be 5 to 25 25-fold higher than those observed in normal liver. The data on LDL receptors indicate that human steatohepatitis and atherosclerosis are linked diseases [Shimano H et al 1996]. Further studies have proven this biological mechanism to fit both alcoholic and non-alcoholic steatohepatitis within in vitro and in vivo studies [Horton JD et al 2002; Ji C et al 2006; You M et al 2002; Browning JD et al 2004].

4.3 Origins of fatty liver disease through hepatic liquid-crystals

In human, liver contains 4~6% fat, mostly made up of phospholipids, glycerides and cholesterol. When these fats accumulate in the liver, patients suffer from steatohepatitis or Hepatic steatosis (Fatty Liver Disease), which in turn causes liver enlargement, and abnormal liver function. Though Fatty Liver Disease is most common in overweight and diabetic patients [Hickman IJ et al 2007], a number of pathologic conditions such as excessive alcohol consumption or genetic disorder triggers this accumulation and will be followed by fibrosis and cirrhosis [Ban CR et al 2008; Preiss D et al 2008; Wilfred de Alwis NM et al 2008]. These associated illnesses make steatohepatitis a high mortality disease [El-Zayadi AR 2008; Xirouchakis E et al 2008]. Massive maltese-cross liquid-crystal droplets, like those found in embryonic liver, are observed in the biopsies of a large number of steatohepatitic patients. These hepatic liquid-crystal droplets (HLCDs) have been detected in large numbers of steatohepatitic patients through biopsy examination. Although their data has not yet been published, the phenomenon has been developed as a clinical examines procedure and filed for US documentation and invention disclosure.

4.4 Embryonic-like liquid-crystal linking steatohepatitis to atherosclerosis

Since the 1970s, liquid-crystalline structures have been observed in atherosclerotic lesions [Lang PD et al 1970, Saul S et al 1976]. The first investigation carried out on 95 individual atherosclerotic lesions obtained from 26 patients' classified thelesions into three groups, fatty streaks, fibrous plaques, and gruel (atheromatous) plaques. Using chromatography, the lipid composition of these legions was determined to be cholesterol, cholesterol ester, and phospholipid. Using polarizing microscopy and X-ray diffraction, these lesion lipids were revealed to accumulate as liquid-crystals in lesions composed of special smooth foam cells [Saul S et al 1976, Kruth HS et al 2001]. This phenomenon was further confirmed by another group [Goldstein JL *et al* 1977] and later mimicked *in vitro* [Joseph B et al 1984]. Analysis of familial hypercholesterolemia through *in vitro* fibroblast overloading experiments demonstrated that lipid deposition in the vascular wall is accomplished via low density lipoprotein-cholesteryl esters (LDL). These complexes are mediated by LDL receptors distributed on cell surfaces [Goldstein JL *et al* 2008; Brown MS *et al* 1974, 1975; Goldstein JL 1977 and 1979].

As discussed above, gain-of-function mutations containing overexpression of a truncated form of SREBP-1a links human steatohepatitis to atherosclerosis through LDL receptors. At this mutation results in liquid crystal depositions as part of the disease pathology, the finding directs a new prospect to exploring the biological function of liquid crystal. As a structure normally only found during embryogenesis, the existence of liquid crystals during disease biology suggests a new method approaching pathology from an embryological point of view. Further understanding of the role liquid crystals play in embryogenesis would doubtless reveal its role in pathogenesis of human diseases and help develop early diagnostics biophysics marker and more effective treatments.

5. Conclusion

Based on current discoveries obtained via XRD, SAXS, confocol microscope, and polarization microscopy in combination with cryo-section, push-release procedure for fluidity measurement, and thermal stage for phase transition progress has been made in the field of liquid crystal function in embryogenesis and pathogenesis of human diseases. With this methodology, the research has proved that, during the embryo development, liquid crystals are readily identifiable in the embryo through their Maltese Crosse birefringence texture. Liquid crystals with this configuration display strong fluidity accompanied with shape-changing properties under direct pressure conditions. XRD and SAXS analysis display a single-peak patterncorresponding to the Bragg distance of liquid crystal. Liquid crystals almost identical to those found in the developing embryo have been found in the affected tissue of multiple diseases.

Liquid crystal configuration within the embryo, animal disease models, and diseased human tissues are all cytoplasmic with Maltese-Crosses situated in cells of various tissues, especially in the luminal portion of kidney during diseases and embryonic blood. Further investigation into liquid crystal involvement in disease through its embryonic mechanisms is expected to generate new diagnostic protocols for liquid crystal related diseases, such as ARMD, Steatohepatitis, Atherosclerotic lesions, and Fabry-Anderson.

6. References

[1] Amico L, Visconti G, Amato A, Azzolina V, Sessa A, Li Vecchi M. Anderson-Fabry disease: a protean clinical behavior and a chance diagnosis. J Nephrol. 18(6):770-2 (2005)
[2] Ban CR, Twigg SM. Fibrosis in diabetes complications: pathogenic mechanisms and circulating and urinary markers. *Vasc Health Risk Manag.* 4(3):575-96 (2008)
[3] Brown GH and Wolken JJ, Liquid-crystals and Biological Structures. Academic Press, Inc. New York. 1979

[4] Brown MS, Goldstein JL. Familial hypercholesterolemia: defective binding of lipoproteins to cultured fibroblasts associated with impaired regulation of 3-hydroxy-3-methylglutaryl coenzyme A reductase activity. *Proc Natl Acad Sc. USA.* 71:788-92 (1974)

[5] Brown MS, Faust JR, Goldstein JL. Role of the low density lipoprotein receptor in regulating the content of free and esterified cholesterol in human fibroblasts. *J Clin Invest.* 55:783-93 (1975)

[6] Cho ME, Kopp JB. Fabry disease in the era of enzyme replacement therapy: a renal perspective. Pediatr Nephrol. 19(6):583-93 (2004)

[7] El-Zayadi AR. Hepatic steatosis: a benign disease or a silent killer. *World J Gastroenterol.* 14(26):4120-6 (2008)

[8] Fehér G. Yolk sac stones in domestic fowl. Anat Histol Embryol. 8(4):360-4 (1979)

[9] Ginsberg L, Manara R, Valentine AR, Kendall B, Burlina AP. Magnetic resonance imaging changes in Fabry disease. *Acta Paediatrica Suppl.* 451:57–62 (2006)

[10] Goldstein JL, Anderson RG, Buja LM, Basu SK, Brown MS. Overloading human aortic smooth muscle cells with low density lipoprotein-cholesteryl esters reproduces features of atherosclerosis in vitro. *J Clin Invest.* 59(6):1196–1202 (1977)

[11] Goldstein JL, Anderson RG, Brown MS. Coated pits, coated vesicles, and receptor-mediated endocytosis. *Nature.* 279:679-85 (1979)

[12] Goldstein JL, Brown, MS. The clinical investigator: bewitched, bothered, and bewildered--but still beloved *J Clin Invest,* 99, 2803-12 (1997)

[13] Goldstein JL, Brown MS. From fatty streak to fatty liver: 33 years of joint publications in the JCI. *J Clin Invest.* 118:1220-2 (2008)

[14] Haimovici R, Gantz DL, Rumelt S, Freddo TF, Small DM. The lipid composition of drusen, Bruch's membrane, and sclera by hot stage polarizing light microscopy. *IOVS.* 42:1592-9 (2001)

[15] He H, Zhou H, Wang G, Wu X. Liquid crystalline in yolk sac during chicken development. *J Wuhan Univ (Nature Science Ed.)* 4:32-46 (1978)

[16] Hickman IJ, Macdonald GA. Impact of diabetes on the severity of liver disease. Am J Med. 120(10):829-34 (2007)

[17] Horton JD, Goldstein JL, Brown MS. SREBPs: activators of the complete program of cholesterol and fatty acid synthesis in the liver. *J Clin Invest.* 109:1125-31 (2002)

[18] Ji C, Chan C, Kaplowitz N. Predominant role of sterol response element binding proteins (SREBP) lipogenic pathways in hepatic steatosis in the murine intragastric ethanol feeding model.Kruth HS. Lipoprotein cholesterol and atherosclerosis. *Cur Mol Med.* 1: 633-3 (2001)

[19] Kruth H. Lipoprotein cholesterol and atherosclerosis. *Curr Mol Med.* 1(6): 633-53 (2001)

[20] Lang PD, Insull W Jr. Lipid droplets in atherosclerotic fatty streaks of human aorta. *J Clin Invest.* 49:1479-88 (1970)

[21] Li M, Chao L. Structural study of lipid liquid crystal droplets in chicken development. *Acta Biophysica Sinica,* 12, 299-305(1988)

[22] Lidove O, Klein I, Lelièvre JD, Lavallée P, Serfaty JM, Dupuis E, Papo T, Laissy JP.Imaging features of Fabry disease. Am J Roentgenol. 186(4):1184-91 (2006)

[23] Linhart A, Kampmann C, Zamorano JL, Sunder-Plassmann G, Beck M, Mehta A, Elliott PM; European FOS Investigators. Cardiac manifestations of Anderson-Fabry

disease: results from the international Fabry outcome survey. *Eur Heart J.* 28(10):1228-35 (2007)

[24] Maruyama M*, BY Li*, HY Chen*. Xuehong Xu*, Long-Sheng Song, Wuqiang Zhu, Weidong Yong, Wenjun Zhang, Gui-Xue Bu, Shien-Fong Lin, Michael C. Fishbein, W. Jonathan Lederer, John H. Schild Loren J. Field, Michael Rubart, Peng-Sheng Chen, Weinian Shou. FKBP12 is a Critical Regulator of the Heart Rhythm and the Cardiac Voltage-Gated Sodium Current in Mice. *Circulation Research*, 108(9):1042-52 (2011) (*co-first author)

[25] Moore DF, Ye F, Schiffmann R, Butman JA. Increased signal intensity in the pulvinar on T1-weighted images: a pathognomonic MR imaging sign of Fabry disease. *Am J Neuroradiol.* 24: 1096–101 (2003).

[26] Pieroni M, Chimenti C, De Cobelli F, Morgante E, Del Maschio A, Gaudio C, Russo MA, Frustaci A.Pieroni M. Fabry's disease cardiomyopathy: echocardiographic detection of endomyocardial glycosphingolipid compartmentalization. *J Am Coll Cardiol.* 47(8):1663-71 (2006)

[27] Preiss D, Sattar N. Non-alcoholic fatty liver disease: an overview of prevalence, diagnosis, pathogenesis and treatment considerations. *Clin Sci (Lond).* 15(5):141-50 (2008)

[28] Robert S. Hillman, Kenneth A. Ault, Henry M. Rinder Hematology in clinical practice: a guide to diagnosis and management Chapter 26 McGraw-Hill Professional (2005)

[29] Shimano H, Horton JD, Hammer RE, Shimomura I, Brown MS, Goldstein JL. Overproduction of cholesterol and fatty acids causes massive liver enlargement in transgenic mice expressing truncated SREBP-1a. *J Clin Invest.* 98:1575–84 (1996)

[30] Small DM. Surface Chemistry of Biologic Systems. Plenum Press: New York, US. (1970)

[31] Small DM. The Physical Chemistry of Lipids. Plenum Press: New York, US. (1986)

[32] Small DM. George Lyman Duff memorial lecture. Progression and regression of atherosclerotic lesions. Insights from lipid physical biochemistry. Arteriosclerosis. 8(2):103-29 (1988).

[33] Utsumi K, Mitsuhashi F, Asahi K, Sakurazawa M, Arii K, Komaba Y, Katsumata T, Katsura K, Kase R, Katayama Y. Enzyme replacement therapy for Fabry disease: morphologic and histochemical changes in the urinary sediments. Clin Chim Acta.360(1-2):103-7 (2005)

[34] Wang X, Sato R, Brown MS, Hua X, Goldstein JL. SREBP-1, a membrane-bound transcription factor released by sterol-regulated proteolysis. *Cell.* 77:53–62 (1994).

[35] Wilfred de Alwis NM, Day CP. Genes and nonalcoholic fatty liver disease. *Curr Diab Rep.* 8(2):156-63 (2008)

[36] Xu MM, Xu XH, Cao G, Pan Y, Jones O, Bryant JL, Anthony DD, He H, Yan G, Zhang C. The liquid crystalline in normal renal development amplifies the comprehension for Anderson-Fabry disease. *Molecular Crystals and Liquid Crystals*, 508: 52-66 (2009)

[37] Xu MM, Jones OD, Chen X, Li YF , Yan G, Pan Y, Davis HG, Anthony DD, Xu Y, Zheng S, Bryant JL, Xu XH. Cytoplasmic Accumulation of Liquid-Crystal Like Droplets in Post-Infection Sputum Generated by Gram-Positive Bacteria. *Molecular Crystals and Liquid Crystals*, 547: 173-180 (2011)

[38] Xu MM, Jones OD, Cao G, Yan G, He H, Zhang C, Xu XH. Crystallization of calcium carbonate vaterite involves with another mechnisim associated with liquid crystal in embryonic yolk sacs. *Key Engineering Materials*, 428-9: 349-55 (2010)

[39] Xirouchakis E, Sigalas A, Manousou P, Calvaruso V, Pleguezuelo M, Corbani A, Maimone S, Patch D, Burroughs AK. Models for non-alcoholic fatty liver disease: a link with vascular risk. *Curr Pharm Des*. 14(4):378-84 (2008)

[40] Xu XH, Tang C, He H, Zhang X, The lipid components of hepatic liquid crystal lipoid droplets during developing chick embryo, *Acta Biochimica et Biophysica Sinica*, 24(4): 339-343 (1992)

[41] Xu XH, Tang C, Tang Z, Tong H, He H, The studies on otolith CaCO3 crystal structure of three species of fishes in *Sciaenidae*, *Acta Biophysica Sinica*, 9(1): 41-45, 1993

[42] Xu XH, Wang C, Ai X, Confirmation of liquid state of the trophocyte protein granules in the fat body of Chinese honeybee (*Apis cerana*) chrysalis, *Acta Biochimica et Biophysica Sinica*, 26(1): 105-110 (1994)

[43] Xu XH, Wang C, Wu X, He H. Comparative studies on the hepatic liquid crystal lipoid droplets of newborn ducks and newborn pigeons, *Molecular Crystal and Liquid Crystal*, 265: 659-668 (1995a)

[44] Xu XH, Wang C, Wang X, Lei X, He H, Comparative studies on liquid crystal lipoid droplets of livers from newborn birds (Domestic Fowl), *Acta Biochimica et Biophysica Sinica*, 27(5): 551-558 (1995b)

[45] Xu XH, Zhang C, He H, Lu Qand, Ji H, Construction of imitated liquid crystal lipoid droplets of chicken embryo liver, *Acta Biophysica Sinica*, 13(1): 29-34 (1997)

[46] Xu XH, Dong C, Vogel B, Hemicentin assemblies on diverse epithelia in the mouse, *Journal of Histochemistry and Cytochemistry*, 55(2):119-126 (2007)

[47] Xu XH, Xu MM, Cao G, Jones O, Zhao C, Cao L, Yan G, He H, Zhang C. Co-subsistence of liquid crystal droplets and calcium carbonate vaterite crystals reveals a molecular mechanism of calcium preservation in embryogenesis. *Molecular Crystals and Liquid Crystals*, 508:77–90 (2009)

[48] Xu XH, Xu MM, Jones O, Chen X, Li Y, Yan G, Pan Y, Davis HG, Xu Y, Bryant JL, Zheng S, Anthony DD, Liquid Crystal in Lung Development and Chicken Embryogenesis. *Molecular Crystals and Liquid Crystals*, 547: 164-172 (2011a)

[49] Xu XH, Vogel B. A Secreted Protein Promotes Cleavage Furrow Maturation during Cytokinesis. *Current Biology*, 21(2):114-119 (2011b)

[50] You M, Fischer M, Deeg MA, Crabb DW. Ethanol induces fatty acid synthesis pathways by activation of sterol regulatory element-binding protein (SREBP). *J Biol Chem*. 9;277(32):29342-7 (2002)

[51] Zachary D. Goodman and Hala R. Makhlouf,,Chapter 9, Hepatic Histopathology in Schiff's diseases of the liver Volume 2 (edited by Eugene R. Schiff), Lippincott Williams & Wilkins (2009)

Permissions

The contributors of this book come from diverse backgrounds, making this book a truly international effort. This book will bring forth new frontiers with its revolutionizing research information and detailed analysis of the nascent developments around the world.

We would like to thank Dr. Ken-ichi Sato, for lending his expertise to make the book truly unique. He has played a crucial role in the development of this book. Without his invaluable contribution this book wouldn't have been possible. He has made vital efforts to compile up to date information on the varied aspects of this subject to make this book a valuable addition to the collection of many professionals and students.

This book was conceptualized with the vision of imparting up-to-date information and advanced data in this field. To ensure the same, a matchless editorial board was set up. Every individual on the board went through rigorous rounds of assessment to prove their worth. After which they invested a large part of their time researching and compiling the most relevant data for our readers. Conferences and sessions were held from time to time between the editorial board and the contributing authors to present the data in the most comprehensible form. The editorial team has worked tirelessly to provide valuable and valid information to help people across the globe.

Every chapter published in this book has been scrutinized by our experts. Their significance has been extensively debated. The topics covered herein carry significant findings which will fuel the growth of the discipline. They may even be implemented as practical applications or may be referred to as a beginning point for another development. Chapters in this book were first published by InTech; hereby published with permission under the Creative Commons Attribution License or equivalent.

The editorial board has been involved in producing this book since its inception. They have spent rigorous hours researching and exploring the diverse topics which have resulted in the successful publishing of this book. They have passed on their knowledge of decades through this book. To expedite this challenging task, the publisher supported the team at every step. A small team of assistant editors was also appointed to further simplify the editing procedure and attain best results for the readers.

Our editorial team has been hand-picked from every corner of the world. Their multi-ethnicity adds dynamic inputs to the discussions which result in innovative outcomes. These outcomes are then further discussed with the researchers and contributors who give their valuable feedback and opinion regarding the same. The feedback is then collaborated with the researches and they are edited in a comprehensive manner to aid the understanding of the subject.

Apart from the editorial board, the designing team has also invested a significant amount of their time in understanding the subject and creating the most relevant covers. They scrutinized every image to scout for the most suitable representation of the subject and create an appropriate cover for the book.

The publishing team has been involved in this book since its early stages. They were actively engaged in every process, be it collecting the data, connecting with the contributors or procuring relevant information. The team has been an ardent support to the editorial, designing and production team. Their endless efforts to recruit the best for this project, has resulted in the accomplishment of this book. They are a veteran in the field of academics and their pool of knowledge is as vast as their experience in printing. Their expertise and guidance has proved useful at every step. Their uncompromising quality standards have made this book an exceptional effort. Their encouragement from time to time has been an inspiration for everyone.

The publisher and the editorial board hope that this book will prove to be a valuable piece of knowledge for researchers, students, practitioners and scholars across the globe.

List of Contributors

Xinchun Lin and Wei Fang
The Nurturing Station for the State Key Laboratory of Subtropical Silviculture, Zhejiang Agriculture and Forestry University, Zhejiang

Lichun Huang
Institute of Plant and Microbial Biology, Academia Sinica, Taibei, Taiwan, China

J. Malá and P. Máchová
Forestry and Game Management Research Institute, Jíloviště

M. Cvikrová and L. Gemperlová
Institute of Experimental Botany, Academy of Sciences of the Czech Republic, Praha, Czech Republic

Katarzyna Nawrot-Chorabik
Department of Forest Pathology, Faculty of Forestry, University of Agriculture in Kraków, Kraków, Poland

Tsuyoshi E. Maruyama and Yoshihisa Hosoi
Forestry and Forest Products Research Institute, Department of Molecular and Cell Biology, Tsukuba, Japan

Na Xu
Department of Cell and Developmental Biology, Weill Medical, College of Cornell University, New York
Department of Cell Biology, Albert Einstein College of Medicine, New York, USA

Carolyn Pirraglia, Unisha Patel and Monn Monn Myat
Department of Cell and Developmental Biology, Weill Medical, College of Cornell University, New York, USA

A.K.M. Mahbub Hasan
Laboratory of Gene Biology, Department of Biochemistry and Molecular Biology, University of Dhaka, Dhaka, Bangladesh

Takashi Matsumoto, Shigeru Kihira, Junpei Yoshida and Ken-ichi Sato
Laboratory of Cell Signaling and Development, Department of Molecular Biosciences, Faculty of Life Sciences, Kyoto Sangyo University, Kamigamo-Motoyama, Kita-ku, Kyoto, Japan

A.K.M. Mahbub Hasan
Laboratory of Gene Biology, Department of Biochemistry and Molecular Biology, University of Dhaka, Dhaka, Bangladesh

Takashi Matsumoto,Shigeru Kihira, Junpei Yoshida and Ken-ichi Sato
Laboratory of Cell Signaling and Development, Department of Molecular Biosciences, Faculty of Life Sciences, Kyoto Sangyo University, Kamigamo-Motoyama, Kita-ku, Kyoto, Japan

Elena A. Severtsova, David R. Aguillón Gutiérrez and Aleksey S. Severtsov
Lomonosov Moscow State University, Russian Federation

Tara D. Silva
University of Colombo, Sri Lanka

Shingo Toyoshima
NARO Institute of Vegetable and Tea Science, Japan

Hiroshi Amano
Graduate School of Agriculture, Kyoto University, Japan

Laura Yesenia Solís-Ramos
Unidad de Bioquímica y Biología Molecular de Plantas, Centro de Investigación Científica de Yucatán, (CICY), Mérida, Yucatán, Studies on habanero chilli biotechnology, Mexico
Escuela de Biología, Universidad de Costa Rica, San Pedro, Genetic Plant Transformation and Biotechnology, Costa Rica

Antonio Andrade-Torres
Unidad de Biotecnología, Centro de Investigación Científica de Yucatán, (CICY), Mérida, Yucatán, Biotechnological Studies on Coconut, Mexico
INBIOTECA- Instituto de Biotecnología y Ecología Aplicada, Universidad Veracruzana, Xalapa, Veracruz, México, Genetic Plant Transformation and Biotechnology, Mexico

Luis Alfonso Sáenz Carbonell and Carlos M. Oropeza Salín
Unidad de Biotecnología, Centro de Investigación Científica de Yucatán, (CICY), Mérida, Yucatán, Biotechnological studies on coconut, Mexico

Enrique Castaño de la Serna
Unidad de Bioquímica y Biología Molecular de Plantas,Centro de Investigación Científica de Yucatán, (CICY), Mérida, Yucatán,Studies on habanero chilli biotechnology, Mexico

MengMeng Xu
Medical Scientist Training Program, Duke University School of Medicine, USA

Xuehong Xu
Department of Physiology Center for Biomedical Engineering Technology, USA
Center for Stem Cell Biology and Regenerative Medicine, University of Maryland School
of Medicine, USA
Shaanxi Normal University School of Life Science, China

Printed in the USA
CPSIA information can be obtained
at www.ICGtesting.com
JSHW011503221024
72173JS00005B/1184